INTRODUCTION TO LINEAR ALGEBRA

INTRODUCTION TO LINEAR ALGEBRA

FRANZ E. HOHN

UNIVERSITY OF ILLINOIS, URBANA

Under the Editorship of
Carl B. Allendoerfer

THE MACMILLAN COMPANY, NEW YORK

COLLIER-MACMILLAN LIMITED, LONDON

Printed in the United States of America

The Macmillan Company
866 Third Avenue, New York, New York 10022

Collier-Macmillan Canada, Ltd., Toronto, Ontario

Library of Congress catalog card number: 73–161431

First Printing

PREFACE

The purpose of this book is to make essential ideas and techniques available to the increasing range of students who need to know of linear algebra early in their academic careers. The plan is therefore to provide an unsophisticated but mathematically sound and reasonably comprehensive introduction to the subject. No calculus background is required. Throughout, notation is kept as simple as possible. Explanations and proofs are intended to be full. There are many illustrative examples. There are many different types of exercises and they range from routine practice problems to important extensions of the theory.

To allow space for full exposition, no applications are presented in this book. However, all the material included is "practical" for students in computer science, the engineering sciences, the physical sciences, and the social sciences. For those going on in mathematics, experience shows that the study of a book like this should make a later, mature study of linear algebra more rewarding.

The purposes of the individual chapters are as follows:

Chapter 1. Linear Equations. The object of this chapter is to make it possible for the reader to solve a given system of linear equations, homogeneous or nonhomogeneous, for the complete solution when it exists, by sweepout procedures. This in turn permits extensive use of concrete examples and exercises in following chapters, so that the new abstract concepts that appear are more readily learned.

Chapter 2. Matrices. This chapter presents the basic laws of the algebra of matrices. The material is related to Chapter 1 wherever possible in order to motivate the algebra of matrices and to provide a suitable notation for Chapter 3 on vector geometry in $\mathscr{E}^2$ and $\mathscr{E}^3$. Computation of inverses by sweepout is introduced here to complete the set of computational tools needed for the study of vector spaces.

Chapter 3. Vector Geometry in $\mathscr{E}^2$ and $\mathscr{E}^3$. The purpose here is to provide needed geometrical ideas, techniques, and terminology as well as to lay intuitive foundations for the abstract concepts of vector space, linear dependence, and so on. The reader who has already studied the analytic geometry of three dimensions will be able to dispose of this chapter very rapidly.

Chapter 4. Vector Geometry in $\mathscr{E}^n$. Here the main purpose is to introduce ideas and terminology that are common in the literature of applications. A second goal is further intuitive preparation for Chapter 5.

Chapter 5. Vector Spaces. The treatment here is simple but fully general. Many algebraic and geometric examples are given, to make the abstract ideas more comprehensible. Since the reader can solve systems of linear equations and invert matrices, much useful computational exercise material is possible.

Chapter 6. The Rank of a Matrix. In this chapter, the notion of rank is applied to much that precedes. All computations are based on the sweepout process. The structure of the solution space of a system of linear equations is treated completely and is interpreted geometrically.

Chapter 7. Determinants. In preceding chapters, the deliberate avoidance of determinantal computations has served to place emphasis on the basic concepts and methods of linear algebra. It is now appropriate, for use in following chapters, to present the most basic facts about determinants. In addition, determinants are employed as an alternative computational tool in the solution of many of the preceding algebraic and geometric problems so that the chapter provides valuable review.

Chapter 8. Linear Transformations. This chapter begins with some simple geometrical examples which are designed to provide a strong intuitive base for the algebraic concepts that follow. Linear transformations are interpreted both as linear operators (vector or point mappings) and as transformations of coordinate systems. The concept of change of basis is then applied to the transformation of an operator. Orthogonal transformations are treated in detail. A strong geometrical theme is maintained throughout the chapter.

Chapter 9. The Characteristic Value Problem. This chapter exploits and unifies all the material that precedes. The characteristic value problem is defined and the simpler general theorems are presented. Then symmetric matrices and quadratic forms are treated in considerable detail. Definiteness receives special emphasis because of its importance in applications. Again, geometrical interpretations are emphasized throughout the chapter.

The author hopes that this book will provide a pleasant and rewarding learning experience for beginning students of linear algebra. Suggestions for additions, corrections, and revisions will be gratefully received.

The author is indebted to Professor Carl B. Allendoerfer, editor of this series, to Professor John E. Wetzel, who taught from a preliminary draft of this book, and to Professors Ralph E. Walde and C. R. B. Wright for able critical assistance. As always, Mrs. Carolyn Bloemker was an expert and helpful technical typist. The editors of The Macmillan Company, in particular Mr. Leo Malek, performed most competently and cooperatively throughout so that it was a pleasure to work with them. The compositors, William Clowes & Sons, Limited, provided virtually flawless typesetting. Finally, my wife, Marian, was patient and understanding during the many evenings and weekends that had to be reserved for this project. I am deeply grateful to all of these people for their parts in the making of this book.

F. E. H.

CONTENTS

CHAPTER 1

Linear Equations

1.1 LINEAR EQUATIONS

A natural beginning for the study of linear algebra is the study of systems of linear equations, that is, systems of equations of the form

$$
\begin{aligned}
a_{11}x_1 + a_{12}x_2 + \cdots + a_{1n}x_n &= b_1 \\
a_{21}x_1 + a_{22}x_2 + \cdots + a_{2n}x_n &= b_2 \\
&\vdots \\
a_{m1}x_1 + a_{m2}x_2 + \cdots + a_{mn}x_n &= b_m,
\end{aligned} \tag{1.1.1}
$$

where the a_{ij}'s and the b_j's represent numerical coefficients (real or complex numbers) and where $x_1, x_2, \ldots, x_n$ represent unknowns. Such systems arise in virtually every area in which mathematics is applied.

Most readers will be familiar with the solution of such systems of equations by elimination or by determinants. The latter method employs specific formulas that are easy to understand but the computation is prohibitively tedious except in very simple cases. The elimination procedures have the advantage that they can often benefit from clever inspection. However, the use of special tricks is not always possible or useful. Except in very simple cases and even when tricks cannot be used, elimination involves much less computation than does the determinant method. Hence we begin our study of linear algebra by developing some systematic elimination procedures and investigating their consequences. These methods apply equally well whether or not the number of equations is equal to the number of unknowns.

An elementary understanding of linear equations and of their solutions, which this first chapter is designed to develop, will help us to construct a theory of vector spaces, which will in turn enable us to study, with greater insight, the set of all solutions of an arbitrary system of equations. In some cases such a system will have no solution; that is, the solution set will be empty. In other cases, there will be a unique solution. In still other cases, the number of solutions will be infinite. We shall see that for systems of equations with numerical coefficients, these three cases represent the only possibilities.

When a system has no solution, it is said to be **inconsistent**. When it has at least one solution, it is said to be **consistent**. Our objective in this chapter is to develop methods of determining whether or not a given system is consistent and of computing any solutions that may exist. Fortunately, the same procedure may be used to accomplish both purposes.

1.2 THREE EXAMPLES

These examples illustrate both inconsistent and consistent systems of equations. They employ only operations which the reader has used before. That these familiar operations on systems of equations are valid will be proved in Section 1.4.

Example 1: Find all solutions of the system of equations

$$\begin{aligned} 3x_1 + 2x_2 - x_3 &= -2 \\ x_1 - x_2 + 2x_3 &= 3 \\ 4x_1 + x_2 + x_3 &= 0. \end{aligned} \tag{1.2.1}$$

A clever trick is too useful to ignore here. Adding the first two equations, we obtain

$$4x_1 + x_2 + x_3 = 1,$$

whereas the third equation says that

$$4x_1 + x_2 + x_3 = 0.$$

These last two equations imply that $0 = 1$, which is false. Since both equations cannot simultaneously be true, the given system cannot hold for any set of values (x_1, x_2, x_3). That is, there is no solution and the system is inconsistent.

Example 2: Find all solutions of

$$\begin{aligned} 3x_1 + 2x_2 - x_3 &= -2 \\ x_1 + x_2 + x_3 &= 0 \\ -x_1 + x_2 + 2x_3 &= 3 \\ 3x_1 + 4x_2 + 2x_3 &= 1. \end{aligned} \tag{1.2.2}$$

This system of equations illustrates the fact that there may well be more equations than unknowns. Also, we illustrate a more systematic elimination procedure for the first time.

It is convenient to begin by exchanging the first two equations so that we have

$$\begin{aligned} x_1 + x_2 + x_3 &= 0 \\ 3x_1 + 2x_2 - x_3 &= -2 \\ -x_1 + x_2 + 2x_3 &= 3 \\ 3x_1 + 4x_2 + 2x_3 &= 1. \end{aligned}$$

This yields an initial equation with leading coefficient 1. (If no equation of the system had had a leading coefficient 1, we would have begun by dividing the first equation by its leading coefficient, as in the next example.)

We now use the first equation to eliminate x_1 from the other three equations. Subtracting three times the first equation from the second and the fourth equations, and adding the first equation to the third, we obtain the new system of equations

$$\begin{aligned} x_1 + x_2 + x_3 &= 0 \\ -x_2 - 4x_3 &= -2 \\ 2x_2 + 3x_3 &= 3 \\ x_2 - x_3 &= 1. \end{aligned}$$

Now we multiply the second equation by -1 so as to get a leading coefficient 1. This makes it easy to eliminate x_1 from all equations but the second. (We could also have interchanged the second and fourth equations.) Then adding -1 times the resulting second equation once to the first, twice to the third, and once to the fourth, we obtain

$$\begin{aligned} x_1 \qquad - 3x_3 &= -2 \\ x_2 + 4x_3 &= 2 \\ -5x_3 &= -1 \\ -5x_3 &= -1. \end{aligned}$$

Multiplying the third equation by $-\frac{1}{5}$ and employing the resulting equation to eliminate x_3 from all the other equations, we obtain the system

$$\begin{aligned} x_1 &= -\tfrac{7}{5} \\ x_2 &= \tfrac{6}{5} \\ x_3 &= \tfrac{1}{5} \\ 0 &= 0. \end{aligned}$$

It is not hard to check, by substitution in the original set of equations, that this is, in fact, a solution, so that the system is consistent. Moreover, the development here makes clear that this is the only possible solution—in short, that it is *unique*. Indeed, we have shown that if (x_1, x_2, x_3) represents a solution of (1.2.2), that solution must be $(-\frac{7}{5}, \frac{6}{5}, \frac{1}{5})$.

Putting it another way, we have reduced the given system of equations to the simpler system

$$\begin{aligned} x_1 &= -\tfrac{7}{5} \\ x_2 &= \tfrac{6}{5} \\ x_3 &= \tfrac{1}{5}, \end{aligned} \tag{1.2.3}$$

which we call a solution.

The basic idea of the above procedure is to eliminate from its column all appearances but one of a given variable. For this reason, the method is called the **sweepout process**.

Example 3: Find all solutions of the system of equations

$$\begin{aligned} 2x_1 + 3x_2 + x_3 &= 1 \\ 3x_1 - x_2 + 2x_3 &= 5. \end{aligned} \tag{1.2.4}$$

First let us multiply the leading equation by $\frac{1}{2}$:

$$\begin{aligned} x_1 + \tfrac{3}{2}x_2 + \tfrac{1}{2}x_3 &= \tfrac{1}{2} \\ 3x_1 - x_2 + 2x_3 &= 5. \end{aligned}$$

Now subtract three times the new first equation from the second:

$$\begin{aligned} x_1 + \tfrac{3}{2}x_2 + \tfrac{1}{2}x_3 &= \tfrac{1}{2} \\ -\tfrac{11}{2}x_2 + \tfrac{1}{2}x_3 &= \tfrac{7}{2}. \end{aligned}$$

Multiplying the second of these equations by $-\frac{2}{11}$, we obtain

$$\begin{aligned} x_1 + \tfrac{3}{2}x_2 + \tfrac{1}{2}x_3 &= \tfrac{1}{2} \\ x_2 - \tfrac{1}{11}x_3 &= -\tfrac{7}{11}. \end{aligned}$$

Adding $-\frac{3}{2}$ times the second equation to the first, we obtain

$$\begin{aligned} x_1 \qquad + \tfrac{7}{11}x_3 &= \tfrac{16}{11} \\ x_2 - \tfrac{1}{11}x_3 &= -\tfrac{7}{11}. \end{aligned} \tag{1.2.5}$$

Finally, we rewrite these equations as follows:

$$\begin{aligned} x_1 &= -\tfrac{7}{11}x_3 + \tfrac{16}{11} \\ x_2 &= \tfrac{1}{11}x_3 - \tfrac{7}{11}. \end{aligned} \tag{1.2.6}$$

We have shown that if equations (1.2.4) are true, then equations (1.2.6) are true, that is, that every solution of (1.2.4) must also satisfy (1.2.6). On the other hand, if we substitute these last expressions for x_1 and x_2 into the original equations, then, independently of the value of x_3, they reduce respectively to the identities $1 = 1$ and $5 = 5$. (Check this.) Hence we can choose any particular value for x_3 that we please, compute x_1 and x_2 from (1.2.6), and obtain a particular solution of the original pair. This will be one solution of an infinite set of solutions so obtainable, so this system, too, is consistent. For example, if we put $x_3 = 0$ in (1.2.6), we have the solution $(\frac{16}{11}, -\frac{7}{11}, 0)$ of (1.2.4). If we put $x_3 = 7$, we have the solution $(-3, 0, 7)$. If we put $x_3 = \frac{16}{7}$, we have the solution $(0, -\frac{3}{7}, \frac{16}{7})$, and so on.

The complete set of solutions of this system may be represented in another way. Suppose we put $x_3 = t$, where t represents an arbitrary parameter, that is, an independent variable. Then we have

$$\begin{aligned} x_1 &= -\tfrac{7}{11}t + \tfrac{16}{11} \\ x_2 &= \tfrac{1}{11}t - \tfrac{7}{11} \\ x_3 &= t. \end{aligned} \tag{1.2.7}$$

For each value assigned to t, we get a solution of the system. Moreover, every solution of the system may be obtained by appropriate choice of t, since choosing t is equivalent to choosing x_3.

Such a parametric solution is often useful. Other parametric representations of the set of all solutions are possible. Thus, we could put $x_3 = at + b$, where a and b are any convenient real numbers ($a \neq 0$), and still have x_1, x_2, and x_3 expressed as linear functions of t. For example, if we put

$$x_3 = 11t + 18,$$

we obtain the parametric solution

$$\begin{aligned} x_1 &= -7t - 10 \\ x_2 &= t + 1 \\ x_3 &= 11t + 18, \end{aligned} \tag{1.2.8}$$

the correctness of which is easily checked by substitution.

These examples illustrate the fact that *a given system of equations may have no solution, exactly one solution, or an infinite number of solutions.* Which of these cases occurs depends on certain relations among the coefficients of the system of equations being solved. Just how this depends on the coefficients is made clear in following chapters.

The technique employed in Examples 2 and 3 illustrates the following general procedure. Use the first equation to eliminate the first variable from all the other equations. Then use the second equation to eliminate another variable (usually but not necessarily the second) from all the other equations, and so on. This results in a system in which each equation contains one selected variable appearing in no other equation. If the system is consistent, the solution is then completed by transposing all remaining terms containing variables other than those selected in the elimination process. Sometimes it is necessary to permute equations at some stage in order to proceed with the elimination in the prescribed manner. We shall show in Section 1.5 how to make this process completely systematic and completely general.

1.3 EXERCISES

1. Solve these systems of equations by the method employed in Example 2:

(a) $$\begin{aligned} x_1 - x_2 + x_3 &= 1 \\ 2x_1 + 3x_2 - x_3 &= 4 \\ -x_1 - 2x_2 + 5x_3 &= 2, \end{aligned}$$

(b) $$\begin{aligned} 2x_1 + 3x_2 + 4x_3 &= 8 \\ x_1 - x_2 + 2x_3 &= 9 \\ -3x_1 + 2x_2 + x_3 &= -4, \end{aligned}$$

(c) $$\begin{aligned} x_1 + x_2 - x_3 + x_4 &= 3 \\ x_1 \qquad + x_3 + x_4 &= 2 \\ x_2 + x_3 - x_4 &= -1 \\ 2x_1 - x_2 + x_3 \qquad &= 2. \end{aligned}$$

2. Obtain expressions for the complete solutions of

(a) $$\begin{aligned} x_1 - 2x_2 + x_3 &= 0 \\ 2x_1 + 3x_2 - x_3 &= 4, \end{aligned}$$

(b) $$\begin{aligned} 3x - 4y + 2z &= 12 \\ x + y - 5z &= -4 \\ -2x + 3y + 8z &= 1, \end{aligned}$$

(c) $$\begin{aligned} -3x_1 + x_2 + x_3 + 5x_4 &= 6 \\ 2x_2 - 3x_3 + 5x_4 &= 4 \\ - x_3 + 5x_4 &= 4, \end{aligned}$$

(d) $$\begin{aligned} 2x_1 + 3x_2 - x_3 - x_4 &= 3 \\ x_1 + 2x_2 + x_3 + 4x_4 &= 8 \\ -x_1 - x_2 + 3x_3 + x_4 &= 2. \end{aligned}$$

3. Show that the system of equations

$$\begin{aligned} 2x_1 - 3x_2 &= 1 \\ x_1 + 2x_2 &= 3 \\ 5x_1 - 7x_2 &= 0 \end{aligned}$$

is inconsistent. There are several ways to do this: Solve two of the equations and show that the solution does not satisfy the third or, better, just start the solution process, as in Example 2, and show that it leads to a contradiction.

4. Determine whether or not each system is consistent:

(a) $$\begin{aligned} x_1 + 2x_2 + 3x_3 &= 4 \\ 2x_1 + 3x_2 + 4x_3 &= 5 \\ 3x_1 + 4x_2 + 5x_3 &= 7, \end{aligned}$$

(b) $$\begin{aligned} x_1 - x_2 + x_3 &= 1 \\ 2x_1 + x_2 - 2x_3 &= 1 \\ 3x_1 - 2x_2 - x_3 &= 0 \\ 4x_1 + x_2 + 3x_3 &= 2. \end{aligned}$$

5. Given that

$$\begin{aligned} X &= x + y - z \\ Y &= x - y + z \\ Z &= -x + y + z, \end{aligned}$$

solve for x, y, and z as functions of X, Y, and Z and show that $X + Y + Z = x + y + z$.

6. Express in parametric form the complete solution of the system consisting of the single equation $2x - 5y = 7$. Then do the same for the system consisting of the single equation $5x_1 - 2x_2 + 3x_3 = 4$. In the latter case you will need to use *two* parameters; for example, let $x_2 = s$, $x_3 = t$. Then solve for x_1.

7. Obtain the solution of this system of equations by inspection:

$$\begin{aligned} x_1 - 2x_2 + 3x_3 + x_4 &= 3 \\ x_2 - x_3 + 2x_4 &= 2 \\ 3x_3 + x_4 &= 4 \\ x_4 &= 1. \end{aligned}$$

8. Find the complete solution in parametric form:

(a) $$\begin{aligned} x_1 - x_2 - x_3 &= 1 \\ 2x_1 - 2x_2 - 3x_3 &= 5, \end{aligned}$$

(b) $$\begin{aligned} 2x - y + z &= 0 \\ x + 2y - 3z &= 0. \end{aligned}$$

9. Obtain the complete solution of the system

$$\begin{aligned} a + 2b - 3c &= 0 \\ -3a + b + 2c &= 0 \\ 2a - 3b + c &= 0. \end{aligned}$$

10. What must be the values of b_1, b_2, and b_3 if $(1, 1, 1)$ is to be a solution of the following system?

$$\begin{aligned} 2x + 3y &= b_1 \\ 5x - 4y &= b_2 \\ -x + 4y &= b_3. \end{aligned}$$

11. How must you choose a, b, and c in order that $(1, 1, 1)$ will be a solution of this system?

$$\begin{aligned} ax_1 + 2x_2 + 3x_3 &= 6 \\ 3x_1 + bx_2 - 5x_3 &= 0 \\ 5x_1 - 4x_2 + cx_3 &= 7. \end{aligned}$$

Show that $(0, 1, 1)$ is not a solution of the system, no matter how a, b, and c are chosen.

12. Show that the system of equations

$$\begin{aligned} a_{11}x_1 + a_{12}x_2 &= b_1 \\ a_{21}x_1 + a_{22}x_2 &= b_2 \end{aligned}$$

has a unique solution if and only if $a_{11}a_{22} - a_{12}a_{21} \neq 0$.

13. For what values of k will each of the following systems fail to have a *unique* solution? Will they have *any* solutions in these cases? Obtain the solutions in all cases where they exist.

(a) $\begin{aligned} x + ky &= 1 \\ kx + y &= 1, \end{aligned}$ (b) $\begin{aligned} x + 2y - z &= 1 \\ 2x - 3y + z &= 2 \\ kx + 9ky - 4z &= 0, \end{aligned}$

(c) $\begin{aligned} x_1 + 2x_2 + 3x_3 &= 4 \\ 2x_1 + 3x_2 + 4x_3 &= 5 \\ 3x_1 + 4x_2 + 5x_3 &= 2k. \end{aligned}$

14. Under what conditions on t will these systems be consistent? What are the solutions in these cases?

(a) $\begin{aligned} 3x_1 + x_2 + x_3 &= t \\ x_1 - x_2 + 2x_3 &= 1 - t \\ x_1 + 3x_2 - 3x_3 &= 1 + t, \end{aligned}$ (b) $\begin{aligned} tx_1 + x_2 \qquad &= 0 \\ x_1 + tx_2 - x_3 &= 1 \\ - x_2 + tx_3 &= 0. \end{aligned}$

15. Given that a parametric solution of

$$\begin{aligned} a_{11}x_1 + a_{12}x_2 + a_{13}x_3 &= b_1 \\ a_{21}x_1 + a_{22}x_2 + a_{23}x_3 &= b_2 \end{aligned}$$

is

$$\begin{aligned} x_1 &= \alpha_1 t + \beta_1 \\ x_2 &= \alpha_2 t + \beta_2 \\ x_3 &= \alpha_3 t + \beta_3, \end{aligned}$$

show that a parametric solution of

$$\begin{aligned} a_{11}x_1 + a_{12}x_2 + a_{13}x_3 &= 0 \\ a_{21}x_1 + a_{22}x_2 + a_{23}x_3 &= 0 \end{aligned}$$

is

$$\begin{aligned} x_1 &= \alpha_1 t \\ x_2 &= \alpha_2 t \\ x_3 &= \alpha_3 t \end{aligned}$$

and that $(\beta_1, \beta_2, \beta_3)$ is a solution of the first system.

16. Illustrate Exercise 15 by comparing the solutions of

(a) $\begin{aligned} x_1 + 2x_2 - x_3 &= 2 \\ 2x_1 + x_2 + x_3 &= 4 \end{aligned}$ and (b) $\begin{aligned} x_1 + 2x_2 - x_3 &= 0 \\ 2x_1 + x_2 + x_3 &= 0. \end{aligned}$

17. The parametric solution (1.2.8) of (1.2.4) has only integer coefficients. Can you discover by what means a and b were chosen in the expression $x_3 = at + b$ so as to cause this to happen? Are other parametric representations with integer coefficients possible? Can you tell how to find them all? Given two consistent equations in three unknowns, with integer coefficients, is a parametric representation of the solution having only integer coefficients always possible? If your answer is no, given an example that proves your claim. If your answer is yes, prove it.

1.4 EQUIVALENT SYSTEMS OF EQUATIONS

The examples of Section 1.2 appealed to previous experience and were designed to illustrate the basic manipulations used in solving a given system of linear equations by elimination. It is time to make the ideas involved precise.

Consider any system of m linear equations in n unknowns:

$$\begin{aligned} a_{11}x_1 + a_{12}x_2 + \cdots + a_{1n}x_n - b_1 &= 0 \\ a_{21}x_1 + a_{22}x_2 + \cdots + a_{2n}x_n - b_2 &= 0 \\ &\vdots \\ a_{m1}x_1 + a_{m2}x_2 + \cdots + a_{mn}x_n - b_m &= 0, \end{aligned} \tag{1.4.1}$$

where the coefficients a_{ij} and the constant terms b_i, $i = 1, 2, \ldots, m$; $j = 1, 2, \ldots, n$, are real or complex numbers.

Any set of numerical values of $x_1, x_2, \ldots, x_n$ which simultaneously satisfy these equations is called a **particular solution** of the system. Any set of expressions [for example (1.2.6)] which yields all solutions and only solutions of a given system of equations is called a **complete solution** of that system.

Two systems of equations in the variables $x_1, x_2, \ldots, x_n$ are said to be **equivalent** if and only if every particular solution of either one is also a solution of the other. The process of "solving" a system of equations amounts to deducing from the given system an equivalent system of a prescribed form. Thus (1.2.2) and (1.2.3) are equivalent, as are (1.2.4), (1.2.5), and (1.2.6).

In finding the solutions (1.2.3) and (1.2.5), we performed operations that fall into three categories:

(a) The interchange of two equations of a system.
(b) The multiplication of an equation by an arbitrary nonzero constant.
(c) The addition of an arbitrary multiple of one equation of a system to another equation of the system.

For the proof of the validity of the usual procedure for solving systems of equations, it is only necessary to show that any such operation, applied to a given system, always leads to an equivalent system.

For the purpose of providing this proof, it is convenient to abbreviate equations (1.4.1) thus:

$$
\begin{aligned}
f_1 &= 0\\
&\vdots\\
f_i &= 0\\
&\vdots\\
f_j &= 0\\
&\vdots\\
f_m &= 0,
\end{aligned}
\tag{1.4.2}
$$

where f_i denotes the expression which constitutes the left member of the ith equation; that is,

$$f_i = a_{i1}x_1 + a_{i2}x_2 + \cdots + a_{in}x_n - b_i.$$

In this notation, replacing $f_i = 0$ by

$c_i f_i = 0 \quad (c_i \neq 0)$ (multiplication of the ith equation by a nonzero constant)

or by

$f_i + c_j f_j = 0 \quad (j \neq i)$ (addition of c_j times the jth equation to the ith equation)

is how one accomplishes operations of the types (b) and (c). The result of substituting a set of values $(a_1, a_2, \ldots, a_n)$ into the expression f_k is $f_k(a_1, a_2, \ldots, a_n)$ and $(a_1, a_2, \ldots, a_n)$ is a solution of (1.4.2) if and only if $f_k(a_1, a_2, \ldots, a_n) = 0$ for $k = 1, 2, \ldots, m$.

The effects of the operations (a), (b), and (c) on the system (1.4.2) may be represented, respectively, by the three systems

$$
\text{(a)}\ \begin{cases}
f_1 = 0\\
\vdots\\
f_j = 0\\
\vdots\\
f_i = 0\\
\vdots\\
f_m = 0,
\end{cases}
\quad
\text{(b)}\ \begin{cases}
f_1 = 0\\
\vdots\\
c_i f_i = 0 \quad (c_i \neq 0)\\
\vdots\\
f_j = 0\\
\vdots\\
f_m = 0,
\end{cases}
\quad
\text{(c)}\ \begin{cases}
f_1 = 0\\
\vdots\\
f_i + c_j f_j = 0 \quad (j \neq i)\\
\vdots\\
f_j = 0\\
\vdots\\
f_m = 0.
\end{cases}
\tag{1.4.3}
$$

Now, suppose we know any solution $(a_1, a_2, \ldots, a_n)$ of (1.4.2). Then we have $f_k(a_1, a_2, \ldots, a_n) = 0$, $k = 1, 2, \ldots, m$. This implies that every equation of the three systems (a), (b), (c) is also satisfied by $(a_1, a_2, \ldots, a_n)$. Conversely, if $(a_1, a_2, \ldots, a_n)$ satisfies each equation of system (a), then each equation of system (1.4.2) is satisfied. If it satisfies each equation of system (b), then, because $c_i \neq 0$, $f_i(a_1, a_2, \ldots, a_n) = 0$, so again each equation of (1.4.2) is satisfied. If it satisfies each equation of system (c), then, because $f_j(a_1, a_2, \ldots, a_n) = 0$ and $f_i(a_1, a_2, \ldots, a_n) + c_j f_j(a_1, a_2, \ldots, a_n) = 0$, it follows that $f_i(a_1, a_2, \ldots, a_n) = 0$, so each equation of (1.4.2) is satisfied. This shows that each of the systems (a), (b), and (c) is equivalent to (1.4.2). We have therefore proved

Theorem 1.4.1: *Interchanging two equations of a system, multiplying an equation of a system by a nonzero constant, and adding any constant multiple of one equation of a system to another equation of the same system all lead to an equivalent system of equations.*

Now, by the definition of equivalence of systems of equations, if a system A is equivalent to B, and B to C, then A must be equivalent to C. Hence repeated applications of these three basic operations necessarily also lead to equivalent systems of equations. Either by this observation or by a direct proof, in the manner of the proof of the preceding theorem, one can establish

Corollary 1.4.2: *The system of linear equations* (1.4.2) *is equivalent to the following system of equations in which the coefficients* $c_1, c_2, \ldots, c_m$ *are arbitrary except that the particular coefficient* c_i *must not be zero:*

$$\begin{aligned} f_1 &= 0 \\ f_2 &= 0 \\ &\vdots \\ f_{i-1} &= 0 \\ c_1f_1 + c_2f_2 + \cdots + c_{i-1}f_{i-1} + c_if_i + c_{i+1}f_{i+1} + \cdots + c_mf_m &= 0 \quad (c_i \neq 0) \\ f_{i+1} &= 0 \\ &\vdots \\ f_m &= 0. \end{aligned} \tag{1.4.4}$$

That is, we may replace the ith equation by any nonzero multiple thereof plus arbitrary multiples of the remaining equations. Any or all of the coefficients other than c_i may be zero.

If we can choose the coefficients $c_1, c_2, \ldots, c_m(c_i \neq 0)$ so that $\sum_{j=1}^{m} c_jf_j$ is identically zero in $x_1, x_2, \ldots, x_n$, then we will have reduced the given system to an equivalent system with one less equation in it. This is often a useful device.

In the same way, one can prove

Corollary 1.4.3: *If* $i_1, i_2, \ldots, i_m$ *is any permutation of the integers* $1, 2, \ldots, m$, *then the system* (1.4.2) *is equivalent to the system*

$$\begin{aligned} f_{i_1} &= 0 \\ f_{i_2} &= 0 \\ &\vdots \\ f_{i_m} &= 0. \end{aligned} \tag{1.4.5}$$

That is, one may rearrange the equations of a given system in any order one finds convenient.

These two corollaries justify many of the clever tricks one commonly uses in solving particularly simple systems of equations.

1.5 THE ECHELON FORM FOR SYSTEMS OF EQUATIONS

We now describe formally the systematic elimination process used in Example 3 of Section 1.2. Consider an arbitrary system of equations represented in the form (1.5.1):

$$\begin{aligned} a_{11}x_1 + a_{12}x_2 + \cdots + a_{1n}x_n &= b_1 \\ a_{21}x_1 + a_{22}x_2 + \cdots + a_{2n}x_n &= b_2 \\ &\vdots \\ a_{m1}x_1 + a_{m2}x_2 + \cdots + a_{mn}x_n &= b_m. \end{aligned} \tag{1.5.1}$$

We rearrange the order of the equations, if necessary, so that the variable x_1 appears with a nonzero coefficient in the first position of the first equation. Then we divide the first equation by the coefficient of x_1 and eliminate x_1 from equations 2 through m by the sweepout process.

Now let x_{i_2} be the first variable (usually x_2) actually appearing in equations 2 through m, and again rearrange the order of the equations, if necessary, so that x_{i_2} appears in the second row. Then we divide the second equation by the coefficient of x_{i_2} and eliminate x_{i_2} from equations $1, 3, 4, \ldots, m$.

Continuing in this fashion, *we eventually stop because we have swept columns corresponding to m of the variables or because we have no more variable terms in the remaining equations.*

We illustrate this observation with an example:

$$\begin{aligned} x_1 - 2x_2 + x_3 - x_4 &= -1 \\ 2x_1 - 4x_2 + 3x_3 - 3x_4 &= 4 \\ -x_1 + 2x_2 + x_3 - x_4 &= b_3. \end{aligned}$$

First we use x_1 to sweep the first column:

$$\begin{aligned} x_1 - 2x_2 + x_3 - x_4 &= -1 \\ x_3 - x_4 &= 6 \\ 2x_3 - 2x_4 &= b_3 - 1. \end{aligned}$$

Because of the proportionality of the first two columns of coefficients, this eliminates x_2 also. Hence we must go next to x_3 and use it to sweep the third column:

$$\begin{aligned} x_1 - 2x_2 \qquad\qquad &= -7 \\ x_3 - x_4 &= 6 \\ 0 &= b_3 - 13. \end{aligned}$$

This eliminates x_4 also from the third equation.

If now $b_3 \neq 13$, the equations imply a contradiction and are inconsistent. On the other hand, if $b_3 = 13$, the final equation is $0 = 0$, the system is consistent, and we can write the complete solution thus:

$$\begin{aligned} x_1 &= 2x_2 - 7 \\ x_3 &= x_4 + 6. \end{aligned}$$

From these last two equations, by assigning values to x_2 and x_4, then computing x_1 and x_3, one can obtain any particular solution whatsoever of the original system.

The example illustrates the fact that in sweeping a column, we may well eliminate more than one variable. It also illustrates the fact that we may eliminate all variables from some of the equations. Because of this, although there were three equations, we were able to employ only two variables for the purpose of sweeping columns. If there are m equations, one can employ at most m variables for this purpose. When this is possible, one can express these m variables in terms of the remaining $n - m$ variables, just as we have done in previous examples.

A system of linear equations will be said to be in **echelon form** (pronounced esh′-uh-lon) if and only if

1. the first nonzero coefficient in each equation is 1, and
2. the number of leading zero coefficients in any equation beyond the first is greater than the number of leading zero coefficients in the preceding equation.

The system of equations is said to be in **reduced echelon form** if and only if, in addition to 1 and 2,

3. the variable appearing in the leading term of any equation appears in no other equation.

The above-described computational process is the process of *deriving a system of equations in reduced echelon form equivalent to a given system.*

The process can be programmed for a computer just as described and is often called, after its inventors, **Gauss–Jordan elimination.**

Let $x_1, x_{i_2}, \ldots, x_{i_r}$ denote the variables used for sweeping columns and let $x_{i_{r+1}}, x_{i_{r+2}}, \ldots, x_{i_n}$ denote the remaining variables. For purposes of further discussion, it is convenient to group all terms containing these latter variables at the right. We thus obtain, from the reduced echelon form, the following system of equations, which, because of the operations performed to obtain it, is equivalent to the original system:

$$\begin{array}{lllllll} x_1 & & & + \alpha_{1,r+1}x_{i_{r+1}} + \alpha_{1,r+2}x_{i_{r+2}} + \cdots + \alpha_{1,n}x_{i_n} & = & \beta_1 \\ & x_{i_2} & & + \alpha_{2,r+1}x_{i_{r+1}} + \alpha_{2,r+2}x_{i_{r+2}} + \cdots + \alpha_{2,n}x_{i_n} & = & \beta_2 \\ & & & & & \vdots \\ & & x_{i_r} & + \alpha_{r,r+1}x_{i_{r+1}} + \alpha_{r,r+2}x_{i_{r+2}} + \cdots + \alpha_{r,n}x_{i_n} & = & \beta_r \\ & & & & 0 = & \beta_{r+1} \\ & & & & & \vdots \\ & & & & 0 = & \beta_m. \end{array} \tag{1.5.2}$$

Necessarily, $r \leq m$. If we have employed m of the variables to sweep columns, $r = m$ and there are no equations of the form $0 = \beta_j$. The quantities $\beta_{r+1}, \ldots, \beta_m$ may or may not all be zero in the event that $r < m$.

If at least one of the quantities $\beta_{r+1}, \beta_{r+2}, \ldots, \beta_m$ is *not* zero, then the given equations imply a contradiction and hence are inconsistent. If all the β's just named are zero, then the equations are consistent and we can solve for $x_1, x_{i_2}, \ldots, x_{i_r}$ by transposing all other terms to the right, thereby obtaining what we have called the "complete solution" of the system. *Thus the same procedure gives us a practical test for consistency and also gives the complete solution, when there is one.* The example with which we began illustrates both situations.

In a given case, when something clever can be done, one may find it useful to depart from the strict algorithm outlined here and solve for other variables, rearrange the equations, and so on, always still adhering to the basic pattern of sweeping columns.

1.6 SYNTHETIC ELIMINATION

In solving a system of linear equations, we actually operate only on the coefficients. The variables simply serve to keep the coefficients properly aligned in columns. If we copy the coefficients in columns and keep them carefully aligned, we don't need to copy the variables at all and hence can save a great deal of needless writing. The process is called **synthetic elimination** or the **method of detached coefficients**, because no variables appear in it.

For example, suppose we want to solve, by the sweepout process, the system of equations

$$\begin{aligned} x_1 + x_2 - x_3 + x_4 &= 3 \\ x_1 \qquad + x_3 + x_4 &= 2 \\ x_2 + x_3 - x_4 &= -1 \\ 2x_1 - x_2 + x_3 \qquad &= 2. \end{aligned}$$

We copy down the coefficients, taking care to locate all zero coefficients in their proper places and then operate on rows of coefficients as though they were the equations they represent, following closely the rules for obtaining the reduced echelon form, but not recopying equations needlessly.

In Table 1.1, R_1, R_2, R_3, and R_4 identify the rows that represent the given equations. The extra column, headed Sum, is used for checking purposes only. The sum entry in any row is the sum of the coefficients and the constant term recorded in that same row. In operating on the rows, we perform the same operations on the sum entries also. Thus when we "subtract row 1 from row 2" to get the fifth row ($R_2' = R_2 - R_1$), we subtract every entry of row 1 from the corresponding entry of row 2, and also subtract the sum entry of row 1 from that of row 2. The resulting entry in the sum column and row R_2' will agree with

TABLE 1.1
EXAMPLE OF SYNTHETIC ELIMINATION

Key to Operations	Coefficients				Constant Terms	Sum (for checking)
R_1	1	1	−1	1	3	5
R_2	1	0	1	1	2	5
R_3	0	1	1	−1	−1	0
R_4	2	−1	1	0	2	4
$R_2' = R_2 - R_1$	0	−1	2	0	−1	0
$R_4' = R_4 - 2R_1$	0	−3	3	−2	−4	−6
$R_1'' = R_1 - R_2''$	1	0	1	1	2	5
$*R_2'' = \quad - R_2$	0	1	−2	0	1	0
$R_3'' = R_3 - R_2''$	0	0	3	−1	−2	0
$R_4'' = R_4' + 3R_2''$	0	0	−3	−2	−1	−6
$R_1''' = R_1'' - R_3'''$	1	0	0	$\frac{4}{3}$	$\frac{8}{3}$	5
$R_2''' = R_2'' + 2R_3'''$	0	1	0	$-\frac{2}{3}$	$-\frac{1}{3}$	0
$*R_3''' = \quad \frac{1}{3}R_3''$	0	0	1	$-\frac{1}{3}$	$-\frac{2}{3}$	0
$R_4''' = R_4'' + 3R_3'''$	0	0	0	−3	−3	−6
$R_1^{(iv)} = R_1''' - \frac{4}{3}R_4^{(iv)}$	1	0	0	0	$\frac{4}{3}$	$\frac{7}{3}$
$R_2^{(iv)} = R_2''' + \frac{2}{3}R_4^{(iv)}$	0	1	0	0	$\frac{1}{3}$	$\frac{4}{3}$
$R_3^{(iv)} = R_3''' + \frac{1}{3}R_4^{(iv)}$	0	0	1	0	$-\frac{1}{3}$	$\frac{2}{3}$
$*R_4^{(iv)} = \quad -\frac{1}{3}R_4'''$	0	0	0	1	1	2

the sum of the coefficients and the constant term in row R_2' if no errors have been made. Thus the sum column provides a step-by-step check on accuracy, unless one makes compensating errors, of course.

Sweeping out the x_1-column, we need alter only rows 2 and 4, so we do not recopy rows 1 and 3. R_2' and R_4' denote the new second and fourth rows: $R_2' = R_2 - R_1$ and $R_4' = R_4 - 2R_1$.

Now we compute $*R_2'' = -R_2'$ in the next set of rows, then use it to compute R_1'', R_3'', and R_4'' in the fashion indicated in the leftmost column of the table. This amounts to sweeping the x_2-column in the standard way. $*R_2''$ is marked with an asterisk to indicate that it is the first row computed in this set.

Next we compute $*R_3'''$ (marked with an asterisk) and use it to get R_1''', R_2''', and R_4''' by the indicated computations, thus sweeping the x_3-column.

Finally, we compute $*R_4^{(iv)}$ and sweep the x_4-column with the result. This yields the last four rows of the table, from which, by recalling the meanings of the locations of the row entries, we can read off the solution:

$$x_1 = \frac{4}{3}$$
$$x_2 = \frac{1}{3}$$
$$x_3 = -\frac{1}{3}$$
$$x_4 = 1.$$

The successive major stages of the computation are indicated by the primes on the R's. When a row at one such stage is the same as the corresponding row at the previous stage, it need not be recopied. This explains why only two rows appear at the second level. There is, of course, no objection to writing a full set of rows at each level. Except in simple cases, full sets would occur anyhow.

When the number of variables is small and the coefficients are simple, solving by the usual elimination procedure is often easier than using this tabular array. In any other case, the tabular array is economical and helpful. Moreover, it is closely similar to the computer procedure used in solving such problems.

The method of synthetic elimination works equally well when the solution is not unique and it will also detect inconsistent systems of equations. One need only take careful note of the meanings of the entries in the table in order to apply the procedure in these cases. We illustrate with two examples.

Consider first the system of equations

$$\begin{aligned} 2x_1 - x_2 + 3x_3 &= 4 \\ 4x_1 + 3x_2 + 2x_3 &= 9 \\ -x_1 + x_2 + 4x_3 &= 4 \\ x_1 + 5x_2 + 3x_3 &= 2. \end{aligned}$$

We proceed in Table 1.2, this time dispensing with the sum column for the sake of brevity. Also, we copy down the rows of coefficients in a more convenient order.

The procedure need not be continued to the very end, for the last two lines of the table assert respectively that $\frac{59}{6}x_3 = 11$ and $\frac{59}{6}x_3 = 18$, so the equations are clearly inconsistent.

As a final example, consider the system of three equations in four unknowns:

$$\begin{aligned} x_1 + x_2 - x_3 + x_4 &= 2 \\ 2x_1 - x_2 + x_3 - x_4 &= 1 \\ x_1 + 4x_2 - 4x_3 + 4x_4 &= 5. \end{aligned}$$

TABLE 1.2
AN INCONSISTENT SYSTEM

Key to Operations	Coefficients			Constant Terms
R_1	1	5	3	2
R_2	-1	1	4	4
R_3	2	-1	3	4
R_4	4	3	2	9
$R_2' = R_2 + R_1$	0	6	7	6
$R_3' = R_3 - 2R_1$	0	-11	-3	0
$R_4' = R_4 - 4R_1$	0	-17	-10	1
$R_1'' = R_1 - 5R_2''$	1	0	$-\frac{17}{6}$	-3
$R_2'' = \frac{1}{6}R_2'$	0	1	$\frac{7}{6}$	1
$R_3'' = R_3' + 11R_2''$	0	0	$\frac{59}{6}$	11
$R_4'' = R_4' + 17R_2''$	0	0	$\frac{59}{6}$	18

TABLE 1.3
A SYSTEM WITH INFINITELY MANY SOLUTIONS

Key to Operations	Coefficients				Constant Terms
R_1	1	1	−1	1	2
R_2	2	−1	1	−1	1
R_3	1	4	−4	4	5
$R_2' = R_2 - 2R_1$	0	−3	3	−3	−3
$R_3' = R_3 - R_1$	0	3	−3	3	3
$R_1'' = R_1 - R_2''$	1	0	0	0	1
$*R_2'' = -\frac{1}{3}R_2'$	0	1	−1	1	1
$R_3'' = R_3' - 3R_2''$	0	0	0	0	0

Here the arithmetic proceeds as shown in Table 1.3. From the last two nonzero rows of Table 1.3 we read the equations

$$x_1 = 1$$
$$x_2 = x_3 - x_4 + 1.$$

That is, x_3 and x_4 may be chosen arbitrarily and their choice determines x_2, but x_1 is inevitably 1.

If we put $x_3 = t$, $x_4 = s$, we may write the solution in parametric form thus:

$$\begin{aligned} x_1 &= \qquad\quad 1 \\ x_2 &= t - s + 1 \\ x_3 &= t \\ x_4 &= \qquad s. \end{aligned}$$

The parametric form of the solution is often particularly useful.

1.7 SYSTEMS OF HOMOGENEOUS LINEAR EQUATIONS

A system of linear equations of the form

$$\begin{aligned} a_{11}x_1 + a_{12}x_2 + \cdots + a_{1n}x_n &= 0 \\ a_{21}x_1 + a_{22}x_2 + \cdots + a_{2n}x_n &= 0 \\ &\vdots \\ a_{m1}x_1 + a_{m2}x_2 + \cdots + a_{mn}x_n &= 0, \end{aligned} \tag{1.7.1}$$

in which the constant terms are all zero, is called a **system of homogeneous linear equations**. Since $(0, 0, \ldots, 0)$ is a solution of every such system, *all* *systems of homogeneous linear equations are consistent.* The solution $(0, 0, \ldots, 0)$ is called the **trivial solution** because it is inevitably present.

A solution *not* consisting entirely of 0's is called **nontrivial**. The problems of interest are to determine whether or not nontrivial solutions exist and to find them if they do. This is readily accomplished by transforming the system into reduced echelon form. If there are n variables and if the reduced echelon form has $r < n$ nonzero equations, then we can solve for r variables in terms of the remaining $n - r$ variables. By assigning a nonzero value to at least one of these $n - r$ independent variables, we can obtain a particular nontrivial solution. Moreover, since the final system is equivalent to the original system, all solutions are obtainable by assigning values to these $n - r$ variables and every such assignment yields a solution.

If $r = n$, since the constant terms are all 0, the original system is equivalent to

$$\begin{aligned} x_1 \qquad\qquad &= 0 \\ x_2 \qquad &= 0 \\ &\vdots \\ x_n &= 0, \end{aligned}$$

so the trivial solution is the *only* solution.

We illustrate with two examples:

$$\text{(a)}\quad \begin{aligned} x_1 + x_2 + x_3 &= 0 \\ 3x_1 + 2x_2 - 4x_3 &= 0 \\ 2x_1 - x_2 + x_3 &= 0. \end{aligned}$$

The computation

R_1:	1	1	1	0
R_2:	3	2	−4	0
R_3:	2	−1	1	0
R_2':	0	−1	−7	0
R_3':	0	−3	−1	0
R_3'':	0	0	20	0

shows that $x_3 = 0$, so $x_2 = 0$ and $x_1 = 0$ also; that is, the only solution is the trivial solution.

$$\text{(b)}\quad \begin{aligned} x_1 - x_2 + x_3 &= 0 \\ 2x_1 + 3x_2 - x_3 &= 0 \\ x_1 + 4x_2 - 2x_3 &= 0 \\ x_1 + 14x_2 - 8x_3 &= 0. \end{aligned}$$

In this case, the computation proceeds as follows:

R_1:	1	-1	1	0
R_2:	2	3	-1	0
R_3:	1	4	-2	0
R_4:	1	14	-8	0
R_2':	0	5	-3	0
R_3':	0	5	-3	0
R_4':	0	15	-9	0
R_1'':	1	0	$\frac{2}{5}$	0
R_2'':	0	1	$-\frac{3}{5}$	0

Since R_3'' and R_4'' consist of all 0's, they have been omitted.

From this we have the complete solution

$$\begin{aligned} x_1 &= -\tfrac{2}{5}x_3 \\ x_2 &= \tfrac{3}{5}x_3 \end{aligned} \qquad (n = 3, r = 2, n - r = 1).$$

Let us put $x_3 = 5t$ so that the solution becomes

$$\begin{aligned} x_1 &= -2t \\ x_2 &= 3t \\ x_3 &= 5t \end{aligned}$$

in parametric form. In this case there are infinitely many nontrivial solutions, each of which is obtainable by assigning an appropriate nonzero value to t.

1.8 EXERCISES

1. Solve each system of equations by deriving the equivalent reduced echelon form, or else show that it is inconsistent.

(a) $$\begin{aligned} x_1 - x_2 + x_3 &= 4 \\ 2x_1 + x_2 - 3x_3 &= -3 \\ -3x_1 + 2x_2 + x_3 &= -6, \end{aligned}$$

(b) $$\begin{aligned} 2x_1 - 3x_2 + x_3 - x_4 &= -1 \\ 4x_1 - 6x_2 + 3x_3 + 2x_4 &= 3, \end{aligned}$$

(c) $$\begin{aligned} 2x_1 + x_2 &= 5 \\ x_1 - 3x_2 &= -1 \\ 3x_1 + 4x_2 &= 6, \end{aligned}$$

(d) $$\begin{aligned} x_1 - 2x_2 + x_3 &= 2 \\ 2x_1 - x_2 - x_3 &= 7 \\ 4x_1 - 2x_2 + x_3 &= 0. \end{aligned}$$

2. Solve by the method of synthetic elimination:

(a) $$\begin{aligned} 2x_1 - x_2 + x_3 &= 1 \\ 3x_1 - 2x_2 + x_3 &= 0 \\ 5x_1 + x_2 + 2x_3 &= 9, \end{aligned}$$

(b) $$\begin{aligned} x_1 - x_2 + x_3 &= 0 \\ 2x_1 - 3x_2 + 4x_3 &= 0 \\ 3x_1 + x_2 - 5x_3 &= 0, \end{aligned}$$

(c)
$$\begin{aligned} x_1 - x_2 + 2x_3 + x_4 &= 1 \\ 2x_1 - x_2 - x_3 + 3x_4 &= 2 \\ -3x_1 + 2x_2 - x_3 - 4x_4 &= -3, \end{aligned}$$

(d)
$$\begin{aligned} x_1 + 2x_2 \qquad\qquad &= 3 \\ 2x_2 + 3x_3 \qquad &= 5 \\ 3x_3 + 4x_4 &= 7 \\ x_1 \qquad\qquad + 4x_4 &= 5. \end{aligned}$$

3. For what choice of k will the system

$$\begin{aligned} 2x_1 + x_2 &= 5 \\ x_1 - 3x_2 &= -1 \\ 3x_1 + 4x_2 &= k \end{aligned}$$

be consistent?

4. For what pairs of variables can one solve this system of equations?

$$\begin{aligned} x_1 + 2x_2 - 3x_3 + 6x_4 &= 6 \\ 2x_1 + 4x_2 + 2x_3 - 4x_4 &= 4. \end{aligned}$$

5. Find the complete solution of the system

$$\begin{aligned} x_1 + 2x_2 + 3x_3 + 6x_4 &= 0 \\ x_1 - 2x_2 + x_3 - 2x_4 &= 0 \end{aligned}$$

and represent the solution in parametric form.

6. Compare the solution found in Exercise 5 with the parametric form of the solution of

$$\begin{aligned} x_1 + 2x_2 + 3x_3 + 6x_4 &= 12 \\ x_1 - 2x_2 + x_3 - 2x_4 &= -2. \end{aligned}$$

What general theorem does this suggest?

7. For what values of k will the system

$$\begin{aligned} x_1 + x_2 + kx_3 &= 0 \\ x_1 + kx_2 + x_3 &= 0 \\ kx_1 + x_2 + x_3 &= 0 \end{aligned}$$

have nontrivial solutions? What are these solutions in each case?

8. Show that any one equation of a system (1.4.1) may be replaced by the sum of this equation and any subset of the remaining equations, the result being a system equivalent to (1.4.1). Use this fact to solve for x_1 by inspection, where

$$\begin{aligned} x_1 + 2x_2 - x_3 &= 3 \\ 2x_1 - x_2 + 3x_3 &= 2 \\ 3x_1 - x_2 - 2x_3 &= 1. \end{aligned}$$

9. Under what conditions on m, n, r will the reduced echelon form indicate that a system of equations has a unique solution?

10. Under what conditions on the coefficients will a system of equations of this form have a unique solution?

$$\begin{aligned} a_{11}x_1 + a_{12}x_2 \qquad\qquad\qquad\qquad &= b_1 \\ a_{22}x_2 + a_{23}x_3 \qquad\qquad &= b_2 \\ a_{33}x_3 + a_{34}x_4 &= b_3 \\ a_{44}x_4 &= b_4. \end{aligned}$$

What is this solution? (Many systems of equations appearing in applications have large blocks of coefficients that are zero.)

11. Use a desk calculator to solve this system:

$$\begin{aligned} x_1 + 0.43210x_2 + 0.61257x_3 &= 1 \\ 0.43210x_1 + \qquad\quad x_2 + 0.94761x_3 &= 2 \\ 0.61257x_1 + 0.94761x_2 + \qquad\quad x_3 &= 3. \end{aligned}$$

12. Solve this system of equations:

$$\begin{aligned} x_1 + (1 - i)x_2 + \qquad\quad ix_3 &= 1 \\ (1 + i)x_1 + \qquad x_2 + (1 + i)x_3 &= 1 + i \\ -ix_1 + (1 - i)x_2 + \qquad\quad x_3 &= 4 + 3i. \end{aligned}$$

1.9 NUMBER FIELDS

All coefficients in systems of equations treated in preceding sections have been assumed to represent real or complex numbers. Real, rational, or complex coefficients have always led, respectively, to real, rational, or complex solutions or expressions that define the set of all solutions. To examine carefully the relation between the nature of the coefficients and the nature of the solutions we need the concept of a number field.

A nonempty collection or set $\mathscr{F}$ of real or complex numbers that does not consist of the number zero alone will be called a **number field** if and only if the sum, difference, product, and quotient of any two numbers of $\mathscr{F}$ are again numbers of $\mathscr{F}$, division by zero being excepted. According to this definition, each of the following familiar sets of numbers constitutes a field:

1. The set of all rational numbers, that is, the set of all quotients of the form a/b, where a and b are integers but $b \neq 0$.
2. The set of all real numbers.
3. The set of all complex numbers, that is, the set of all numbers of the form $a + bi$, where a and b are real and $i^2 = -1$.

These fields are called, respectively, the **rational number field**, the **real number field**, and the **complex number field**. They are the most important number fields as far as applications are concerned. Most statistical work employs the real and rational number fields. Much work in physics and engineering employs the complex number field.

The operations of addition, subtraction, multiplication, and division involved in the definition of a field are known as the **four rational operations**.

Whenever the numbers of one field are all members of another field, the first field is called a **subfield** of the second. In the case of the three fields just listed, the rational field is a subfield of the real field, and the real field is a subfield of the field of complex numbers. It is customary to regard any field as a subfield of itself.

Since every number field $\mathscr{F}$ contains $a - a$ for each number a in $\mathscr{F}$, it contains the number 0. Since it contains a/a for some $a \neq 0$, it also contains 1. Since it contains 1, it contains $0 - 1 = -1$, $1 + 1 = 2$, $0 - 2 = -2$, $2 + 1 = 3$, $0 - 3 = -3, \ldots$; that is, it contains all integers. Since it contains all integers and hence all quotients a/b, where a and b are integers ($b \neq 0$), it contains all rational numbers. Thus *the rational number field is the smallest possible number field and is a subfield of every number field.*

The concept of what a number field is may be clarified by examples of sets of numbers that are *not* fields. The set $\mathscr{Z}$ of all integers is not a number field because the quotient of one integer by another integer (not zero) is not always an integer, that is, it is not always a number of $\mathscr{Z}$. For example, $\frac{3}{5}$ is not an integer. The set of all nonnegative real numbers is not a field because not every difference of two such numbers, for example, $\pi - 13$, is a nonnegative real number. The set of all complex numbers of the form bi, where b is real, that is, the set of pure imaginary numbers, is not a field because the product of two of these is not a pure imaginary number: $bi \cdot ci = -bc$, which is real.

It is worthwhile to note that there are number fields other than the three familiar ones just mentioned. For example, the set of all numbers of the form $a + b\sqrt{2}$, where a and b are rational numbers, is a number field. In fact, if $\alpha + \beta\sqrt{2}$ and $\gamma + \delta\sqrt{2}$ are any two numbers of this kind, we have

$$(\alpha + \beta\sqrt{2}) \pm (\gamma + \delta\sqrt{2}) = (\alpha \pm \gamma) + (\beta \pm \delta)\sqrt{2}$$

$$(\alpha + \beta\sqrt{2}) \cdot (\gamma + \delta\sqrt{2}) = (\alpha\gamma + 2\beta\delta) + (\beta\gamma + \alpha\delta)\sqrt{2}.$$

Now let $\gamma + \delta\sqrt{2} \neq 0$. Because $\sqrt{2}$ is not rational and $\gamma + \delta\sqrt{2} \neq 0$, we have $\gamma - \delta\sqrt{2} \neq 0$ and $\gamma^2 - 2\delta^2 \neq 0$. Hence

$$\frac{\alpha + \beta\sqrt{2}}{\gamma + \delta\sqrt{2}} = \frac{(\alpha + \beta\sqrt{2})(\gamma - \delta\sqrt{2})}{(\gamma + \delta\sqrt{2})(\gamma - \delta\sqrt{2})} = \frac{\alpha\gamma - 2\beta\delta}{\gamma^2 - 2\delta^2} + \frac{\beta\gamma - \alpha\delta}{\gamma^2 - 2\delta^2}\sqrt{2}.$$

Since the set of rational numbers itself forms a number field, the expression on the right is of the form $a + b\sqrt{2}$, with a and b rational numbers. We have thus shown that, when the four rational operations are applied to any two numbers of the form $a + b\sqrt{2}$ with a and b rational, the result is a number of the same kind. Hence the set of all such numbers constitutes a field.

There are, of course, infinitely many other number fields of various types, but in this book our attention is directed almost exclusively to the three fields defined in 1, 2, and 3 above.

Now consider a system of linear equations whose coefficients and constant terms all belong to a particular number field $\mathscr{F}$. Such a system is called a **system of linear equations over** $\mathscr{F}$. The sweepout process involves only applying certain of the four rational operations to the coefficients and constant terms at any stage. When a constant multiplier or divisor is used, it is one of the coefficients. Hence every operation leads to a system of equations not only equivalent to the first but also over $\mathscr{F}$. Thus, if there exists a unique solution, it will be over $\mathscr{F}$. When there are infinitely many solutions, the complete solution obtained by sweepout will be over $\mathscr{F}$. One could, possibly, assign values from other fields to the independent variables in this case, thus obtaining solutions over other number fields than $\mathscr{F}$. This possibility is of no importance in this book.

There are other fields than number fields: finite fields, rational function fields, fields of matrices, and so on. Many of the operations we develop and many of the conclusions we reach are valid in such fields; others are not. We leave a detailed discussion of such matters to more advanced books on linear algebra.

1.10 EXERCISES

1. Show that if α $(\neq 0)$ belongs to a number field $\mathscr{F}$, so do $n\alpha$ and α^n, where n is any integer.

2. Show that the set of all complex numbers $a + bi$, where a and b are real and where $i^2 = -1$, actually constitutes a number field, as stated in the text.

3. Show that the set of all numbers of the form $a + b\sqrt{p}$, where a and b are arbitrary rational numbers and where p is a prime number, constitutes a number field.

4. Show that if a, b, c belong to a number field $\mathscr{F}$, if $a \neq 0$, and if $ax + b = c$, then x also belongs to $\mathscr{F}$.

5. Show that if $a_1, b_1, c_1, a_2, b_2, c_2$ belong to a number field $\mathscr{F}$, if $a_1b_2 - a_2b_1 \neq 0$, and if

$$a_1x + b_1y = c_1$$
$$a_2x + b_2y = c_2,$$

then x and y also belong to $\mathscr{F}$.

CHAPTER 2

Matrices

2.1 MATRICES

Consider again an arbitrary system of m equations in n unknowns:

$$\begin{aligned} a_{11}x_1 + a_{12}x_2 + \cdots + a_{1n}x_n &= b_1 \\ a_{21}x_1 + a_{22}x_2 + \cdots + a_{2n}x_n &= b_2 \\ &\vdots \\ a_{m1}x_1 + a_{m2}x_2 + \cdots + a_{mn}x_n &= b_m. \end{aligned} \tag{2.1.1}$$

The coefficients of the variables, the variables themselves, and the constant terms can be conveniently arranged in rectangular arrays:

$$\begin{bmatrix} a_{11} & a_{12} & \cdots & a_{1n} \\ a_{21} & a_{22} & \cdots & a_{2n} \\ \vdots & & & \\ a_{m1} & a_{m2} & \cdots & a_{mn} \end{bmatrix}, \quad \begin{bmatrix} x_1 \\ x_2 \\ \vdots \\ x_n \end{bmatrix}, \quad \begin{bmatrix} b_1 \\ b_2 \\ \vdots \\ b_m \end{bmatrix}. \tag{2.1.2}$$

Any such rectangular array of numbers, variables, or functions is called a **matrix**. We shall see that there are many ways to operate usefully with matrices.

Let

$$(2.1.3) \qquad A = \begin{bmatrix} a_{11} & a_{12} & \cdots & a_{1n} \\ a_{21} & a_{22} & \cdots & a_{2n} \\ \vdots \\ a_{m1} & a_{m2} & \cdots & a_{mn} \end{bmatrix}$$

denote any matrix with m rows and n columns. The expressions a_{ij} of this matrix are called its **elements** or its **entries**. If the elements all come from a common number field $\mathscr{F}$, the matrix is said to be a matrix over $\mathscr{F}$. The subscripts i and j of a_{ij} denote, respectively, the **row** and the **column** of A in which the element a_{ij} is located. The rows and columns are also called **lines** of the matrix.

If a matrix A has m rows and n columns, we say it is an $m \times n$ (read "m by n") matrix or is of **order** (m, n). When $m = n$, so that A is square, A is said to be of **order** n or to be an n**-square matrix**. When A is of order n, the elements $a_{11}, a_{22}, \ldots, a_{nn}$ constitute its **main diagonal**, and the elements $a_{n1}, a_{n-1,2}, \ldots, a_{2,n-1}, a_{1n}$ constitute its **secondary diagonal**. A matrix of the form $[a_1, a_2, \ldots, a_n]$ is called a **row vector** and one of the form

$$\begin{bmatrix} a_1 \\ a_2 \\ \vdots \\ a_n \end{bmatrix}$$

is called a **column vector**. A column vector will be called an n**-vector** when the number of components needs to be identified.

It is often convenient to abbreviate (2.1.3) thus:

$$A = [a_{ij}]_{m \times n}$$

where the expression on the right means "the m by n matrix whose elements are the a_{ij}'s." If the order is clearly understood, we write simply $A = [a_{ij}]$.

2.2 EQUALITY OF MATRICES

The reader will recall that two complex numbers $a + bi$ and $c + di$, where a, b, c, and d are real, are defined to be equal if and only if $a = c$ and $b = d$, and that two quadratic polynomials $a_1x^2 + a_2x + a_3$ and $b_1x^2 + b_2x + b_3$ are equal if and only if $a_1 = b_1$, $a_2 = b_2$, $a_3 = b_3$. There are many other instances in which two mathematical objects of the same kind are equal if and only if certain corresponding parts are equal. In analogous fashion, we define two matrices $[a_{ij}]_{m \times n}$ and $[b_{ij}]_{m \times n}$ to be equal if and only if $a_{ij} = b_{ij}$ for each pair of subscripts i and j, $1 \le i \le m$, $1 \le j \le n$. That is, *two matrices are equal if and only if they have the same order and have equal corresponding elements throughout.*

To illustrate,

$$\begin{bmatrix} 1 & 1-t \\ t & 1+t \end{bmatrix} = \begin{bmatrix} u & 3+t \\ v-3 & w \end{bmatrix}$$

if and only if

$$1 = u, \qquad 1 - t = 3 + t,$$
$$t = v - 3, \qquad 1 + t = w,$$

that is, if and only if

$$u = 1, \qquad t = -1,$$
$$v = 2, \qquad w = 0.$$

2.3 ADDITION OF MATRICES

The reader will recall that complex numbers are added by adding their corresponding parts:

$$(a + bi) + (c + di) = (a + c) + (b + d)i.$$

Also, polynomials of the same degree are added by adding corresponding coefficients, for example,

$$(a_1x^2 + a_2x + a_3) + (b_1x^2 + b_2x + b_3)$$
$$= (a_1 + b_1)x^2 + (a_2 + b_2)x + (a_3 + b_3).$$

These examples suggest the analogous definition of the **sum of two matrices** of the same order: If $A = [a_{ij}]_{m \times n}$ and $B = [b_{ij}]_{m \times n}$, then

$$A + B = [a_{ij}]_{m \times n} + [b_{ij}]_{m \times n} = [(a_{ij} + b_{ij})]_{m \times n}.$$

In words, *the matrix $A + B$, which is the sum of two matrices A and B of the same order, is found by adding corresponding elements of A and B throughout.*

For example,

$$\begin{bmatrix} 2+3t & 1 \\ 4 & 1-t \end{bmatrix} + \begin{bmatrix} t-2 & t \\ -4 & 1+t \end{bmatrix} = \begin{bmatrix} 4t & 1+t \\ 0 & 2 \end{bmatrix}.$$

Two matrices of the same order are said to be **conformable for addition.**

Since the sum of two $m \times n$ matrices over a given number field $\mathscr{F}$ is again an $m \times n$ matrix over $\mathscr{F}$, we say that the set of all such matrices is **closed under addition.** This same closure property holds for matrices over many other sets of elements.

2.4 THE COMMUTATIVE AND ASSOCIATIVE LAWS OF ADDITION FOR MATRICES

Throughout this book, the numbers of the field $\mathscr{F}$ over which we are working will be called **scalars** to distinguish them from the arrays that are called matrices.

At times, functions of variables with domains $\mathscr{F}$ are used. These are also called scalars. In scalar algebra, the fact that $a + b = b + a$ for any two scalars a and b is known as the **commutative law of addition**. The fact that $a + (b + c) = (a + b) + c$ for any three scalars a, b, and c is known as the **associative law of addition**. It is not hard to see that these laws extend to matrix addition. For example,

$$\begin{bmatrix} a & b \\ c & d \end{bmatrix} + \begin{bmatrix} 2 & 1 \\ -1 & 3 \end{bmatrix} = \begin{bmatrix} a+2 & b+1 \\ c-1 & d+3 \end{bmatrix} = \begin{bmatrix} 2+a & 1+b \\ -1+c & 3+d \end{bmatrix}$$
$$= \begin{bmatrix} 2 & 1 \\ -1 & 3 \end{bmatrix} + \begin{bmatrix} a & b \\ c & d \end{bmatrix}$$

and

$$\begin{aligned}([2, -1, 3] + [a, b, c]) + [\alpha, \beta, \gamma] &= [2 + a, -1 + b, 3 + c] + [\alpha, \beta, \gamma] \\ &= [(2 + a) + \alpha, (-1 + b) + \beta, (3 + c) + \gamma] \\ &= [2 + (a + \alpha), -1 + (b + \beta), 3 + (c + \gamma)] \\ &= [2, -1, 3] + [a + \alpha, b + \beta, c + \gamma] \\ &= [2, -1, 3] + ([a, b, c] + [\alpha, \beta, \gamma]).\end{aligned}$$

The principles used in these examples apply in general. Let A, B, C be arbitrary matrices of the same order. Then, using the definition of the sum of two matrices and the commutative law of addition of scalars, we have, in the abbreviated notation,

$$A + B = [a_{ij} + b_{ij}] = [b_{ij} + a_{ij}] = B + A.$$

Similarly, applying the associative law for the addition of scalars, we have

$$A + (B + C) = [a_{ij} + (b_{ij} + c_{ij})] = [(a_{ij} + b_{ij}) + c_{ij}] = (A + B) + C.$$

We have thus proved

Theorem 2.4.1: *The addition of matrices is both commutative and associative; that is, if A, B, and C are conformable for addition,*

$$A + B = B + A, \tag{2.4.1}$$

$$A + (B + C) = (A + B) + C. \tag{2.4.2}$$

The reader who finds the above notation a little too condensed should write out the details in full for matrices of order, say, (2, 3).

These two laws, applied repeatedly if necessary, enable us to arrange the terms of a sum in any order we wish, and to group them in any fashion we wish. In particular, they justify the absence of parentheses in an expression like $A + B + C$, which (for given A, B, and C of the same order) has a uniquely defined meaning.

Another important property of matrix addition is given in

Theorem 2.4.2: $A + C = B + C$ *if and only if* $A = B$.

The fact that $A + C = B + C$ implies that $A = B$ is called the **cancellation law for the addition of matrices**.

Indeed, $A + C = B + C$ if and only if $a_{ij} + c_{ij} = b_{ij} + c_{ij}$ in every case. But $a_{ij} + c_{ij} = b_{ij} + c_{ij}$ if and only if $a_{ij} = b_{ij}$, by the cancellation law of addition in any number field. This implies that $A + C = B + C$ if and only if $A = B$.

2.5 ZERO, NEGATIVES, AND SUBTRACTION OF MATRICES

A matrix all of whose elements are 0 is called a **zero matrix** or **null matrix** and is denoted by $0_{m \times n}$, by 0_n (when it is square and of order n), or simply by 0, when the order will not be misunderstood. The basic property of the matrix $0_{m \times n}$ is that, for all $m \times n$ matrices A, it is an **additive identity element**; that is,

$$A + 0 = A. \tag{2.5.1}$$

The **negative** of an $m \times n$ matrix A is defined to be a matrix "$-A$" such that

$$A + (-A) = 0. \tag{2.5.2}$$

Since the scalar equation $a_{ij} + x = 0$ has the unique solution $x = -a_{ij}$, it follows that the matrix equation $A_{m \times n} + X_{m \times n} = 0_{m \times n}$ has the unique solution $X = [-a_{ij}]_{m \times n}$; that is, the unique negative of A is given by

$$-A = [-a_{ij}]_{m \times n}. \tag{2.5.3}$$

The matrix $-A$ is also called the **additive inverse** of A.

We may now define subtraction of matrices of the same order thus:

$$A - B = A + (-B) = [a_{ij} - b_{ij}]. \tag{2.5.4}$$

That is, *the difference* $A - B$ *may be found by subtracting the elements of* B *from the corresponding elements of* A. For example,

$$\begin{bmatrix} 1 & -5 & 3 \\ -3 & -1 & 0 \end{bmatrix} - \begin{bmatrix} 2 & -5 & 2 \\ -1 & 1 & -4 \end{bmatrix} = \begin{bmatrix} -1 & 0 & 1 \\ -2 & -2 & 4 \end{bmatrix}$$

and

$$\begin{bmatrix} 2b_1 \\ b_2 \\ -3b_3 \end{bmatrix} - \begin{bmatrix} b_1 \\ b_2 \\ b_3 \end{bmatrix} = \begin{bmatrix} b_1 \\ 0 \\ -4b_3 \end{bmatrix}.$$

An important consequence of the preceding definitions is that, if X, A, and B are all of the same order, then the unique solution of the equation

$$X + A = B \tag{2.5.5}$$

is

$$X = B - A. \tag{2.5.6}$$

In fact, replacement of X by $B - A$ in $X + A$ yields

$$\begin{aligned}(B - A) + A &= (B + (-A)) + A\\ &= B + ((-A) + A)\\ &= B + 0\\ &= B,\end{aligned}$$

so $B - A$ is indeed a solution. Moreover, it is the *only* solution, for if Y is any solution, then

$$Y + A = B,$$

and adding $-A$ to each member yields

$$(Y + A) + (-A) = B + (-A),$$

which reduces to

$$Y = B - A.$$

That is, $B - A$ is the unique solution of (2.5.5), as claimed. In particular, this proves that the equations

$$X + A = A$$

and

$$X + A = 0$$

have, respectively, the unique solutions 0 and $-A$; that is, *for a given order* (m, n) *there is precisely one additive identity element, namely the matrix* $0_{m \times n}$, *and each matrix* A *has precisely one inverse with respect to addition, namely the matrix* $-A$ *defined above.*

In summary, with respect to addition, the set of all $m \times n$ matrices has the same properties as the set of scalars. This is because matrix addition is effected by mn independent scalar additions—one in each of the mn positions.

2.6 SCALAR MULTIPLES OF MATRICES

If $A = [a_{ij}]$ and if α is a scalar, we define $\alpha A = A\alpha = [\alpha a_{ij}]$. In words, *to multiply a matrix* A *by a scalar* α, *multiply every element of* A *by* α. This definition is, of course, suggested by the fact that, if we add n A's, we obtain a matrix whose elements are those of A each multiplied by n. For example,

$$\begin{bmatrix} a & b \\ c & d \end{bmatrix} + \begin{bmatrix} a & b \\ c & d \end{bmatrix} = \begin{bmatrix} 2a & 2b \\ 2c & 2d \end{bmatrix} = 2\begin{bmatrix} a & b \\ c & d \end{bmatrix}.$$

The operation of multiplying a matrix by a scalar has these basic properties for all scalars α, β and for all $m \times n$ matrices A, B:

$$\begin{aligned} 1 \cdot A &= A \\ (-1) \cdot A &= -A \\ (\alpha + \beta)A &= \alpha A + \beta A \\ \alpha(A + B) &= \alpha A + \alpha B \\ \alpha(\beta A) &= (\alpha\beta)A. \end{aligned} \tag{2.6.1}$$

All five are readily proved by appealing to the definition. In the case of the third, we have

$$(\alpha + \beta)A = [(\alpha + \beta)a_{ij}] = [\alpha a_{ij} + \beta a_{ij}] = [\alpha a_{ij}] + [\beta a_{ij}] = \alpha A + \beta A.$$

We leave it as an exercise for the reader to prove the other laws in a similar fashion and to illustrate each by a concrete example.

As an example of how these ideas are used, we note that the rules of equality of matrices, of addition, and of multiplication by scalars permit us to represent the system of linear equations (2.1.1) as a single vector equation thus:

$$x_1 \begin{bmatrix} a_{11} \\ a_{21} \\ \vdots \\ a_{m1} \end{bmatrix} + x_2 \begin{bmatrix} a_{12} \\ a_{22} \\ \vdots \\ a_{m2} \end{bmatrix} + \cdots + x_n \begin{bmatrix} a_{1n} \\ a_{2n} \\ \vdots \\ a_{mn} \end{bmatrix} = \begin{bmatrix} b_1 \\ b_2 \\ \vdots \\ b_n \end{bmatrix}. \tag{2.6.2}$$

This is easily seen to be equivalent to (2.1.1) by multiplying and adding, as indicated. The reader should do this.

2.7 MULTIPLICATION OF MATRICES BY MATRICES

We first define the product of a row vector and a column vector, each of which has n entries, to be a scalar, namely the **scalar product** of the two vectors:

$$[a_1, a_2, \ldots, a_n] \cdot \begin{bmatrix} b_1 \\ b_2 \\ \vdots \\ b_n \end{bmatrix} = a_1 b_1 + a_2 b_2 + \cdots + a_n b_n. \tag{2.7.1}$$

(If a row and a column do not have the same number of entries, their scalar product is not defined.) For example,

$$[3, 1, 1] \cdot \begin{bmatrix} 4 \\ 0 \\ 2 \end{bmatrix} = 14 \qquad \text{and} \qquad [3, 1, 1] \cdot \begin{bmatrix} -1 \\ 1 \\ 2 \end{bmatrix} = 0.$$

Next we define the product of an $m \times n$ matrix A and a column vector B with n rows to be the column vector whose entry in the ith row is the scalar product of the ith row of A and the column B, thus:

(2.7.2)
$$AB = \begin{bmatrix} a_{11} & a_{12} & \cdots & a_{1n} \\ a_{21} & a_{22} & \cdots & a_{2n} \\ \vdots \\ a_{m1} & a_{m2} & \cdots & a_{mn} \end{bmatrix} \cdot \begin{bmatrix} b_1 \\ b_2 \\ \vdots \\ b_n \end{bmatrix} = \begin{bmatrix} (a_{11}b_1 + a_{12}b_2 + \cdots + a_{1n}b_n) \\ (a_{21}b_1 + a_{22}b_2 + \cdots + a_{2n}b_n) \\ \vdots \\ (a_{m1}b_1 + a_{m2}b_2 + \cdots + a_{mn}b_n) \end{bmatrix}.$$

For example:

$$\begin{bmatrix} 2 & 3 & -1 \\ 0 & 4 & 2 \end{bmatrix} \begin{bmatrix} 1 \\ 2 \\ -1 \end{bmatrix} = \begin{bmatrix} 9 \\ 6 \end{bmatrix} \quad \text{and} \quad \begin{bmatrix} 2 & -1 \\ 3 & 2 \end{bmatrix} \begin{bmatrix} x_1 \\ x_2 \end{bmatrix} = \begin{bmatrix} 2x_1 - x_2 \\ 3x_1 + 2x_2 \end{bmatrix}.$$

An immediate consequence of the definition of equality of matrices and of this rule of multiplication is that we can represent the system of m linear equations in n unknowns (2.1.1) in matrix form as follows:

(2.7.3)
$$\begin{bmatrix} a_{11} & a_{12} & \cdots & a_{1n} \\ a_{21} & a_{22} & \cdots & a_{2n} \\ \vdots \\ a_{m1} & a_{m2} & \cdots & a_{mn} \end{bmatrix} \begin{bmatrix} x_1 \\ x_2 \\ \vdots \\ x_n \end{bmatrix} = \begin{bmatrix} b_1 \\ b_2 \\ \vdots \\ b_m \end{bmatrix}.$$

If we know a set of values for the x's which makes this a true equation, we know a solution of (2.1.1). In this case we call the vector

$$\begin{bmatrix} x_1 \\ x_2 \\ \vdots \\ x_n \end{bmatrix}$$

a solution of the system.

The complete solution may also be written in matrix form. For example, the system of equations

$$\begin{aligned} 2x_1 - x_2 + 3x_3 + x_4 &= 8 \\ -x_1 + x_2 \qquad + 2x_4 &= 2 \end{aligned} \quad \text{or} \quad \begin{bmatrix} 2 & -1 & 3 & 1 \\ -1 & 1 & 0 & 2 \end{bmatrix} \begin{bmatrix} x_1 \\ x_2 \\ x_3 \\ x_4 \end{bmatrix} = \begin{bmatrix} 8 \\ 2 \end{bmatrix}$$

has the parametric solution

$$\begin{aligned} x_1 &= -3s - 3t + 10 \\ x_2 &= -3s - 5t + 12 \\ x_3 &= s \\ x_4 &= t \end{aligned} \quad \text{or} \quad \begin{bmatrix} x_1 \\ x_2 \\ x_3 \\ x_4 \end{bmatrix} = s \begin{bmatrix} -3 \\ -3 \\ 1 \\ 0 \end{bmatrix} + t \begin{bmatrix} -3 \\ -5 \\ 0 \\ 1 \end{bmatrix} + \begin{bmatrix} 10 \\ 12 \\ 0 \\ 0 \end{bmatrix}.$$

Note that, in accord with Exercise 15, Section 1.5,

$$\begin{bmatrix} 2 & -1 & 3 & 1 \\ -1 & 1 & 0 & 2 \end{bmatrix} \begin{bmatrix} -3 \\ -3 \\ 1 \\ 0 \end{bmatrix} = \begin{bmatrix} 0 \\ 0 \end{bmatrix}, \quad \begin{bmatrix} 2 & -1 & 3 & 1 \\ -1 & 1 & 0 & 2 \end{bmatrix} \begin{bmatrix} -3 \\ -5 \\ 0 \\ 1 \end{bmatrix} = \begin{bmatrix} 0 \\ 0 \end{bmatrix},$$

and

$$\begin{bmatrix} 2 & -1 & 3 & 1 \\ -1 & 1 & 0 & 2 \end{bmatrix} \begin{bmatrix} 10 \\ 12 \\ 0 \\ 0 \end{bmatrix} = \begin{bmatrix} 8 \\ 2 \end{bmatrix}.$$

We shall find this representation extremely useful in what follows.

We are now ready to define matrix multiplication in general. The product of an $m \times n$ matrix A and an $n \times p$ matrix B is defined to be the $m \times p$ matrix whose jth column is the product of A and the jth column of B, computed as in (2.7.2):

$$AB = \begin{bmatrix} a_{11} & a_{12} & \cdots & a_{1n} \\ a_{21} & a_{22} & \cdots & a_{2n} \\ \vdots & & & \\ a_{m1} & a_{m2} & \cdots & a_{mn} \end{bmatrix} \begin{bmatrix} b_{11} & b_{12} & \cdots & b_{1p} \\ b_{21} & b_{22} & \cdots & b_{2p} \\ \vdots & & & \\ b_{n1} & b_{n2} & \cdots & b_{np} \end{bmatrix}$$

(2.7.4)

$$= \begin{bmatrix} (a_{11}b_{11} + a_{12}b_{21} + \cdots + a_{1n}b_{n1}) & \cdots & (a_{11}b_{1p} + a_{12}b_{2p} + \cdots + a_{1n}b_{np}) \\ (a_{21}b_{11} + a_{22}b_{21} + \cdots + a_{2n}b_{n1}) & \cdots & (a_{21}b_{1p} + a_{22}b_{2p} + \cdots + a_{2n}b_{np}) \\ \vdots & & \vdots \\ (a_{m1}b_{11} + a_{m2}b_{21} + \cdots + a_{mn}b_{n1}) & \cdots & (a_{m1}b_{1p} + a_{m2}b_{2p} + \cdots + a_{mn}b_{np}) \end{bmatrix}$$

For example,

(a)
$$\begin{bmatrix} 1 & -1 & 0 \\ 2 & 3 & 1 \end{bmatrix} \begin{bmatrix} 4 & 2 \\ -1 & 0 \\ 2 & 5 \end{bmatrix}$$

$$= \begin{bmatrix} (1 \cdot 4 + (-1) \cdot (-1) + 0 \cdot 2) & (1 \cdot 2 + (-1) \cdot 0 + 0 \cdot 5) \\ (2 \cdot 4 + 3 \cdot (-1) + 1 \cdot 2) & (2 \cdot 2 + 3 \cdot 0 + 1 \cdot 5) \end{bmatrix}$$

$$= \begin{bmatrix} 5 & 2 \\ 7 & 9 \end{bmatrix},$$

(b)
$$\begin{bmatrix} 4 & 2 \\ -1 & 0 \\ 2 & 5 \end{bmatrix} \begin{bmatrix} 1 & -1 & 0 \\ 2 & 3 & 1 \end{bmatrix} = \begin{bmatrix} 8 & 2 & 2 \\ -1 & 1 & 0 \\ 12 & 13 & 5 \end{bmatrix},$$

(c) $$[1, 2, -1, 3]\begin{bmatrix} 1 & 2 \\ 1 & 0 \\ 4 & -3 \\ 1 & 0 \end{bmatrix} = [2, 5].$$

There are some important things to note here:

1. The product AB has the same number of rows as the first factor A and the same number of columns as the second factor B.
2. The number of *columns* of A must be the same as the number of *rows* of B so that the scalar products of rows of A and columns of B are defined. When this condition holds, we say that the matrix A is **conformable** to the matrix B for multiplication. If A is conformable to B, it does not necessarily follow that B is conformable to A, as the third of the three preceding examples shows:

$$\begin{bmatrix} 1 & 2 \\ 1 & 0 \\ 4 & -3 \\ 1 & 0 \end{bmatrix}[1, 2, -1, 3]$$

   does not define a product because the required scalar products cannot be formed.
3. The definition implies that *the ij-entry of a product AB is the scalar product of the* ith *row of A and the* jth *column of B*. Let us denote the ith row of $A_{m \times n}$ by $A^{(i)}$, the jth column of $B_{n \times p}$ by B_j, and the product AB by C. Then $A_{m \times n}B_{n \times p} = C_{m \times p}$, where the ij-entry of C is given by

(2.7.5) $$c_{ij} = A^{(i)}B_j = a_{i1}b_{1j} + a_{i2}b_{2j} + \cdots + a_{in}b_{nj}.$$

The notation for rows and columns introduced in 3 permits us to represent $A_{m \times n}$ as a matrix of rows

$$A = \begin{bmatrix} A^{(1)} \\ A^{(2)} \\ \vdots \\ A^{(m)} \end{bmatrix}$$

or as a matrix of columns: $A = [A_1, A_2, \ldots, A_n]$. Then the product $A_{m \times n} \cdot B_{n \times p}$ can be written in the form

(2.7.6) $$AB = \begin{bmatrix} A^{(1)} \\ A^{(2)} \\ \vdots \\ A^{(m)} \end{bmatrix}[B_1, B_2, \ldots, B_p] = [A^{(i)}B_j]_{m \times p}.$$

(The parentheses on the superscripts are intended to distinguish them from exponents, which will be introduced shortly.)

The definition of the product $A_{m\times n}B_{n\times p}$ just given is a natural extension of the definition of the product $A_{m\times n}B_{n\times 1}$. However, the same rule is suggested by other considerations as well. For example, suppose that

$$\begin{aligned} z_1 &= a_{11}y_1 + a_{12}y_2 \\ z_2 &= a_{21}y_1 + a_{22}y_2 \end{aligned} \qquad \text{and} \qquad \begin{aligned} y_1 &= b_{11}x_1 + b_{12}x_2 \\ y_2 &= b_{21}x_1 + b_{22}x_2. \end{aligned}$$

Then, substituting the expressions for y_1 and y_2 into the first pair of equations, expanding, and collecting, we get

$$\begin{aligned} z_1 &= (a_{11}b_{11} + a_{12}b_{21})x_1 + (a_{11}b_{12} + a_{12}b_{22})x_2, \\ z_2 &= (a_{21}b_{11} + a_{22}b_{21})x_1 + (a_{21}b_{12} + a_{22}b_{22})x_2. \end{aligned}$$

These three pairs of equations can be written in matrix form thus:

$$\begin{bmatrix} z_1 \\ z_2 \end{bmatrix} = \begin{bmatrix} a_{11} & a_{12} \\ a_{21} & a_{22} \end{bmatrix}\begin{bmatrix} y_1 \\ y_2 \end{bmatrix}, \qquad \begin{bmatrix} y_1 \\ y_2 \end{bmatrix} = \begin{bmatrix} b_{11} & b_{12} \\ b_{21} & b_{22} \end{bmatrix}\begin{bmatrix} x_1 \\ x_2 \end{bmatrix},$$

$$\begin{aligned} \begin{bmatrix} z_1 \\ z_2 \end{bmatrix} &= \begin{bmatrix} a_{11}b_{11} + a_{12}b_{21} & a_{11}b_{12} + a_{12}b_{22} \\ a_{21}b_{11} + a_{22}b_{21} & a_{21}b_{12} + a_{22}b_{22} \end{bmatrix}\begin{bmatrix} x_1 \\ x_2 \end{bmatrix} \\ &= \begin{bmatrix} a_{11} & a_{12} \\ a_{21} & a_{22} \end{bmatrix}\begin{bmatrix} b_{11} & b_{12} \\ b_{21} & b_{22} \end{bmatrix}\begin{bmatrix} x_1 \\ x_2 \end{bmatrix}, \end{aligned}$$

which shows that we can substitute from the second matrix equation into the first and then compute the new coefficient matrix by our rule for matrix multiplication.

In general, because of the definition of matrix multiplication that we have chosen, if

$$Z = A_{m\times n}Y \qquad \text{and} \qquad Y = B_{n\times p}X,$$

then

$$Z = (A_{m\times n}B_{n\times p})X.$$

This sort of substitution from one set of linear expressions into another is of frequent occurrence and of great importance. If we had made any other definition of matrix multiplication, the theory of these linear transformations, as they are called, could not be treated so conveniently.

2.8 PROPERTIES OF MATRIX MULTIPLICATION

In the product AB, we say that B is **premultiplied** by A and that A is **postmultiplied** by B. This terminology is essential since ordinarily $AB \neq BA$, as is illustrated by commuting the factors in examples (a) and (b) of the preceding section. In general, if A is $m \times n$ and B is $n \times p$, and if $m \neq p$, then AB is defined but BA is not, as is illustrated by example (c) of the preceding section. Even if A and B are both of order n, we need not have $AB = BA$. For example,

$$\begin{bmatrix} 1 & 2 \\ 2 & 1 \end{bmatrix} \begin{bmatrix} 1 & 2 \\ 3 & 4 \end{bmatrix} = \begin{bmatrix} 7 & 10 \\ 5 & 8 \end{bmatrix}$$

but

$$\begin{bmatrix} 1 & 2 \\ 3 & 4 \end{bmatrix} \begin{bmatrix} 1 & 2 \\ 2 & 1 \end{bmatrix} = \begin{bmatrix} 5 & 4 \\ 11 & 10 \end{bmatrix}.$$

Thus *matrix multiplication is not in general commutative*. This does not, however, mean that we *never* have $AB = BA$. There are important special cases when this equality holds. For example, for all numbers a, b, c, and d we have

$$\begin{bmatrix} a & b \\ -b & a \end{bmatrix} \begin{bmatrix} c & d \\ -d & c \end{bmatrix} = \begin{bmatrix} c & d \\ -d & c \end{bmatrix} \begin{bmatrix} a & b \\ -b & a \end{bmatrix},$$

as the reader should verify. He should also note that these matrices have a very special form which accounts for the fact that they commute.

The familiar rule of scalar algebra that if a product is zero, then at least one of the factors must be zero, also fails to hold for matrix multiplication. An example is the product

$$\begin{bmatrix} 1 & -1 \\ -2 & 2 \end{bmatrix} \begin{bmatrix} 1 & 3 \\ 1 & 3 \end{bmatrix} = \begin{bmatrix} 0 & 0 \\ 0 & 0 \end{bmatrix}.$$

Here neither factor is a zero matrix, although the product is.

When a product $AB = 0$ but neither A nor B is 0, then the factors A and B are called **divisors of zero**. Thus, in the algebra of matrices, there exist divisors of zero, whereas in the algebra of real or complex numbers there do not.

Now let

$$A = \begin{bmatrix} 1 & -1 \\ -2 & 2 \end{bmatrix}, \quad B = \begin{bmatrix} 2 & 1 \\ 2 & 1 \end{bmatrix}, \quad C = \begin{bmatrix} -1 & 2 \\ -1 & 2 \end{bmatrix}.$$

Then

$$AB = \begin{bmatrix} 0 & 0 \\ 0 & 0 \end{bmatrix} = AC.$$

Thus we can have $AB = AC$ without having $B = C$. In other words, we cannot ordinarily cancel A from $AB = AC$, even if $A \neq 0$. Similarly, we cannot ordinarily cancel A from $BA = CA$. However, there is an important special case when such a cancellation is possible, as we shall see later.

In summary, then, three fundamental properties of multiplication in scalar algebra do not carry over to matrix algebra:

1. The commutative law $AB = BA$ does not hold true generally.
2. From $AB = 0$, we cannot conclude that at least one of A and B must be zero; that is, there exist divisors of zero.

3. From $AB = AC$ or $BA = CA$ we cannot in general conclude that $B = C$, even if $A \neq 0$; that is, the cancellation law does not hold in general in multiplication.

These facts might make one wonder whether matrix multiplication is not a nearly useless operation. This is, of course, not the case, for, as we shall prove, the most vital properties—the associative and the distributive laws—still remain. However, it should be clear at this point why we have been, and must continue to be, so careful to prove the validity of the matrix operations which we employ.

Theorem 2.8.1: *The multiplication of matrices is associative.*

Let

$$A = [a_{ij}]_{m \times n}, \qquad B = [b_{jk}]_{n \times p}, \qquad C = [c_{kr}]_{p \times q}.$$

Then the theorem says that

$$(AB)C = A(BC).$$

Applying the definition of multiplication, we see first that

$$AB = \left[\sum_{j=1}^{n} a_{ij}b_{jk}\right]_{m \times p}.$$

Here i ranges from 1 to m and denotes the row of the element in parentheses, whereas k ranges from 1 to p and denotes its column.

We apply the definition now to AB and C. The new summation will be on the column subscript k of AB, which is also the row subscript of C, so

$$(AB)C = \left[\sum_{k=1}^{p}\left(\sum_{j=1}^{n} a_{ij}b_{jk}\right)c_{kr}\right]_{m \times q}.$$

Multiplying the factor c_{kr} into each sum in parentheses, we obtain

$$(AB)C = \left[\sum_{k=1}^{p}\left(\sum_{j=1}^{n} a_{ij}b_{jk}c_{kr}\right)\right]_{m \times q},$$

in which the row subscript i ranges from 1 to m while the column subscript r ranges from 1 to q.

In the same way we find

$$A(BC) = \left[\sum_{j=1}^{n}\left(\sum_{k=1}^{p} a_{ij}b_{jk}c_{kr}\right)\right]_{m \times q}.$$

Since the order of summation is arbitrary in a finite sum, we have

$$\sum_{k=1}^{p}\left(\sum_{j=1}^{n} a_{ij}b_{jk}c_{kr}\right) = \sum_{j=1}^{n}\left(\sum_{k=1}^{p} a_{ij}b_{jk}c_{kr}\right)$$

for each pair of values of i and r, so $(AB)C = A(BC)$.

If the uses made of the $\sum$ sign in this proof are unfamiliar to the reader, he may refer to an explanation of these matters in Appendix I. It would also help to write out the proof in full for 2×2 matrices.

Theorem 2.8.2: *Matrix multiplication is distributive with respect to addition.*

To make this explicit, let

$$A = [a_{ik}]_{m \times n}, \qquad B = [b_{kj}]_{n \times p}, \qquad C = [c_{kj}]_{n \times p}.$$

Here A is conformable to B and also to C for multiplication, and B is conformable to C for addition. Then the theorem says that

$$A(B + C) = AB + AC.$$

Indeed

$$\begin{aligned} A(B + C) &= [a_{ik}]_{m \times n}[(b_{kj} + c_{kj})]_{n \times p} \\ &= \left[\sum_{k=1}^{n} a_{ik}(b_{kj} + c_{kj})\right]_{m \times p} \\ &= \left[\sum_{k=1}^{n} a_{ik}b_{kj} + \sum_{k=1}^{n} a_{ik}c_{kj}\right]_{m \times p} \\ &= \left[\sum_{k=1}^{n} a_{ik}b_{kj}\right]_{m \times p} + \left[\sum_{k=1}^{n} a_{ik}c_{kj}\right]_{m \times p} \\ &= AB + AC. \end{aligned}$$

The theorem also says that, assuming conformability,

$$(D + E)F = DF + EF.$$

This second distributive law is distinct from the first because matrix multiplication is not in general commutative. It is proved in the same manner as the first, however, and details are left to the reader.

The proofs of the last two theorems involve a detailed examination of the elements of the matrices involved. They are thus essentially scalar in nature. As the theory develops, we shall increasingly employ proofs involving only manipulations with matrices. Such proofs are typically more compact than are scalar-type proofs of the same results and hence are to be preferred. The reader's progress in learning matrix algebra will be accelerated if in the exercises he avoids the use of the scalar-type proof whenever this is possible. For example, to prove that, for conformable matrices,

$$(A + B)(C + D) = AC + BC + AD + BD,$$

we do not again resort to a scalar type of proof. We simply note that, by the first distributive law above, $(A + B)(C + D) = (A + B)C + (A + B)D$, so that, by the second distributive law and the associative law,

$$(A + B)(C + D) = AC + BC + AD + BD.$$

2.9 EXERCISES

Throughout this book, exercises marked with an asterisk (*) develop an important part of the theory and should not be overlooked.

In many problems, conditions of conformability for addition or multiplication must be satisfied for the problem to have meaning. These conditions are usually rather obvious, so that we shall frequently omit statement of them. The reader is then expected to make the necessary assumptions in working the exercises.

In these exercises, the words "prove" and "show" are to be taken as synonymous.

1. Prove that $(A + B) - C = A + (B - C)$ and give the reason for each step in the proof. Why is $(A - B) + C \neq A - (B + C)$ in general?

2. In each case find all solutions (x, y) of the given equation:

$$\text{(a)} \begin{bmatrix} x+y & 2 \\ 1 & x-y \end{bmatrix} = \begin{bmatrix} 3 & 2 \\ 1 & 5 \end{bmatrix}, \qquad \text{(b)} \begin{bmatrix} x+y & 2 \\ 1 & x-y \end{bmatrix} = \begin{bmatrix} 3 & x+y \\ x-y & 4 \end{bmatrix},$$

$$\text{(c)} \begin{bmatrix} x+y & 2 \\ 1 & 2x+2y \end{bmatrix} = \begin{bmatrix} 2 & 2 \\ 1 & 4 \end{bmatrix}.$$

***3.** Prove that for all conformable matrices A and B and for all scalars α and β,

(a) $\alpha A \cdot \beta B = \alpha\beta \cdot AB$,
(b) $(-1)A = -A$,
(c) $(-A)(-B) = AB$,
(d) $A(\alpha B) = \alpha(AB)$.

4. Given that

$$A_1 = \begin{bmatrix} 1 & 0 \\ 0 & 1 \end{bmatrix}, \quad A_2 = \begin{bmatrix} 0 & 1 \\ 1 & 0 \end{bmatrix}, \quad A_3 = \begin{bmatrix} 1 & 0 \\ 0 & -1 \end{bmatrix}, \quad A_4 = \begin{bmatrix} 0 & 1 \\ -1 & 0 \end{bmatrix},$$

(a) compute $2A_1 - 3A_2 + 4A_3 - A_4$,
(b) solve for x_1, x_2, x_3, and x_4:

$$x_1A_1 + x_2A_2 + x_3A_3 + x_4A_4 = \begin{bmatrix} 1 & 1 \\ 1 & 1 \end{bmatrix}.$$

5. (a) Solve for α_1, α_2, and α_3:

$$\alpha_1 \begin{bmatrix} 1 \\ 0 \\ 0 \end{bmatrix} + \alpha_2 \begin{bmatrix} 1 \\ 1 \\ 0 \end{bmatrix} + \alpha_3 \begin{bmatrix} 1 \\ 1 \\ 1 \end{bmatrix} = \begin{bmatrix} 2 \\ -3 \\ 0 \end{bmatrix}.$$

(b) Rewrite this system of equations as a single matrix equation of the form (2.7.3):

$$\begin{aligned} x_1 + x_2 - 2 &= 2x_1 - x_2 + 4 \\ 3x_1 - 2x_2 + 1 &= x_1 + 3x_2 - 5. \end{aligned}$$

Then write it in the form (2.6.2).

(c) Interpret the matrix equation

$$\begin{bmatrix} 1 & 3 & 4 \\ 2 & 1 & 6 \\ -1 & 1 & 5 \end{bmatrix}\begin{bmatrix} 1 \\ 1 \\ 1 \end{bmatrix} = \begin{bmatrix} 8 \\ 9 \\ 5 \end{bmatrix}$$

as a fact about a system of linear equations in three unknowns.

(d) Write this system of equations as a single vector equation in the manner of (2.6.2):

$$\begin{aligned} 2x_1 - 3x_2 + 4x_3 &= 0 \\ x_1 + 4x_2 \qquad &= 5 \\ x_2 - 4x_3 &= -7. \end{aligned}$$

6. Perform the indicated matrix multiplications ($\bar{z}$ denotes the complex conjugate of z and i denotes $\sqrt{-1}$):

(a) $\begin{bmatrix} 1 & 0 & 2 \\ -1 & 4 & 3 \\ 2 & 1 & 0 \end{bmatrix}\begin{bmatrix} 1 & -1 \\ 2 & 3 \\ 0 & 4 \end{bmatrix}$, (b) $[a, b, c, d]\begin{bmatrix} a \\ b \\ c \\ d \end{bmatrix}$,

(c) $[\bar{z}_1, \bar{z}_2, \bar{z}_3, \bar{z}_4]\begin{bmatrix} z_1 \\ z_2 \\ z_3 \\ z_4 \end{bmatrix}$, (d) $[1 \quad 4 \quad x]\begin{bmatrix} 1 & -1 & 0 \\ 0 & 1 & -1 \\ -1 & 0 & 1 \end{bmatrix}\begin{bmatrix} 1 \\ 4 \\ x \end{bmatrix}$,

(e) $\begin{bmatrix} 1 & 0 & 0 \\ 0 & \alpha & 0 \\ 0 & 0 & 1 \end{bmatrix}\begin{bmatrix} a_1 & b_1 & c_1 & d_1 \\ a_2 & b_2 & c_2 & d_2 \\ a_3 & b_3 & c_3 & d_3 \end{bmatrix}\begin{bmatrix} 1 & 0 & 0 & 0 \\ 0 & \beta & 0 & 0 \\ 0 & 0 & 1 & 0 \\ 0 & 0 & 0 & 1 \end{bmatrix}$,

(f) $\begin{bmatrix} 1 & 0 \\ i & 1 \end{bmatrix}\begin{bmatrix} 1 & i \\ -i & 0 \end{bmatrix}\begin{bmatrix} 1 & -i \\ 0 & 1 \end{bmatrix}$, (g) $\begin{bmatrix} 2 \\ -1 \\ 3 \end{bmatrix}[3, 4, 0, -1]$,

(h) $\begin{bmatrix} 1/\sqrt{2} & 1/\sqrt{2} \\ -1/\sqrt{2} & 1/\sqrt{2} \end{bmatrix}\begin{bmatrix} 1 & 2 \\ 2 & 1 \end{bmatrix}\begin{bmatrix} 1/\sqrt{2} & -1/\sqrt{2} \\ 1/\sqrt{2} & 1/\sqrt{2} \end{bmatrix}$.

7. Assuming a, b, c, and d real, what can you say about the result in Exercise 6(b)? What is implied about a, b, c, and d if the product 6(b) is zero? What can you say about the result in 6(c)? What happens in 6(e) if $\alpha = \beta = 1$? If $\alpha = \beta = 0$?

8. Under what conditions is a matrix product $ABCD$ defined? According to the associative law, in how many different ways may it be computed?

***9.** Prove in detail the second distributive law

$$(D + E)F = DF + EF.$$

***10.** Given that A is a square matrix, we define $A^1 = A$, $A^{p+1} = A^p \cdot A$ for $p \geq 1$. Prove by induction on q that, for each positive integer p,

$$A^p A^q = A^{p+q} \qquad \text{and} \qquad (A^p)^q = A^{pq}$$

for all positive integers q. (Note that *neither the definition nor the proof is of the scalar type.*)

11. (a) If

$$A = \begin{bmatrix} 0 & i \\ i & 0 \end{bmatrix},$$

compute A^2, A^3, and A^4. (Here $i^2 = -1$.) Give a general rule for A^n. (You may treat the cases n even and n odd separately.)

(b) Evaluate

$$\begin{bmatrix} 0 & 1 \\ 0 & 0 \end{bmatrix}^2, \quad \begin{bmatrix} 0 & 1 & 0 \\ 0 & 0 & 1 \\ 0 & 0 & 0 \end{bmatrix}^3, \quad \begin{bmatrix} 0 & 1 & 0 & 0 \\ 0 & 0 & 1 & 0 \\ 0 & 0 & 0 & 1 \\ 0 & 0 & 0 & 0 \end{bmatrix}^4.$$

Then state and prove a general rule illustrated by these three examples.

***12.** Show that, if $\alpha A = 0$, where α is a scalar, then either $\alpha = 0$ or $A = 0$. Show also that, if $\alpha A = \alpha B$ and $\alpha \neq 0$, then $A = B$, and that, if $A = B$, then $\alpha A = \alpha B$ for all scalars α.

***13.** (a) Show that $AX = 0$ for all n-vectors X if and only if $A = 0$.

(b) Show that, if A and B are both $m \times n$ matrices, then $A = B$ if and only if $AX = BX$ for all n-vectors X.

(c) Show that $AX = X$ for all n-vectors X if and only if $A = I_n$ (Sec. 2.10).

***14.** (a) Explain why in matrix algebra

$$(A + B)^2 \neq A^2 + 2AB + B^2$$

and

$$(A + B)(A - B) \neq A^2 - B^2,$$

except in special cases. Under what circumstances would equality hold?

(b) Expand $(A + B)^3$.

15. Let $AB = C$, where A and B are of order n. If, in the ith row of A, $a_{ik} = 1$, where i and k are fixed but all other elements of the ith row are zero, what can be said about the ith row of C? What is the analogous fact for columns?

16. Let A and B be of order n and let

$$C_1 = \alpha_1 A + \beta_1 B,$$
$$C_2 = \alpha_2 A + \beta_2 B,$$

where α_1, α_2, β_1, and β_2 are scalars such that $\alpha_1\beta_2 \neq \alpha_2\beta_1$. Show that $C_1C_2 = C_2C_1$ if and only if $AB = BA$.

17. Let

$$A = \begin{bmatrix} 3 & 4 & 2 \\ -2 & -1 & -1 \\ -1 & -3 & -1 \end{bmatrix}, \qquad B = \begin{bmatrix} -1 & -1 & -1 \\ 2 & 2 & 2 \\ 1 & 1 & 1 \end{bmatrix}.$$

Compare the products AB and BA.

18. For what values of x will

$$[x \quad 4 \quad 1] \cdot \begin{bmatrix} 2 & 1 & 0 \\ 1 & 0 & 2 \\ 0 & 2 & 4 \end{bmatrix} \cdot \begin{bmatrix} x \\ 4 \\ 1 \end{bmatrix} = 2? \qquad = 0?$$

***19.** A square matrix of the form

$$D_n = \begin{bmatrix} d_{11} & 0 & \cdots & 0 \\ 0 & d_{22} & \cdots & 0 \\ \vdots & & & \\ 0 & 0 & \cdots & d_{nn} \end{bmatrix},$$

that is, one in which $d_{ij} = 0$ if $i \neq j$, is called a **diagonal matrix** of order n. (Note that this does *not* say $d_{ii} \neq 0$.) Let A be any matrix of order (p, q) and evaluate the products D_pA and AD_q. Describe the results in words. What happens in the special cases $d_{11} = d_{22} = \cdots = \alpha$ and $d_{11} = d_{22} = \cdots = 1$?

***20.** (a) Show that any two diagonal matrices of the same order commute.

(b) Give a formula for D^p, where D is diagonal and p is a positive integer.

(c) Show that if D is diagonal, then $AD^p = D^pA$ for all positive integers p if and only if $AD = DA$.

21. Prove by induction that if B and C are of order n and if $A = B + C$, $C^2 = 0$, and $BC = CB$, then for every positive integer k, $A^{k+1} = B^k(B + (k + 1)C)$.

22. If $AB = BA$, the matrices A and B are said to be **commutative** or to **commute**. Show that for all values of a, b, c, and d, the matrices

$$A = \begin{bmatrix} a & b \\ -b & a \end{bmatrix} \quad \text{and} \quad B = \begin{bmatrix} c & d \\ -d & c \end{bmatrix}$$

commute.

23. What must be true about a, b, c, and d if the matrices

$$\begin{bmatrix} a & b \\ c & d \end{bmatrix} \quad \text{and} \quad \begin{bmatrix} 1 & 1 \\ -1 & 1 \end{bmatrix}$$

are to commute?

24. Show that, if A and B are of order n, and if I_n is an $n \times n$ diagonal matrix all of whose diagonal entries are 1, then A and B commute if and only if $A - \lambda I_n$ and $B - \lambda I_n$ commute for every scalar λ.

25. If $AB = -BA$, the matrices A and B are said to be **anticommutative** or to **anticommute**. Show that each of the matrices

$$\sigma_x = \begin{bmatrix} 0 & 1 \\ 1 & 0 \end{bmatrix}, \quad \sigma_y = \begin{bmatrix} 0 & -i \\ i & 0 \end{bmatrix}, \quad \sigma_z = \begin{bmatrix} 1 & 0 \\ 0 & -1 \end{bmatrix} \quad (i^2 = -1)$$

anticommutes with the others. These are the **Pauli spin matrices**, which are used in the study of electron spin in quantum mechanics.

26. The matrix $AB - BA$ (A and B of order n) is called the **commutator** of A and B. Using Exercise 25, show that the commutators of σ_x and σ_y, σ_y and σ_z, and σ_z and σ_x are, respectively, $2i\sigma_z$, $2i\sigma_x$, and $2i\sigma_y$.

***27.** Show by induction that, if A is square and $AB = \lambda B$, where λ is a scalar, then $A^p B = \lambda^p B$ for every positive integer p.

28. If

$$A_i = \begin{bmatrix} \cos\theta_i & -\sin\theta_i \\ \sin\theta_i & \cos\theta_i \end{bmatrix}, \qquad i = 1, 2,$$

show that A_1 and A_2 commute. What is the connection with transformations used in plane analytic geometry?

***29.** The sum of the main diagonal elements a_{ii}, $i = 1, 2, \ldots, n$, of a square matrix A is called the **trace** of A:

$$\operatorname{tr} A = a_{11} + a_{22} + \cdots + a_{nn}.$$

(a) If A and B are of order n, show that $\operatorname{tr}(A + B) = \operatorname{tr} A + \operatorname{tr} B$.

(b) If C is of order (m, n) and G is of order (n, m), show that $\operatorname{tr} CG = \operatorname{tr} GC$.

30. Prove that, if A has two equal rows and AB is defined, then AB has two equal rows also. If all the rows of A are equal and if BA is also defined, what can you say about the columns of BA?

31. A matrix A such that $A^p = 0$ for some positive integer p is called **nilpotent**. Show that every 2×2 nilpotent matrix A such that $A^2 = 0$ may be written in the form

$$\begin{bmatrix} \lambda\mu & \mu^2 \\ -\lambda^2 & -\lambda\mu \end{bmatrix},$$

where λ and μ are scalars, and that every such matrix is nilpotent. If A is real, must λ and μ also be real?

***32.** Given that $\alpha A = \beta A$ and $A \neq 0$, prove that $\alpha = \beta$ (α and β scalars).

33. Given that A, B, and C all have order n, use Exercise 29(b) to show that

$$\operatorname{tr} ABC = \operatorname{tr} BCA = \operatorname{tr} CAB; \quad \operatorname{tr} ACB = \operatorname{tr} BAC = \operatorname{tr} CBA.$$

What is the generalization of this observation?

34. A matrix A such that $A^2 = A$ is called **idempotent**. Determine all diagonal matrices of order n which are idempotent. How many are there?

35. Given that

$$\begin{bmatrix} 1 & 0 & 0 \\ 0 & 2 & 0 \\ 0 & 0 & -3 \end{bmatrix} \cdot A \cdot \begin{bmatrix} 1 & 0 & 0 \\ 0 & 0 & 1 \\ 0 & 1 & 0 \end{bmatrix} = \begin{bmatrix} 1 & 2 & 3 \\ 4 & 5 & 4 \\ 3 & 2 & 1 \end{bmatrix},$$

find the matrix A.

***36.** Given that $AB = BA$, show that, for all positive integers r and s, $A^r B^s = B^s A^r$.

2.10 DIAGONAL, SCALAR, AND IDENTITY MATRICES

There are many special types of matrices that are particularly useful. For example, a **diagonal matrix** is a square matrix of the form

$$\begin{bmatrix} \alpha_1 & 0 & \cdots & 0 \\ 0 & \alpha_2 & \cdots & 0 \\ \vdots & & & \\ 0 & 0 & \cdots & \alpha_n \end{bmatrix},$$

a useful abbreviation for which is diag $[\alpha_1, \alpha_2, \ldots, \alpha_n]$. Here every element *off* the main diagonal is zero. Only main diagonal elements may be unequal to zero, but this is not required of them. Some 3×3 examples are

$$\begin{bmatrix} 1 & 0 & 0 \\ 0 & -1 & 0 \\ 0 & 0 & 2 \end{bmatrix}, \quad \begin{bmatrix} 1 & 0 & 0 \\ 0 & 0 & 0 \\ 0 & 0 & 0 \end{bmatrix}, \quad \begin{bmatrix} 0 & 0 & 0 \\ 0 & 0 & 0 \\ 0 & 0 & 0 \end{bmatrix}, \quad \begin{bmatrix} 1 & 0 & 0 \\ 0 & 1 & 0 \\ 0 & 0 & 1 \end{bmatrix}.$$

If A is any $n \times p$ matrix,

$$A = \begin{bmatrix} A^{(1)} \\ A^{(2)} \\ \vdots \\ A^{(n)} \end{bmatrix} = [A_1, A_2, \ldots, A_p],$$

then

$$\text{diag}\,[\alpha_1, \alpha_2, \ldots, \alpha_n]A = \begin{bmatrix} \alpha_1 A^{(1)} \\ \alpha_2 A^{(2)} \\ \vdots \\ \alpha_n A^{(n)} \end{bmatrix} \tag{2.10.1}$$

and

$$A\,\text{diag}\,[\alpha_1, \alpha_2, \ldots, \alpha_p] = [\alpha_1 A_1, \alpha_2 A_2, \ldots, \alpha_p A_p], \tag{2.10.2}$$

as the reader may verify in detail by executing the indicated computations. For example,

$$\begin{bmatrix} 1 & 0 & 0 \\ 0 & 2 & 0 \\ 0 & 0 & -3 \end{bmatrix} \begin{bmatrix} a_{11} & a_{12} \\ a_{21} & a_{22} \\ a_{31} & a_{32} \end{bmatrix} = \begin{bmatrix} a_{11} & a_{12} \\ 2a_{21} & 2a_{22} \\ -3a_{31} & -3a_{32} \end{bmatrix}$$

and

$$\begin{bmatrix} a_{11} & a_{12} & a_{13} \\ a_{21} & a_{22} & a_{23} \end{bmatrix} \begin{bmatrix} 1 & 0 & 0 \\ 0 & 2 & 0 \\ 0 & 0 & -3 \end{bmatrix} = \begin{bmatrix} a_{11} & 2a_{12} & -3a_{13} \\ a_{21} & 2a_{22} & -3a_{23} \end{bmatrix}.$$

A diagonal matrix in which all diagonal elements are equal is called a **scalar matrix**. By what precedes, we have

$$(2.10.3) \qquad \begin{bmatrix} \alpha & 0 & \cdots & 0 \\ 0 & \alpha & \cdots & 0 \\ \vdots & & & \\ 0 & 0 & \cdots & \alpha \end{bmatrix}_{n \times n} A_{n \times p} = \alpha A_{n \times p} = A_{n \times p} \begin{bmatrix} \alpha & 0 & \cdots & 0 \\ 0 & \alpha & \cdots & 0 \\ \vdots & & & \\ 0 & 0 & \cdots & \alpha \end{bmatrix}_{p \times p}$$

since every row (column) and hence every element of A is multiplied by α. Thus a scalar matrix has the same effect as a scalar in matrix multiplication, which explains the name.

A scalar matrix of order n in which every diagonal element is equal to 1 is called the **identity matrix** of order n or the **unit matrix** of order n:

$$I_n = \begin{bmatrix} 1 & 0 & \cdots & 0 \\ 0 & 1 & \cdots & 0 \\ \vdots & & & \\ 0 & 0 & \cdots & 1 \end{bmatrix}_{n \times n}.$$

We have

$$(2.10.4) \qquad I_n A_{n \times p} = A_{n \times p} I_p = A_{n \times p}.$$

Thus *identity matrices are analogous to the number* 1 *of arithmetic*, as far as multiplication is concerned, since for all scalars α,

$$1 \cdot \alpha = \alpha \cdot 1 = \alpha.$$

Suppose now that X is any $n \times n$ matrix such that

$$AX = XA = A$$

for *every* $n \times n$ matrix A. Then, in particular, when $A = I_n$,

$$I_n X = X I_n = I_n,$$

so that

$$X = I_n.$$

Thus, for given n, the identity matrix is the *only* matrix of order n which behaves in multiplication like 1 in arithmetic.

2.11 THE INVERSE OF A MATRIX

For any scalar $\alpha \neq 0$, there exists another scalar, $\alpha^{-1} \neq 0$, the reciprocal or multiplicative inverse of α, such that

$$\alpha\alpha^{-1} = \alpha^{-1}\alpha = 1.$$

Analogously, for scalar matrices, if $\alpha \neq 0$,

$$\begin{bmatrix} \alpha & 0 & \cdots & 0 \\ 0 & \alpha & \cdots & 0 \\ \vdots \\ 0 & 0 & \cdots & \alpha \end{bmatrix} \begin{bmatrix} \alpha^{-1} & 0 & \cdots & 0 \\ 0 & \alpha^{-1} & \cdots & 0 \\ \vdots \\ 0 & 0 & \cdots & \alpha^{-1} \end{bmatrix} = \begin{bmatrix} \alpha^{-1} & 0 & \cdots & 0 \\ 0 & \alpha^{-1} & \cdots & 0 \\ \vdots \\ 0 & 0 & \cdots & \alpha^{-1} \end{bmatrix} \begin{bmatrix} \alpha & 0 & \cdots & 0 \\ 0 & \alpha & \cdots & 0 \\ \vdots \\ 0 & 0 & \cdots & \alpha \end{bmatrix} = \begin{bmatrix} 1 & 0 & \cdots & 0 \\ 0 & 1 & \cdots & 0 \\ \vdots \\ 0 & 0 & \cdots & 1 \end{bmatrix}.$$

That is, the product of these two matrices is the unit matrix.

The preceding special case suggests the following definition: Given a matrix A, if there exists a matrix B such that

$$AB = BA = I, \tag{2.11.1}$$

we say B is an **inverse** of A. Since then also

$$BA = AB = I,$$

A is an inverse of B; that is, A and B are inverses of each other.

Since AB and BA have the same order if and only if A and B are square and of the same order, it follows that *only square matrices have inverses.*

Suppose now that we have both $AB = BA = I$ and $AC = CA = I$. Then, premultiplying both sides of $AC = I$ by B, we find $B(AC) = BI = B$. Similarly, from $BA = I$, we obtain $(BA)C = IC = C$. But $B(AC) = (BA)C$ by the associative law. Hence $B = C$. Thus we have

Theorem 2.11.1: *A square matrix A has at most one inverse; that is, the inverse is unique if it exists.*

In view of this result, it is proper to refer to *the* inverse of a matrix A and to denote it by the convenient symbol A^{-1}, when the inverse exists.

In some cases, the inverse is readily found by inspection. For instance, we have, in the case of a diagonal matrix D,

$$D = \begin{bmatrix} d_1 & 0 & \cdots & 0 \\ 0 & d_2 & \cdots & 0 \\ \vdots \\ 0 & 0 & \cdots & d_n \end{bmatrix} \quad \text{and} \quad D^{-1} = \begin{bmatrix} d_1^{-1} & 0 & \cdots & 0 \\ 0 & d_2^{-1} & \cdots & 0 \\ \vdots \\ 0 & 0 & \cdots & d_n^{-1} \end{bmatrix},$$

provided that all the d_j's are different from zero, for it is easily verified that, in this case, $DD^{-1} = D^{-1}D = I_n$.

That an inverse does not always exist is shown by the following example. The product

$$\begin{bmatrix} a & b \\ c & d \end{bmatrix} \cdot \begin{bmatrix} 0 & 1 \\ 0 & 0 \end{bmatrix} = \begin{bmatrix} 0 & a \\ 0 & c \end{bmatrix}$$

shows that there is no way to choose a, b, c, and d so as to make the right member equal to I_2. Hence the matrix $\begin{bmatrix} 0 & 1 \\ 0 & 0 \end{bmatrix}$ has no inverse.

Ordinarily, though, one cannot determine by simple inspection whether or not a given square matrix has an inverse. A general answer to this question rests on the theory of linear dependence and will be given in Chapter 5. We can, however, determine the inverse of a square matrix, if it exists, by the sweepout process. Moreover, if no inverse exists, the failure of the sweepout process to produce a solution will reveal the fact.

To see how the inverse may be computed, consider the system of n equations in n variables represented by

$$AX = B. \tag{2.11.2}$$

If A^{-1} exists, premultiplication by A^{-1} shows that any X that satisfies (2.11.2) must be given by

$$X = A^{-1}B. \tag{2.11.3}$$

Substitution from (2.11.3) into (2.11.2) shows that we have indeed a solution. That is, if A^{-1} exists, the equation (2.11.2) has the *unique* solution (2.11.3). This implies that when we apply synthetic elimination to the $n \times (n + 1)$ matrix formed by augmenting the matrix A by the column B:

$$[A, B],$$

when A^{-1} exists the end result must be the array

$$[I_n, A^{-1}B].$$

More generally, one can treat the systems of equations

$$AX = B_1, AX = B_2, \ldots, AX = B_p$$

all at the same time by applying sweepout to the array

$$[A, B_1, B_2, \ldots, B_p].$$

This observation applies whether or not A^{-1} exists. When the elimination process is complete, A has been replaced by the coefficient matrix corresponding to the reduced echelon form of these systems. The last p columns of the array tell the story concerning the solutions of the p systems of equations. In the particular case when A^{-1} exists, the final array is, necessarily,

$$[I_n, A^{-1}B_1, A^{-1}B_2, \ldots, A^{-1}B_p].$$

For the computation of the inverse by sweepout, we make a special choice of the B_j's. We define the **elementary n-vectors** $E_1, E_2, \ldots, E_n$ as follows:

$$E_1 = \begin{bmatrix} 1 \\ 0 \\ 0 \\ \vdots \\ 0 \\ 0 \end{bmatrix}, E_2 = \begin{bmatrix} 0 \\ 1 \\ 0 \\ \vdots \\ 0 \\ 0 \end{bmatrix}, \ldots, E_n = \begin{bmatrix} 0 \\ 0 \\ 0 \\ \vdots \\ 0 \\ 1 \end{bmatrix}.$$

These are just the columns of I_n:

$$I_n = [E_1, E_2, \ldots, E_n].$$

Because of the isolated 1's in these vectors, for any matrix $C_{k \times n}$ we have $CE_1 = C_1$, $CE_2 = C_2, \ldots, CE_n = C_n$. That is, CE_j is the jth column of C for $j = 1, 2, \ldots, n$.

Returning now to the solution of p systems of equations with a common coefficient matrix A of order n, we let $p = n$ and let $B_j = E_j$, $j = 1, 2, \ldots, n$. Then the initial array is

$$[A, E_1, E_2, \ldots, E_n] = [A, I_n]$$

and, if A^{-1} exists, the final array becomes

$$[I_n, A^{-1}E_1, A^{-1}E_2, \ldots, A^{-1}E_n] = [I_n, A_1^{-1}, A_2^{-1}, \ldots, A_n^{-1}] = [I_n, A^{-1}].$$

That is, to invert A we use the sweepout process to reduce $[A, I_n]$ to $[I_n, A^{-1}]$.

We do not need to know that A^{-1} exists before beginning this process. If A^{-1} does not exist, it will not be possible to reduce the left block to I_n. That is, the sweepout process determines whether or not A has an inverse and produces the inverse when it exists.

We illustrate with an example. Find

$$\begin{bmatrix} 3 & 1 & 2 \\ 1 & -4 & 1 \\ 2 & 3 & 0 \end{bmatrix}^{-1}.$$

The arithmetic is given in Table 2.1, where the starred member of each set of rows is computed first.

Note that the technique is first to get a 1 in the 1,1-position, then use it to sweep the rest of the first column. Then we arrange to get a 1 in the 2,2-position and use it to sweep the rest of the second column, etc. Proceeding thus, we eventually obtain an identity matrix on the left, and A^{-1} on the right, or we eventually obtain a row of zeros on the left, in which case an identity matrix cannot be obtained on the left, so that A^{-1} does not exist.

In paper-and-pencil computation, the sequence of operations one employs in this process need not be unique. One arranges it so as to exploit any obviously

TABLE 2.1

Key to Operations	A to I			I to A^{-1}		
R_1	3	1	2	1	0	0
R_2	1	−4	1	0	1	0
R_3	2	3	0	0	0	1
$*R'_1 = R_2$	1	−4	1	0	1	0
$R'_2 = R_3 - 2R_2$	0	11	−2	0	−2	1
$R'_3 = R_1 - 3R_2$	0	13	−1	1	−3	0
$R''_1 = R'_1 + 4R''_2$	1	0	$\frac{3}{11}$	0	$\frac{3}{11}$	$\frac{4}{11}$
$*R''_2 = \frac{1}{11}R'_2$	0	1	$-\frac{2}{11}$	0	$-\frac{2}{11}$	$\frac{1}{11}$
$R''_3 = R'_3 - 13R''_2$	0	0	$\frac{15}{11}$	1	$-\frac{7}{11}$	$-\frac{13}{11}$
$R'''_1 = R''_1 - \frac{3}{11}R'''_3$	1	0	0	$-\frac{3}{15}$	$\frac{6}{15}$	$\frac{9}{15}$
$R'''_2 = R''_2 + \frac{2}{11}R'''_3$	0	1	0	$\frac{2}{15}$	$-\frac{4}{15}$	$-\frac{1}{15}$
$*R'''_3 = \frac{11}{15}R''_3$	0	0	1	$\frac{11}{15}$	$-\frac{7}{15}$	$-\frac{13}{15}$

advantageous circumstances. This is illustrated in the example by our using R_2 as R'_1 rather than using $\frac{1}{3}R_1$ as R'_1. On the other hand, when this procedure is programmed for a computer, a completely systematic sequence of steps is used.

We give next a result that is often useful.

Theorem 2.11.2: *If the $n \times n$ matrices $A, B, \ldots, M, N$, all have inverses, then their product $AB \cdots MN$ has the inverse $N^{-1}M^{-1} \cdots B^{-1}A^{-1}$; that is, the inverse of a product is the product of the inverses in the reverse order.*

In fact, by the associative law

$$\begin{aligned}(AB \cdots MN)(N^{-1}M^{-1} \cdots B^{-1}A^{-1}) &= (AB \cdots M)(NN^{-1})(M^{-1} \cdots B^{-1}A^{-1}) \\ &= (AB \cdots M)(M^{-1} \cdots B^{-1}A^{-1}) \\ &= \cdots \\ &= AA^{-1} \\ &= I.\end{aligned}$$

The product $(N^{-1}M^{-1} \cdots B^{-1}A^{-1})(AB \cdots MN)$ reduces to I in a similar fashion, so the theorem is proved.

For example, the inverse of the product

$$AB = \begin{bmatrix} 2 & 0 \\ 0 & 3 \end{bmatrix} \cdot \begin{bmatrix} 1 & -1 \\ 0 & 1 \end{bmatrix}$$

is

$$B^{-1}A^{-1} = \begin{bmatrix} 1 & 1 \\ 0 & 1 \end{bmatrix} \cdot \begin{bmatrix} \frac{1}{2} & 0 \\ 0 & \frac{1}{3} \end{bmatrix} = \begin{bmatrix} \frac{1}{2} & \frac{1}{3} \\ 0 & \frac{1}{3} \end{bmatrix},$$

as is readily checked.

Finally, we note that the inverse is useful in solving certain types of matrix equations. Thus, if $A_{n\times n}X_{n\times p} = B_{n\times p}$ and, if A^{-1} exists, then $X = A^{-1}B$.

Again, if A^{-1} and B^{-1} exist, then

$$AWB = C$$

if and only if

$$W = A^{-1}CB^{-1},$$

so that the second equation gives the unique solution for W of the first. Note the importance of distinguishing between pre- and postmultiplication here.

2.12 THE TRANSPOSE OF A MATRIX

The $n \times m$ matrix A^{T} obtained from an $m \times n$ matrix A by interchanging rows and columns is called the **transpose** of A. For example, the transpose of

$$\begin{bmatrix} 3 & -1 & 2 \\ 4 & 0 & 1 \end{bmatrix} \quad \text{is} \quad \begin{bmatrix} 3 & 4 \\ -1 & 0 \\ 2 & 1 \end{bmatrix}.$$

Theorem 2.12.1: *If A^{T} and B^{T} are the transposes of A and B, and if α is any scalar, then*

(a) $(A^{\mathsf{T}})^{\mathsf{T}} = A$.
(b) $(A + B)^{\mathsf{T}} = A^{\mathsf{T}} + B^{\mathsf{T}}$.
(c) $(\alpha A)^{\mathsf{T}} = \alpha A^{\mathsf{T}}$.
(d) $(AB)^{\mathsf{T}} = B^{\mathsf{T}}A^{\mathsf{T}}$.

The first three of these rules are easy to think through. Detailed proofs and illustrative examples are left to the reader. Only (d) will be proved here. Let us first illustrate with an example. Let

$$A = \begin{bmatrix} 1 & 2 & -1 \\ 3 & 0 & -2 \end{bmatrix}, \qquad B = \begin{bmatrix} a & \alpha \\ b & \beta \\ c & \gamma \end{bmatrix}.$$

Then

$$AB = \begin{bmatrix} a + 2b - c & \alpha + 2\beta - \gamma \\ 3a - 2c & 3\alpha - 2\gamma \end{bmatrix},$$

so

$$(AB)^{\mathsf{T}} = \begin{bmatrix} a + 2b - c & 3a - 2c \\ \alpha + 2\beta - \gamma & 3\alpha - 2\gamma \end{bmatrix}.$$

Also

$$B^{\mathsf{T}} = \begin{bmatrix} a & b & c \\ \alpha & \beta & \gamma \end{bmatrix}, \qquad A^{\mathsf{T}} = \begin{bmatrix} 1 & 3 \\ 2 & 0 \\ -1 & -2 \end{bmatrix},$$

so

$$B^{\mathsf{T}}A^{\mathsf{T}} = \begin{bmatrix} a + 2b - c & 3a - 2c \\ \alpha + 2\beta - \gamma & 3\alpha - 2\gamma \end{bmatrix}.$$

Therefore, in this case,

$$(AB)^{\mathsf{T}} = B^{\mathsf{T}}A^{\mathsf{T}}.$$

For the proof, let $A = [a_{ik}]_{(m,n)}$ and $B = [b_{kj}]_{(n,p)}$. Then $AB = [c_{ij}]_{(m,p)}$, where $c_{ij} = \sum_{k=1}^{n} a_{ik}b_{kj}$. Here i, ranging from 1 to m, identifies the row, and j, ranging from 1 to p, identifies the column of the element c_{ij}.

Now B^{T} is of order (p, n) and A^{T} is of order (n, m), so that B^{T} is conformable to A^{T} for multiplication. To compute the element γ_{ji} in the jth row and the ith column of $B^{\mathsf{T}}A^{\mathsf{T}}$, we must multiply the jth row of B^{T} into the ith column of A^{T}. Observing that the second subscript of an element in B or A identifies the row and the first subscript identifies the column in which it appears in B^{T} or A^{T}, we see that

$$B^{\mathsf{T}}A^{\mathsf{T}} = [\gamma_{ji}]_{(p,m)} = \left[\sum_{k=1}^{n} b_{kj}a_{ik}\right]_{(p,m)} = \left[\sum_{k=1}^{n} a_{ik}b_{kj}\right]_{(p,m)}.$$

Here j ranges from 1 to p and identifies the *row* of γ_{ji}, whereas i ranges from 1 to m and identifies the *column* of γ_{ji}. Thus $\gamma_{ji} = c_{ij}$ but with the meanings of i and j for rows and columns just opposite in the two cases, so that $B^{\mathsf{T}}A^{\mathsf{T}} = (AB)^{\mathsf{T}}$. That is, *the transpose of the product of two matrices is the product of their transposes in the reverse order.*

2.13 SYMMETRIC, SKEW-SYMMETRIC, AND HERMITIAN MATRICES

A **symmetric matrix** is a square matrix A such that $A = A^{\mathsf{T}}$. A **skew-symmetric matrix** is a square matrix A such that $A = -A^{\mathsf{T}}$. These definitions may also be stated in terms of the individual elements: A is symmetric if and only if $a_{ij} = a_{ji}$ for all pairs of subscripts; it is skew-symmetric if and only if $a_{ij} = -a_{ji}$ for all pairs of subscripts. The reader should demonstrate the equivalence of the alternative definitions in each case.

The following are examples of symmetric and skew-symmetric matrices, respectively:

$$\text{(a)} \begin{bmatrix} 0 & 1 & 2 \\ 1 & 2 & 3 \\ 2 & 3 & 4 \end{bmatrix}; \quad \text{(b)} \begin{bmatrix} 0 & 1 & 2 \\ -1 & 0 & 3 \\ -2 & -3 & 0 \end{bmatrix}.$$

Example (b) illustrates the fact that *the main diagonal elements of a skew-symmetric matrix must all be zero.* Why is this true?

A matrix is called a **real matrix** if and only if all its elements are real. In the applications, real symmetric matrices occur most frequently. However, matrices of complex elements are also of importance. When the elements of such a matrix A are replaced by their complex conjugates, the resulting matrix is called the **conjugate** of A and is denoted by $\bar{A}$. Evidently a matrix A is real if and only if $A = \bar{A}$. Transposing $\bar{A}$, we obtain the **transposed conjugate** or **tranjugate** $(\bar{A})^{\mathsf{T}}$ of A. This will be denoted by the symbol A^*. (A^* is sometimes called the adjoint of A, but *not in this book*.) For example, if

$$A = \begin{bmatrix} 1 - i & 2 \\ i & 1 + i \end{bmatrix},$$

then

$$\bar{A} = \begin{bmatrix} 1 + i & 2 \\ -i & 1 - i \end{bmatrix}$$

and

$$A^* = \begin{bmatrix} 1 + i & -i \\ 2 & 1 - i \end{bmatrix}.$$

When $A = A^*$, that is, when $a_{ij} = \bar{a}_{ji}$ for all pairs of subscripts, A is called a **Hermitian matrix** (after the French mathematician, Charles Hermite, 1822–1901). When matrices of complex elements appear in the applications, for example, in the theory of atomic physics, they are often Hermitian. The matrices

$$\begin{bmatrix} 0 & i \\ -i & 0 \end{bmatrix} \quad \text{and} \quad \begin{bmatrix} 4 & 1 - i \\ 1 + i & 2 \end{bmatrix}$$

are simple examples of Hermitian matrices, as is readily verified. *The diagonal elements of a Hermitian matrix must all be real numbers.* Why?

If the elements of A are real, $A^{\mathsf{T}} = A^*$, so the property of being real and symmetric is a special case of the property of being Hermitian. The following pages will contain a great many results about Hermitian matrices. The reader interested only in the real case may interpret the word "Hermitian" as "real symmetric" and, in most instances, all will be well.

2.14 POLYNOMIAL FUNCTIONS OF MATRICES

Consider a polynomial

$$f(x) = a_p x^p + a_{p-1} x^{p-1} + \cdots + a_1 x + a_0$$

of degree p in an indeterminate x, with either real or complex coefficients. With any such polynomial, we can associate a polynomial function of an $n \times n$ matrix A:

$$f(A) = a_p A^p + a_{p-1} A^{p-1} + \cdots + a_1 A + a_0 I_n$$

simply by replacing the constant term a_0 by the scalar matrix $a_0 I_n$ and by replacing x by A throughout. For example, if

$$f(x) = x^2 - 3x - 2,$$

then, for any matrix A of order n,

$$f(A) = A^2 - 3A - 2I_n.$$

Such polynomial functions of matrices are frequently useful.

If we can factor $f(x)$ in a certain way, then, because powers of a square matrix A are subject to the same laws of exponents as powers of x and because the commutative, associative, and distributive laws apply in the same way in both cases, we can factor $f(A)$ analogously. For example, if

$$f(x) = x^3 - 3x^2 + 2x = x(x - 1)(x - 2),$$

then

$$f(A) = A^3 - 3A^2 + 2A = A(A - I_n)(A - 2I_n).$$

Now suppose that the solutions of the polynomial equation

$$f(x) = a_p x^p + a_{p-1}x^{p-1} + \cdots + a_1 x + a_0 = 0$$

are $\alpha_1, \alpha_2, \ldots, \alpha_p$ so that, by the factor theorem,

$$f(x) \equiv a_p(x - \alpha_1)(x - \alpha_2)\cdots(x - \alpha_p).$$

The analogous factoring of $f(A)$ is

$$f(A) \equiv a_p(A - \alpha_1 I)(A - \alpha_2 I)\cdots(A - \alpha_p I), \tag{2.14.1}$$

so that the matrix equation $f(A) = 0$ has for solutions the scalar matrices $\alpha_1 I, \ldots, \alpha_p I$. Moreover, it has no other *scalar* matrices as solutions. Indeed, if $\beta \neq \alpha_i$ for all i, $f(\beta I)$ is a nonzero scalar matrix. (What are its diagonal entries?) However, since the product of matrices (2.14.1) may be zero even though no factor is zero, the equation $f(A) = 0$ may also have *nonscalar* matrices as solutions. For example, the equation

$$A^2 + I_2 = 0$$

has for solutions the scalar matrices

$$\begin{bmatrix} i & 0 \\ 0 & i \end{bmatrix}, \quad \begin{bmatrix} -i & 0 \\ 0 & -i \end{bmatrix},$$

corresponding to the solutions i and $-i$ of the equation $x^2 + 1 = 0$. It also has (among others) the nonscalar solution

$$\begin{bmatrix} 1 & 2 \\ -1 & -1 \end{bmatrix},$$

as the reader should verify.

This example illustrates the fact that whereas in scalar algebra a polynomial equation of degree p in a single unknown x has exactly p solutions, the same does not hold true in matrix algebra, where the number of solutions is always *at least p*.

2.15 EXERCISES

***1.** (a) Show that any two diagonal matrices of the same order commute.

(b) Give a formula for D^p, where D is diagonal and where p is a positive integer.

***2.** Given that $d_1 d_2 \cdots d_n \neq 0$ and that $D = \text{diag}\,[d_1, d_2, \ldots, d_n]$, show that D^{-1} exists. Then define D^p, where p is any integer, and give a formula for D^p.

***3.** Show that a matrix A of order n is scalar if and only if it commutes with *every* matrix B of order n.

4. Find an inverse for A, where

$$A = \begin{bmatrix} 1 & 1 & 0 \\ 0 & 1 & 1 \\ 0 & 0 & 1 \end{bmatrix}$$

by solving this equation for the b's:

$$\begin{bmatrix} 1 & 1 & 0 \\ 0 & 1 & 1 \\ 0 & 0 & 1 \end{bmatrix} \begin{bmatrix} b_{11} & b_{12} & b_{13} \\ b_{21} & b_{22} & b_{23} \\ b_{31} & b_{32} & b_{33} \end{bmatrix} = \begin{bmatrix} 1 & 0 & 0 \\ 0 & 1 & 0 \\ 0 & 0 & 1 \end{bmatrix}.$$

5. Find by inspection inverses for the matrices:

$$\text{(a)} \begin{bmatrix} 0 & 0 & \cdots & 0 & d_n \\ 0 & 0 & \cdots & d_{n-1} & 0 \\ \vdots & & & & \\ 0 & d_2 & \cdots & 0 & 0 \\ d_1 & 0 & \cdots & 0 & 0 \end{bmatrix} \qquad (d_1 d_2 \cdots d_n \neq 0).$$

$$\text{(b)} \begin{bmatrix} 1 & 0 & 0 & 0 \\ 0 & 1 & 1 & 0 \\ 0 & 0 & 1 & 0 \\ 0 & 0 & 0 & 1 \end{bmatrix}, \quad \text{(c)} \begin{bmatrix} 0 & 1 & 0 & 0 \\ 0 & 0 & 1 & 0 \\ 0 & 0 & 0 & 1 \\ 1 & 0 & 0 & 0 \end{bmatrix}, \quad \text{(d)} \begin{bmatrix} 1 & 0 & 0 & 0 \\ 0 & 1 & 0 & 0 \\ 0 & 0 & 1 & 0 \\ a & b & c & 1 \end{bmatrix}.$$

(Remember that since $A \cdot A^{-1} = I$, the ith row of A times the jth column of A^{-1} must yield 1 if $i = j$, but 0 otherwise.)

6. Evaluate the product

$$\begin{bmatrix} a_{11} & a_{12} \\ a_{21} & a_{22} \end{bmatrix} \cdot \begin{bmatrix} a_{22} & -a_{12} \\ -a_{21} & a_{11} \end{bmatrix}$$

and hence determine an inverse for the first factor of the product when one exists. Under what conditions is there no inverse?

7. Compute the inverses of these matrices by the sweepout process:

$$\text{(a)} \begin{bmatrix} 1 & 2 & -1 \\ 2 & 3 & 0 \\ -1 & 0 & 4 \end{bmatrix}, \qquad \text{(b)} \begin{bmatrix} 1 & 1 & 0 & -1 \\ 1 & 0 & -1 & 1 \\ 0 & -1 & 1 & 1 \\ -1 & 1 & 1 & 0 \end{bmatrix}.$$

8. Show by the sweepout process that this matrix has no inverse:

$$\begin{bmatrix} 2 & 1 & 0 \\ -1 & 1 & -3 \\ 3 & 2 & -1 \end{bmatrix}.$$

9. Given that

$$\begin{bmatrix} 1 & 0 & 0 \\ 0 & \frac{1}{2} & 0 \\ 0 & 0 & -\frac{1}{3} \end{bmatrix} \cdot W \cdot \begin{bmatrix} 1 & 1 \\ 1 & 2 \end{bmatrix} = \begin{bmatrix} 1 & 0 \\ 0 & 1 \\ 0 & 0 \end{bmatrix},$$

solve for the matrix W.

10. Find the inverse of this product without first computing the product:

$$\begin{bmatrix} 0 & 0 & 1 \\ 0 & 1 & 0 \\ 1 & 0 & 0 \end{bmatrix} \begin{bmatrix} 1 & 0 & 0 \\ 0 & 2 & 0 \\ 0 & 0 & 3 \end{bmatrix} \begin{bmatrix} 0 & 1 & 0 \\ 0 & 0 & 1 \\ 1 & 1 & 1 \end{bmatrix}.$$

***11.** Let $M_1, M_2, \ldots, M_k$ denote any finite sequence of matrices conformable for multiplication in the order given. Prove by induction that

$$(M_1 M_2 \cdots M_k)^\mathsf{T} = M_k^\mathsf{T} \cdots M_2^\mathsf{T} M_1^\mathsf{T}.$$

12. Show by an example that $AA^\mathsf{T} \neq A^\mathsf{T}A$ in general.

***13.** Prove that AA^T and $A^\mathsf{T}A$ are symmetric. Illustrate with an example.

***14.** Prove that if A is real, AA^T and $A^\mathsf{T}A$ have nonnegative principal diagonal elements. Illustrate with an example.

15. What may be said about the main diagonal elements of AA^T when A is a real, skew-symmetric matrix? What about those of A^2 in this case?

16. Prove that if A is symmetric or skew-symmetric, then $AA^\mathsf{T} = A^\mathsf{T}A$. Prove also that, in either case, A^2 is symmetric. Are higher powers of A also symmetric in either case?

17. Given that $A = [a_{\alpha\beta}]_{n \times n}$ and $B = [b_{\alpha\beta}]_{n \times n}$, use the concept of the transpose to represent each of the following as a matrix product (i represents the row and j represents the column of the entry):

$$\left[\sum_{k=1}^{n} a_{ki} b_{jk}\right], \quad \left[\sum_{k=1}^{n} a_{ik} b_{jk}\right], \quad \left[\sum_{k=1}^{n} b_{ik} a_{kj}\right], \quad \left[\sum_{k=1}^{n} b_{kj} a_{ik}\right].$$

18. Show that every square matrix A can be represented uniquely in the form

$$A = A^{(S)} + A^{(SS)},$$

where $A^{(S)}$ is symmetric and $A^{(SS)}$ is skew-symmetric. (*Hint*: If $A = A^{(S)} + A^{(SS)}$, then $A^{\mathsf{T}} = A^{(S)} - A^{(SS)}$, and so on.)

19. Show by means of an example that even though A and B are both symmetric and are of the same order, AB need not necessarily be symmetric.

20. (a) Prove that if A and B of order n are both symmetric or both skew-symmetric and commute, then AB is symmetric.

(b) Given that A, B, and C are all symmetric and that $ABC = CBA$, prove that ABC is symmetric. Generalize.

***21.** Denoting by E_j an n-vector with 1 in the jth row, all other components being 0, interpret each of the products

$$E_j^{\mathsf{T}}A, \quad AE_j, \quad E_i^{\mathsf{T}}AE_j, \quad E_j^{\mathsf{T}}E_k, \quad E_jE_k^{\mathsf{T}},$$

where A is an arbitrary matrix of order n. (The vectors E_j are the elementary n-vectors defined in Section 2.11.)

22. Let $W_n = \sum_{j=1}^{n} E_j$. Then how are AW_n and $W_n^{\mathsf{T}}A$ related to A? Evaluate $W_n^{\mathsf{T}}X$, where X is an arbitrary n-vector.

23. Evaluate each of the following products, then express the results in words. The E's are the elementary n-vectors.

$$(E_{i_1} + E_{i_2} + \cdots + E_{i_k})^{\mathsf{T}}A_{n \times m}, \quad B_{p \times n}(E_{j_1} + E_{j_2} + \cdots + E_{j_r}),$$
$$(E_{i_1} + E_{i_2} + \cdots + E_{i_k})^{\mathsf{T}}C_{n \times n}(E_{j_1} + E_{j_2} + \cdots + E_{j_r}).$$

***24.** (a) Denoting by E_{ij} an $n \times n$ matrix with a 1 in the ith row and jth column and 0's elsewhere, evaluate these products, in which A is an arbitrary matrix of order n:

$$E_{ij}E_{jk}, \quad E_{ij}E_{km}, \quad AE_{ij}, \quad E_{ij}A, \quad E_{ij}AE_{jk}, \quad E_{ij}AE_{km}.$$

(b) Show that

$$E_{ij} = E_iE_j^{\mathsf{T}}, \qquad E_{jj}^2 = E_{jj},$$

and

$$(E_{ij}E_{jk})^{\mathsf{T}} = E_{ki} = E_{kj}E_{ji}.$$

(c) What matrices are defined by

$$(\alpha) \sum_{j=1}^{n} E_jE_j^{\mathsf{T}}, \qquad (\beta) \sum_{i=1}^{n}\sum_{j=1}^{n} E_iE_j^{\mathsf{T}}?$$

25. Show that

$$\sigma_x^2 = \sigma_y^2 = \sigma_z^2 = I_2^2 = I_2,$$

where σ_x, σ_y, and σ_z are the Pauli spin matrices defined in Section 2.9, Exercise 25. What rule about roots in scalar algebra fails to carry over into matrix algebra?

*26. Show that, if α is a scalar, $(\alpha A)^* = \bar{\alpha}A^*$. Show also that $\overline{(\bar{A})} = A$, $(\bar{A})^{\mathsf{T}} = (\overline{A^{\mathsf{T}}})$, $(A^*)^* = A$, $(A + B)^* = A^* + B^*$, $\overline{AB} = \bar{A}\bar{B}$, and $(AB)^* = B^*A^*$.

*27. (a) Show that, for every matrix A, the products A^*A and AA^* are Hermitian matrices. (b) What may be said about the diagonal entries of A^*A and AA^*? (c) Show that, if H is Hermitian, so is B^*HB for every conformable matrix B.

28. If A is square and if $A = -A^*$, that is, if $a_{ji} = -\bar{a}_{ij}$ for all i and j, A is called **skew-Hermitian**. Show that every square matrix A over the complex field can be written in the form $A = A^{(H)} + A^{(SH)}$ where $A^{(H)}$ is Hermitian and $A^{(SH)}$ is skew-Hermitian. Show also that the diagonal elements of a skew-Hermitian matrix are pure imaginaries.

29. Show that every Hermitian matrix H may be written in the form $A + iB$, where A is real and symmetric, B is real and skew-symmetric, and $i^2 = -1$. Then show that H^*H is real if and only if $AB = -BA$, that is, if and only if A and B anticommute.

30. Find all real or complex scalar matrices which satisfy the equations

$$\begin{aligned}
&\text{(a)}\ A^2 - 5A + 6I_2 = 0,\\
&\text{(b)}\ A^3 + 6A^2 + 12A + 8I_2 = 0,\\
&\text{(c)}\ A^3 - A = 0,\\
&\text{(d)}\ A^4 - I_n = 0.
\end{aligned}$$

31. Find all real or complex solutions of the form $\begin{bmatrix} 0 & a \\ b & 0 \end{bmatrix}$ of the equation $A^3 + A = 0$.

32. Determine all diagonal matrices of order n which satisfy a polynomial matrix equation $f(A) = 0$ of degree p over the complex field.

33. Show that any two polynomial functions of a square matrix A are commutative with respect to multiplication.

*34. Show that arbitrary polynomials $f(A)$ and $g(B)$ in fixed matrices A and B of order n commute if and only if A and B commute. (It suffices to show that $A^rB^s = B^sA^r$ for all positive integers r and s if and only if $AB = BA$.)

2.16 PARTITIONING OF MATRICES

It is often useful to subdivide or partition matrices into rectangular blocks of elements and to treat these blocks as matrices. That is, we treat the original matrix as a matrix whose elements are matrices. The blocks are called **sub-matrices** of the original matrix.

For example, we can represent the matrix

$$A = \begin{bmatrix} 1 & -1 & 4 \\ 2 & 3 & 0 \end{bmatrix}$$

as $[A_1, A_2, A_3]$, where

$$A_1 = \begin{bmatrix} 1 \\ 2 \end{bmatrix}, \qquad A_2 = \begin{bmatrix} -1 \\ 3 \end{bmatrix}, \qquad A_3 = \begin{bmatrix} 4 \\ 0 \end{bmatrix},$$

or as

$$\begin{bmatrix} A^{(1)} \\ A^{(2)} \end{bmatrix}$$

where

$$A^{(1)} = [1, -1, 4], \qquad A^{(2)} = [2, 3, 0].$$

This sort of partitioning has already been used in the discussion of the matrix product. We could also write

$$A = [C_1, C_2],$$

where

$$C_1 = \begin{bmatrix} 1 & -1 \\ 2 & 3 \end{bmatrix}, \qquad C_2 = \begin{bmatrix} 4 \\ 0 \end{bmatrix}.$$

The manner in which a matrix is to be partitioned may be indicated by dashed lines. For example,

$$\begin{bmatrix} 4 & 0 & -2 \\ 2 & 1 & 3 \\ -1 & -1 & -1 \end{bmatrix} = \left[\begin{array}{c|cc} 4 & 0 & -2 \\ 2 & 1 & 3 \\ \hline -1 & -1 & -1 \end{array}\right] = \begin{bmatrix} A & B \\ C & D \end{bmatrix},$$

where A, B, C, and D denote the submatrices indicated by the dashed lines. As the example illustrates, two matrices, partitioned or not, are equal if and only if their nonpartitioned forms are equal.

If we wish to compute with the submatrices of partitioned matrices, we must take care to observe all requirements of conformability. For example, we may write the equation

$$\begin{bmatrix} A_1 & B_1 \\ C_1 & D_1 \end{bmatrix} + \begin{bmatrix} A_2 & B_2 \\ C_2 & D_2 \end{bmatrix} = \begin{bmatrix} (A_1 + A_2) & (B_1 + B_2) \\ (C_1 + C_2) & (D_1 + D_2) \end{bmatrix}$$

only if A_1 and A_2 have the same order, and similarly for B_1 and B_2, C_1 and C_2, D_1 and D_2. The condition is also sufficient.

Now let A and B denote arbitrary matrices of the same order. We shall then say that A and B are **identically partitioned** if the resulting matrices *of matrices* contain the same number of rows and the same number of columns and if, in addition, corresponding submatrices have the same order. Thus the matrices

$$\left[\begin{array}{ccc|c} 2 & -1 & 4 & 3 \\ 1 & 2 & -1 & 0 \\ \hline 1 & 1 & 0 & 1 \end{array}\right] \quad \text{and} \quad \left[\begin{array}{ccc|c} a_1 & b_1 & c_1 & d_1 \\ a_2 & b_2 & c_2 & d_2 \\ \hline a_3 & b_3 & c_3 & d_3 \end{array}\right]$$

are identically partitioned. It is not hard to see that *identically partitioned matrices are equal if and only if corresponding submatrices are equal throughout* and that *identically partitioned matrices may be added by adding corresponding submatrices throughout.*

To show how partitioning into submatrices is used in matrix multiplication, we introduce several examples.

1. Let

$$A = \left[\begin{array}{cc|c} 1 & 0 & a \\ 0 & 1 & b \end{array}\right], \qquad B = \left[\begin{array}{cc} 1 & 0 \\ 0 & 1 \\ \hline c & d \end{array}\right].$$

Then, if we treat the submatrices as though they were elements, we have

$$\left[\begin{bmatrix} 1 & 0 \\ 0 & 1 \end{bmatrix}\begin{bmatrix} a \\ b \end{bmatrix}\right]\begin{bmatrix} \begin{bmatrix} 1 & 0 \\ 0 & 1 \end{bmatrix} \\ [c \quad d] \end{bmatrix} = \left[\begin{bmatrix} 1 & 0 \\ 0 & 1 \end{bmatrix}\begin{bmatrix} 1 & 0 \\ 0 & 1 \end{bmatrix} + \begin{bmatrix} a \\ b \end{bmatrix} \cdot [c \quad d]\right]$$

$$= \left[\begin{bmatrix} 1 & 0 \\ 0 & 1 \end{bmatrix} + \begin{bmatrix} ac & ad \\ bc & bd \end{bmatrix}\right] = \begin{bmatrix} 1 + ac & ad \\ bc & 1 + bd \end{bmatrix} = AB.$$

2. Let

$$A = \left[\begin{array}{cc|cc} 1 & 0 & 0 & 1 \\ 0 & 1 & 1 & 0 \\ \hline 0 & 0 & 0 & -1 \\ 0 & 0 & -1 & 0 \end{array}\right], \qquad B = \left[\begin{array}{cc|cc} 0 & 1 & 1 & 0 \\ 1 & 0 & 0 & 1 \\ \hline -1 & 0 & 0 & 0 \\ 0 & -1 & 0 & 0 \end{array}\right].$$

Now, because of the identity submatrices and the zero submatrices, if we again treat submatrices as elements, we have as the product

$$\left[\begin{array}{c|c} \begin{bmatrix} 0 & 1 \\ 1 & 0 \end{bmatrix} + \begin{bmatrix} 0 & -1 \\ -1 & 0 \end{bmatrix} & \begin{bmatrix} 1 & 0 \\ 0 & 1 \end{bmatrix} \\ \hline \begin{bmatrix} 0 & 1 \\ 1 & 0 \end{bmatrix} & \begin{bmatrix} 0 & 0 \\ 0 & 0 \end{bmatrix} \end{array}\right] = \left[\begin{array}{cc|cc} 0 & 0 & 1 & 0 \\ 0 & 0 & 0 & 1 \\ \hline 0 & 1 & 0 & 0 \\ 1 & 0 & 0 & 0 \end{array}\right] = AB.$$

This example illustrates one of the principal advantages of partitioning: Appropriate partitioning often enables one to exploit particularly simple submatrices in the multiplication process.

It is not hard in principle to prove that this procedure of multiplying partitioned matrices is valid in general. The necessary notation is somewhat involved, however, so we content ourselves with an example that is illustrative of the general situation.

Let

$$A = \begin{bmatrix} a_{11} & a_{12} & a_{13} \\ a_{21} & a_{22} & a_{23} \\ a_{31} & a_{32} & a_{33} \end{bmatrix} \quad \text{and} \quad B = \begin{bmatrix} b_{11} & b_{12} \\ b_{21} & b_{22} \\ b_{31} & b_{32} \end{bmatrix}.$$

Let us partition A and B, designating the submatrices with double subscripts:

$$A = \left[\begin{array}{cc|c} a_{11} & a_{12} & a_{13} \\ a_{21} & a_{22} & a_{23} \\ \hline a_{31} & a_{32} & a_{33} \end{array}\right] = \begin{bmatrix} A_{11} & A_{12} \\ A_{21} & A_{22} \end{bmatrix}, \qquad B = \left[\begin{array}{cc} b_{11} & b_{12} \\ b_{21} & b_{22} \\ \hline b_{31} & b_{32} \end{array}\right] = \begin{bmatrix} B_{11} \\ B_{21} \end{bmatrix}.$$

Then

$$\begin{bmatrix} A_{11} & A_{12} \\ A_{21} & A_{22} \end{bmatrix} \cdot \begin{bmatrix} B_{11} \\ B_{21} \end{bmatrix}$$

$$= \begin{bmatrix} (A_{11}B_{11} + A_{12}B_{21}) \\ (A_{21}B_{11} + A_{22}B_{21}) \end{bmatrix}$$

$$= \begin{bmatrix} \begin{bmatrix} (a_{11}b_{11} + a_{12}b_{21}) & (a_{11}b_{12} + a_{12}b_{22}) \\ (a_{21}b_{11} + a_{22}b_{21}) & (a_{21}b_{12} + a_{22}b_{22}) \end{bmatrix} + \begin{bmatrix} a_{13}b_{31} & a_{13}b_{32} \\ a_{23}b_{31} & a_{23}b_{32} \end{bmatrix} \\ [(a_{31}b_{11} + a_{32}b_{21}) \quad (a_{31}b_{12} + a_{32}b_{22})] + [a_{33}b_{31} \quad a_{33}b_{32}] \end{bmatrix}$$

$$= \begin{bmatrix} \begin{bmatrix} a_{11}b_{11} + a_{12}b_{21} + a_{13}b_{31} & a_{11}b_{12} + a_{12}b_{22} + a_{13}b_{32} \\ a_{21}b_{11} + a_{22}b_{21} + a_{23}b_{31} & a_{21}b_{12} + a_{22}b_{22} + a_{23}b_{32} \end{bmatrix} \\ [a_{31}b_{11} + a_{32}b_{21} + a_{33}b_{31} \quad a_{31}b_{12} + a_{32}b_{22} + a_{33}b_{32}] \end{bmatrix}$$

$= AB$ (after the internal brackets are dropped).

The general proof demonstrates the same thing: that portions of the complete entries of the product are computed by the multiplication of the submatrices and that precisely the complete entries are obtained by summing the several portions in the way the multiplication process dictates.

2.17 EXERCISES

1. Compute these products, using the indicated partitioning:

(a) $$\left[\begin{array}{ccc|cc} 0 & 2 & 1 & 0 & 0 \\ -2 & 0 & -1 & 0 & 0 \\ -1 & 1 & 0 & 0 & 0 \\ \hline 0 & 0 & 0 & 1 & 0 \\ 0 & 0 & 0 & 0 & 1 \end{array}\right] \left[\begin{array}{cc|ccc} 0 & 0 & 0 & 2 & 1 \\ 0 & 0 & -2 & 0 & -1 \\ 0 & 0 & -1 & 1 & 0 \\ \hline 1 & 0 & 0 & 0 & 0 \\ 0 & 1 & 0 & 0 & 0 \end{array}\right],$$

(b) $$\left[\begin{array}{cc|cc|c} 1 & 0 & 0 & 0 & 1 \\ 0 & 1 & 0 & 0 & 1 \\ \hline 0 & 0 & 0 & 1 & 0 \\ 0 & 0 & 1 & 0 & 0 \\ \hline 1 & 1 & 0 & 0 & 1 \end{array}\right]^2,$$

(c) $$\left[\begin{array}{c|cc|cc} 1 & 0 & 0 & 0 & 0 \\ \hline 0 & 0 & 1 & 1 & 0 \\ 0 & 1 & 0 & 0 & 1 \\ \hline 0 & 1 & 0 & 0 & 0 \\ 0 & 0 & 1 & 0 & 0 \end{array}\right]^2,$$

(d)
$$\left[\begin{array}{rr|rr|r} -1 & 2 & 0 & 0 & 0 \\ 2 & -1 & 0 & 0 & 0 \\ \hline 0 & 0 & 2 & 1 & 1 \\ 0 & 0 & 1 & 2 & 1 \\ \hline 0 & 0 & 1 & 1 & 2 \end{array}\right] \cdot \left[\begin{array}{rr|rr|r} -1 & -2 & 1 & 0 & 0 \\ -2 & -1 & 0 & 1 & 0 \\ \hline 1 & 0 & 1 & 0 & 0 \\ 0 & 1 & 0 & 1 & 0 \\ \hline 0 & 0 & 0 & 0 & 1 \end{array}\right].$$

***2.** Let A be $m \times n$ and let B be $n \times p$. Then we can partition A into rows and B into columns, thus:

$$A = \begin{bmatrix} A^{(1)} \\ A^{(2)} \\ \vdots \\ A^{(m)} \end{bmatrix}, \qquad B = [B_1, B_2, \ldots, B_p]$$

or A into columns and B into rows, thus:

$$A = [A_1, A_2, \ldots, A_n], \qquad B = \begin{bmatrix} B^{(1)} \\ B^{(2)} \\ \vdots \\ B^{(n)} \end{bmatrix}.$$

Form the product of the partitioned matrices in each case and observe in detail how these products are related to the product AB.

3. Referring to Exercise 2, form these products and observe how they are related to the product AB:

(a) A partitioned into columns, B not partitioned.

(b) A not partitioned, B partitioned into rows.

4. Show that if $A = [A_{ij}]$, where the A_{ij} are submatrices, then $A^{\mathsf{T}} = [B_{ji}]$, where $B_{ji} = A_{ij}^{\mathsf{T}}$. For example,

$$\begin{bmatrix} A_{11} & A_{12} & A_{13} \\ A_{21} & A_{22} & A_{23} \end{bmatrix}^{\mathsf{T}} = \begin{bmatrix} A_{11}^{\mathsf{T}} & A_{21}^{\mathsf{T}} \\ A_{12}^{\mathsf{T}} & A_{22}^{\mathsf{T}} \\ A_{13}^{\mathsf{T}} & A_{23}^{\mathsf{T}} \end{bmatrix}.$$

***5.** Let A_1 and B_1 be square and of the same order. Let the same be true for A_2 and $B_2, \ldots, A_n$ and B_n. Furthermore, let

$$A = \begin{bmatrix} A_1 & 0 & \cdots & 0 \\ 0 & A_2 & \cdots & 0 \\ \vdots & & & \\ 0 & 0 & \cdots & A_n \end{bmatrix}, \qquad B = \begin{bmatrix} B_1 & 0 & \cdots & 0 \\ 0 & B_2 & \cdots & 0 \\ \vdots & & & \\ 0 & 0 & \cdots & B_n \end{bmatrix}.$$

Determine the sum and the product of A and B.

A symmetrically partitioned matrix whose off-diagonal submatrices are all zero matrices is called a **decomposable** or **quasidiagonal** or **pseudodiagonal** matrix. Two decomposable matrices which are identically partitioned are said to be of the **same kind**. This problem then shows that the sum and the product of two decomposable matrices of the same kind are decomposable matrices of the same kind.

If we abbreviate $A = D[A_1, A_2, \ldots, A_n]$, what can be said about $f(A)$, where $f(A)$ is any polynomial function of A?

***6.** Let D denote a quasi-diagonal matrix of order n, and let A denote any matrix of order n which is partitioned identically to D. What is the nature of the products DA and AD?

7. If X_1 is a k_1-vector and X_2 is a k_2-vector, indicate what orders the A_{ij} must have for the product

$$[X_1^T, X_2^T] \cdot \begin{bmatrix} A_{11} & A_{12} \\ A_{21} & A_{22} \end{bmatrix} \cdot \begin{bmatrix} X_1 \\ X_2 \end{bmatrix}$$

to have meaning, and compute the product.

***8.** If $A = [A_1, A_2, \ldots, A_m]$, where the A_j are n-vectors, find a representation of the product AA^T in terms of the A_j's rather than in terms of the elements of A.

9. Given that

$$A = \left[\begin{array}{c|cc|cc} 1 & 0 & 0 & 0 & 0 \\ \hline 0 & 0 & 1 & 1 & 0 \\ 0 & 1 & 0 & 0 & 1 \\ \hline 0 & 1 & 0 & 0 & 1 \\ 0 & 0 & 1 & 1 & 0 \end{array}\right],$$

compute A^2, A^3, and A^4. Then give formulas for A^{2k} and A^{2k+1} and prove them by induction.

CHAPTER 3

Vector Geometry in $\mathscr{E}^2$ and $\mathscr{E}^3$

3.1 GEOMETRIC REPRESENTATION OF VECTORS

In Chapter 2, we defined an n-vector to be a column matrix with n elements. Vectors with two and three components which are real numbers have a ready geometrical interpretation in Euclidean spaces of two and three dimensions. We denote these spaces by $\mathscr{E}^2$ (read "$\mathscr{E}$-two") and $\mathscr{E}^3$ (read "$\mathscr{E}$-three"), respectively.

Given a rectangular coordinate system in $\mathscr{E}^2$, the vector

$$X = \begin{bmatrix} x_1 \\ x_2 \end{bmatrix}$$

determines a unique corresponding directed line segment $\overrightarrow{OP}$ from the origin to the point $P:(x_1, x_2)$. We will also call $\overrightarrow{OP}$ a "vector." When $\overrightarrow{OP}$ is given, its terminal point (x_1, x_2) determines a unique vector

$$X = \begin{bmatrix} x_1 \\ x_2 \end{bmatrix}$$

of which $\overrightarrow{OP}$ is the geometric representation (Figure 3-1a).

The same thing can be done in $\mathscr{E}^3$, where a rectangular coordinate system requires three mutually perpendicular lines, called the **coordinate axes**, which intersect in a common point O, called the **origin**. For ease of generalization, we label the axes 1, 2, and 3. A unit of length is chosen and is used on all three axes, along which distance is measured from the origin. The axes each have a positive sense and a negative sense. The **positive sense** and the **unit of length** on

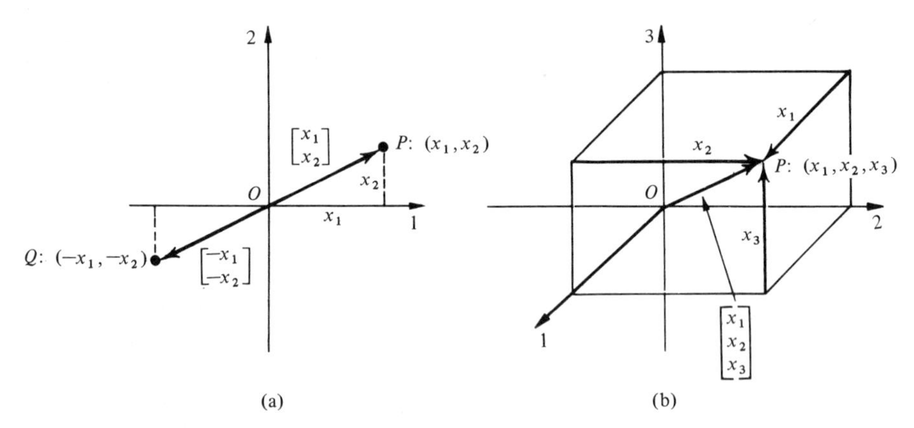

Figure 3-1. *Geometric Representation of Vectors in $\mathscr{E}^2$ and $\mathscr{E}^3$.*

an axis are determined by selection of the point whose directed distance from the origin is $+1$.

Like the axes, every line in $\mathscr{E}^3$, and in $\mathscr{E}^2$ as well, has a direction and two opposite senses. In Section 3.6 we define direction and sense in numerical terms. Often it is desirable to define a positive sense on a line. In the case of lines parallel to (having the same direction as) a coordinate axis, the positive sense is chosen to agree with that of the axis in question.

A coordinate system in $\mathscr{E}^3$ has three mutually perpendicular **coordinate planes**, the 1,2-plane, the 1,3-plane, and the 2,3-plane, each determined by a pair of axes. The point P:(x_1, x_2, x_3) is at a distance x_1 from the 2,3-plane, x_2 from the 1,3-plane, and x_3 from the 1,2-plane (see Figure 3-1b). The numbers x_1, x_2, and x_3 are the **coordinates** of the point. The coordinate planes separate $\mathscr{E}^3$ into eight disjoint regions called **octants**. The octant in which all three of x_1, x_2, x_3 are positive is called the **first octant**.

A **directed line segment** $\overrightarrow{P_1P_2}$ in $\mathscr{E}^3$ is determined by an **initial point** P_1 and a **terminal point**, P_2. A vector

$$X = \begin{bmatrix} x_1 \\ x_2 \\ x_3 \end{bmatrix}$$

is represented by the directed segment $\overrightarrow{OP}$, where P is the point (x_1, x_2, x_3). As in $\mathscr{E}^2$, we also call $\overrightarrow{OP}$ a "vector." The **zero vector**

$$\begin{bmatrix} 0 \\ 0 \\ 0 \end{bmatrix}$$

is represented by the origin O since its initial and terminal points are the same. The **negative** of X is represented by a directed segment or vector $\overrightarrow{OQ}$, where Q is the point $(-x_1, -x_2, -x_3)$. We say that $\overrightarrow{OQ}$ has the same direction as $\overrightarrow{OP}$ but the opposite sense. Given $\overrightarrow{OP}$, its terminal point (x_1, x_2, x_3) determines the unique vector

$$X = \begin{bmatrix} x_1 \\ x_2 \\ x_3 \end{bmatrix}$$

of which $\overrightarrow{OP}$ is the geometric representation.

Since we thus have a one-to-one correspondence between points P and vectors $\overrightarrow{OP}$ in $\mathscr{E}^2$ and similarly in $\mathscr{E}^3$, these spaces may be regarded either as spaces of points or as spaces of vectors. The context will always make clear which interpretation is intended.

3.2 OPERATIONS ON VECTORS

Consider now three directed segments $\overrightarrow{OP}$, $\overrightarrow{OQ}$, and $\overrightarrow{OR}$, corresponding to X, Y, and $X + Y$, respectively. Assume first that O, P, and Q are not in the same straight line. Connect R to each of P and Q to complete a quadrilateral $OPRQ$ (Figure 3-2). Since, in Figure 3-2, the triangles OSQ and PTR are

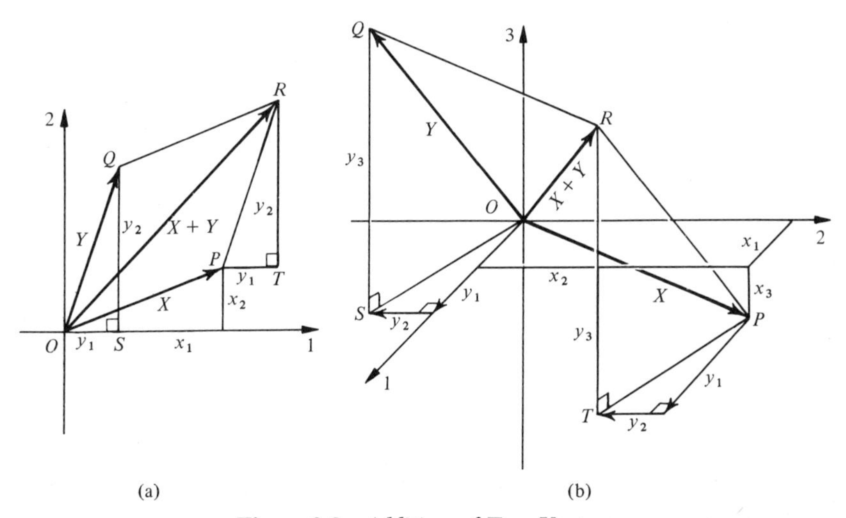

Figure 3-2. *Addition of Two Vectors.*

congruent, have two sides in one parallel to the corresponding sides in the other, and are similarly placed, the line segments OQ and PR are equal and parallel, so that $OPRQ$ is a parallelogram. That is, the directed segment $\overrightarrow{OR}$, representing $X + Y$, is a diagonal of the parallelogram determined by $\overrightarrow{OP}$, representing X, and $\overrightarrow{OQ}$, representing Y. The diagonal $\overrightarrow{OR}$ is therefore called the **sum** of $\overrightarrow{OP}$ and $\overrightarrow{OQ}$ and we write $\overrightarrow{OR} = \overrightarrow{OP} + \overrightarrow{OQ}$. The determination of $\overrightarrow{OR}$ from $\overrightarrow{OP}$ and $\overrightarrow{OQ}$ by completion of the parallelogram $OPRQ$ is called the **parallelogram law of addition** of vectors. Thus *addition of components of real, arithmetic 2-vectors or 3-vectors corresponds to the familiar parallelogram law for the addition of geometric vectors.*

We say that $X, Y, Z, \ldots$ are **coplanar vectors** if and only if the directed segments which represent them all lie on the same plane on the origin. A consequence of the parallelogram law of addition is that the vectors $\overrightarrow{OP}$, $\overrightarrow{OQ}$, and $\overrightarrow{OR}$ are coplanar, that is, that X, Y, *and their sum* $X + Y$ *are always coplanar vectors.*

In the special case when O, P, and Q are collinear, that is, when $\overrightarrow{OP}$ and $\overrightarrow{OQ}$ lie on the same straight line, we say that X and Y are **collinear vectors**. In this case, the sum $X + Y$ is represented by a directed segment $\overrightarrow{OR}$, which is collinear with the segments $\overrightarrow{OP}$ and $\overrightarrow{OQ}$ (Figure 3-3). The reader should provide full details.

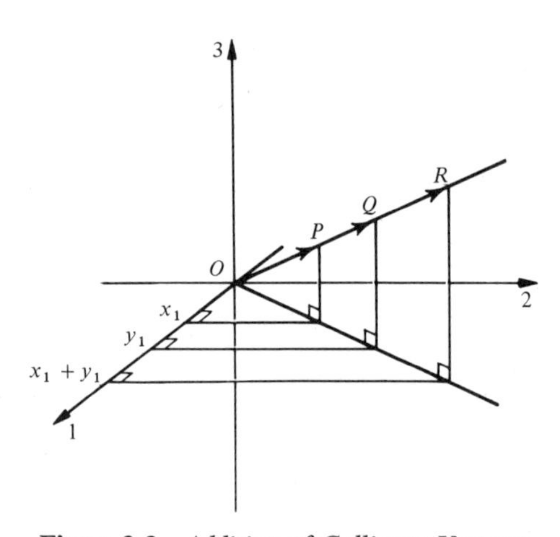

Figure 3-3. *Addition of Collinear Vectors.*

Consider now directed segments $\overrightarrow{OP}$ and $\overrightarrow{OQ}$, corresponding to the vectors X and cX, respectively, neither X nor the real number c being 0. Because triangles OSP and OTQ in Figure 3-4 are similar, O, P, and Q must be collinear and $\overrightarrow{OQ} = c\overrightarrow{OP}$. Conversely, if $\overrightarrow{OQ} = c\overrightarrow{OP}$ for some real number c, similar triangles show that if $\overrightarrow{OP}$ represents X, $\overrightarrow{OQ}$ must represent cX. Thus *multiplica-*

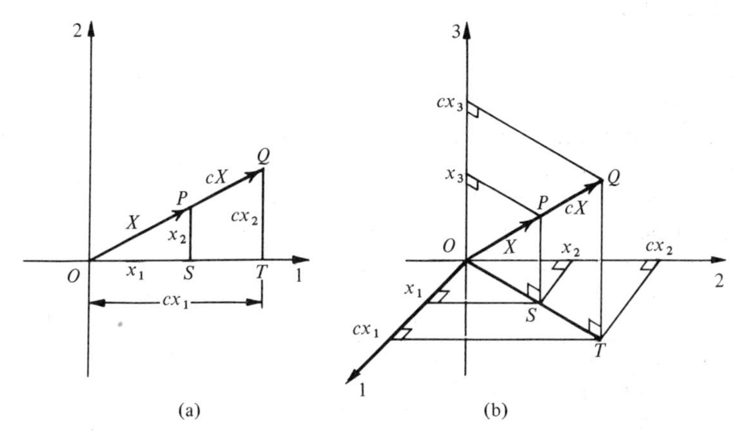

Figure 3-4. *Multiplication of a Vector by a Scalar.*

tion of the vector X by the scalar c corresponds to multiplying the directed segment $\overrightarrow{OP}$ *which represents X by the scalar c.*

If X is 0, cX is also 0 for all c and if c is 0, cX is 0 for all X. In either case, cX is represented by the origin and the preceding conclusion still holds.

Since arithmetic and geometric addition correspond, as do the multiplication of arithmetic and geometric vectors by scalars, other algebraic laws involving only these two operations have easily obtained geometric interpretations. For example, the laws $\alpha X + \alpha Y = \alpha(X + Y)$ and $(X + Y) + Z = X + (Y + Z)$ are illustrated in Figure 3-5.

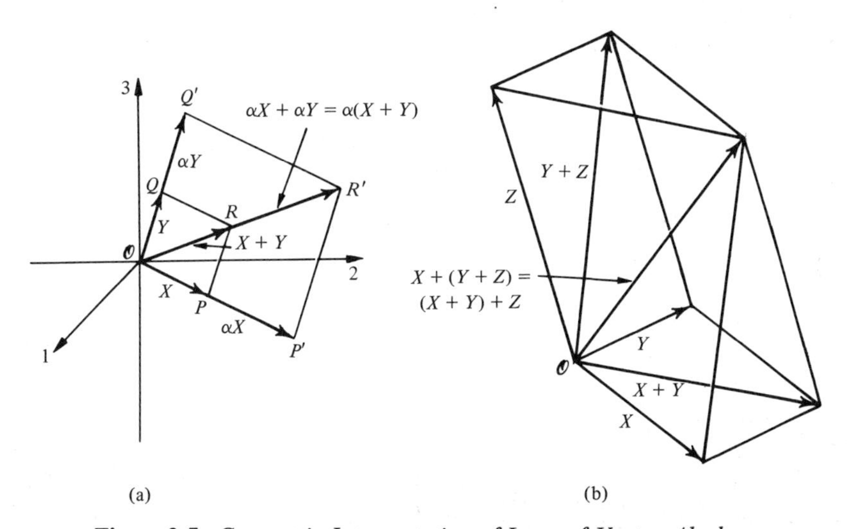

Figure 3-5. *Geometric Interpretation of Laws of Vector Algebra.*

3.3 ISOMORPHISM

Consider again the system of geometric vectors $\overrightarrow{OP}$ in $\mathscr{E}^3$ and the system of 3×1 column matrices with real entries. Each system has an operation of addition and each system has an operation of multiplication by scalars. Moreover, we have seen that we can assign to each geometric vector a unique arithmetic vector with three components. Similarly, to each arithmetic vector we can assign a unique geometric vector. Thus we have a one-to-one correspondence between the elements of the two systems. Moreover, to the sum of two geometric vectors there corresponds the sum of the corresponding arithmetic vectors, and to a scalar multiple of a geometric vector there corresponds the same scalar multiple of its arithmetic counterpart. We can, of course, reverse the roles of geometric and arithmetic vectors in the preceding sentence. We describe this by saying that *the correspondence preserves the operations*: To a sum there corresponds a sum and to a scalar multiple there corresponds a scalar multiple.

The abstract laws governing corresponding operations in the two systems are the same. We identify this fact by calling the systems isomorphic (*iso-morphic* means *same-form*). The only distinction between the two systems is the nature of the representation of their elements: geometric vectors in the one case, column matrices in the other.

Other examples of isomorphic systems will appear later.

3.4 LENGTH AND ANGLE

It is often necessary to make use of the formula for the length of a line segment P_1P_2. We denote this length by $|P_1P_2|$. In Figure 3-6a we have, by the Pythagorean theorem,

$$|P_1P_2| = \sqrt{(y_1 - x_1)^2 + (y_2 - x_2)^2} \qquad (\mathscr{E}^2) \tag{3.4.1}$$

In Figure 3-6b, by the Pythagorean theorem, the length of the segment P_1P is given by $\sqrt{(y_1 - x_1)^2 + (y_2 - x_2)^2}$ and hence, by a second use of the Pythagorean theorem, the length of P_1P_2 is given by

$$|P_1P_2| = \sqrt{(y_1 - x_1)^2 + (y_2 - x_2)^2 + (y_3 - x_3)^2} \qquad (\mathscr{E}^3). \tag{3.4.2}$$

For example, the length of the segment joining P_1:$(1, -1, 0)$ and P_2:$(-2, 2, 4)$ is $\sqrt{(1 + 2)^2 + (-1 - 2)^2 + (0 - 4)^2} = \sqrt{34}$.

Length is not a directed quantity. It is always nonnegative, even when the segment P_1P_2 is directed, and is often called the distance *between* P_1 and P_2. The directed distance *from* P_1 *to* P_2, P_1 and P_2 distinct, is positive or negative, depending on whether the sense of $\overrightarrow{P_1P_2}$ agrees with or is opposite to the positive sense assigned to the line on the points P_1 and P_2.

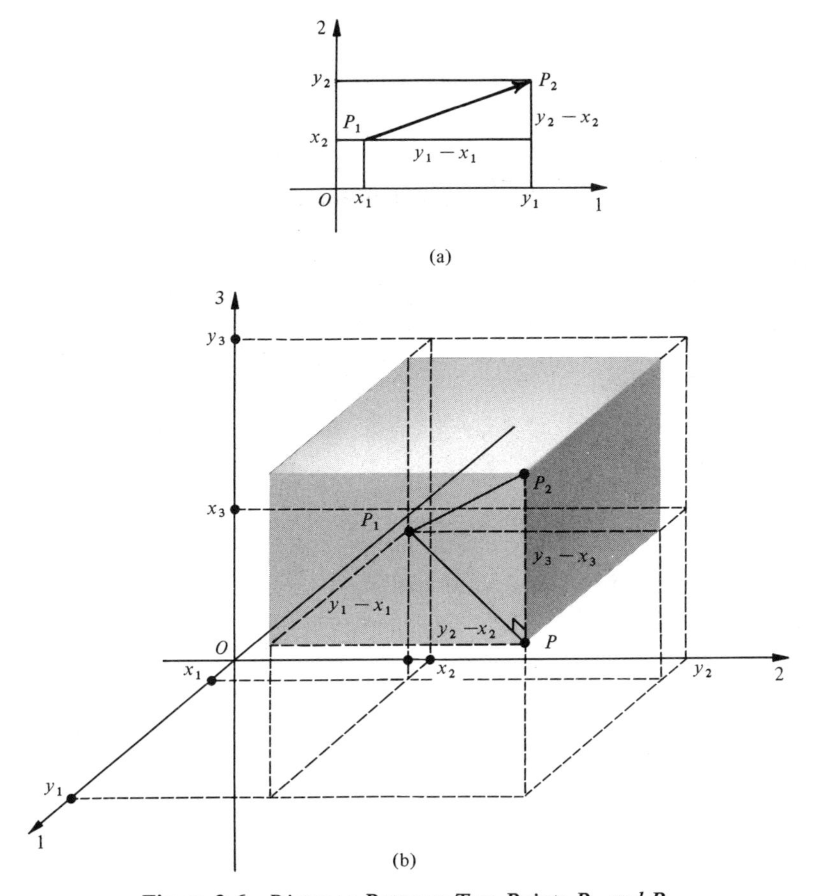

Figure 3-6. *Distance Between Two Points P_1 and P_2.*

The length $|X|$ of a real 2-vector or 3-vector X is defined to be the length of the corresponding directed segment $\overrightarrow{OP}$. Thus we have

(3.4.3)
$$|X| = \begin{cases} \sqrt{x_1^2 + x_2^2} & \text{in } \mathscr{E}^2, \\ \sqrt{x_1^2 + x_2^2 + x_3^2} & \text{in } \mathscr{E}^3. \end{cases}$$

When $|X| = 1$, we call X a **unit vector**. For example, $[\frac{1}{2}, \frac{1}{2}, -1/\sqrt{2}]^\mathsf{T}$ is a unit vector.

The terminal points of the set of directed segments $\overrightarrow{OP}$ representing the set of unit vectors in $\mathscr{E}^2$ constitute a circle with radius 1 and center O. For precisely the points on this circle, we have $|X|^2 = 1$, so the equation of this **unit circle** (Figure 3-7a) is

(3.4.4)
$$x_1^2 + x_2^2 = 1.$$

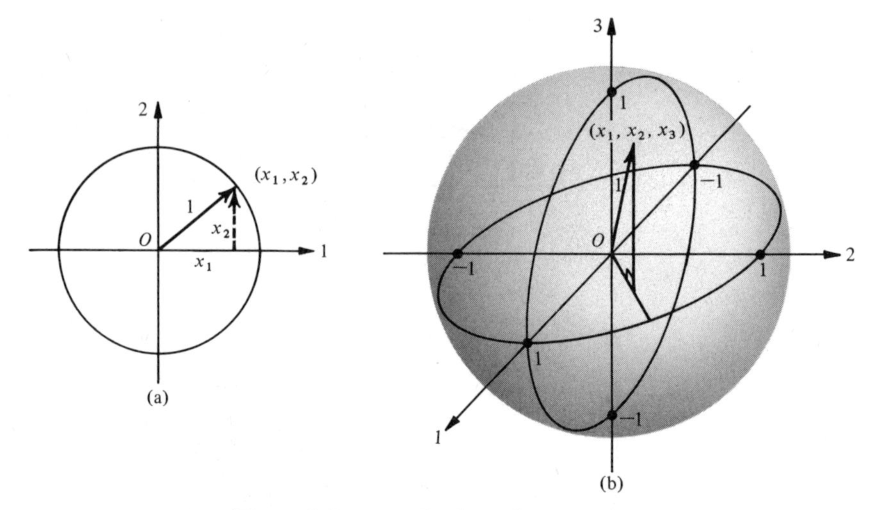

Figure 3-7. *Unit Circle and Unit Sphere.*

In $\mathscr{E}^3$, the corresponding locus is the **unit sphere** (Figure 3-7b) with radius 1 and center O and with equation

(3.4.5) $$x_1^2 + x_2^2 + x_3^2 = 1 \qquad \text{(unit sphere).}$$

Consider now the half-lines parallel to the axes and through P_1, each half-line being sensed in the same way as the corresponding axis. The angles between these half-lines and the directed segment $\overrightarrow{P_1P_2}$ are, respectively, the **direction angles**, α_1, α_2, and α_3, of $\overrightarrow{P_1P_2}$. (See Figure 3-8.) When α_1, α_2, and α_3 are given, both the direction and the sense of any directed segment $\overrightarrow{P_1P_2}$ having these direction angles are determined. If we let $d = |\overrightarrow{P_1P_2}|$, then, also from Figure 3-8,

(3.4.6) $$\cos \alpha_1 = \frac{y_1 - x_1}{d}, \qquad \cos \alpha_2 = \frac{y_2 - x_2}{d}, \qquad \cos \alpha_3 = \frac{y_3 - x_3}{d},$$

and we choose the α's so that $0 \leq \alpha_i \leq \pi$, $i = 1, 2, 3$.

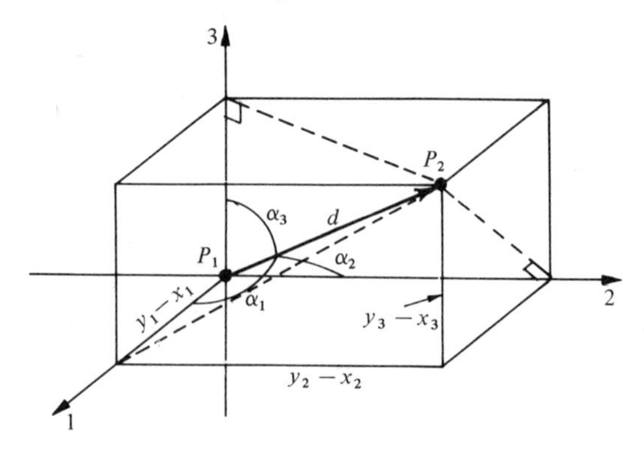

Figure 3-8. *Direction Angles of* $\overrightarrow{P_1P_2}$.

By squaring and adding the formulas in (3.4.6), we find

$$\cos^2 \alpha_1 + \cos^2 \alpha_2 + \cos^2 \alpha_3 = 1, \tag{3.4.7}$$

so *the three direction angles are not independent.* If we choose two, the third is determined as an angle or its supplement. For example, if $\alpha_1 = 45°$ and $\alpha_2 = 60°$, then

$$\left(\frac{1}{\sqrt{2}}\right)^2 + \left(\frac{1}{2}\right)^2 + \cos^2 \alpha_3 = 1.$$

Hence $\cos^2 \alpha_3 = \frac{1}{4}$, so $\cos \alpha_3 = \frac{1}{2}$ or $-\frac{1}{2}$ and hence $\alpha_3 = 60°$ or $120°$. (See Figure 3-9.)

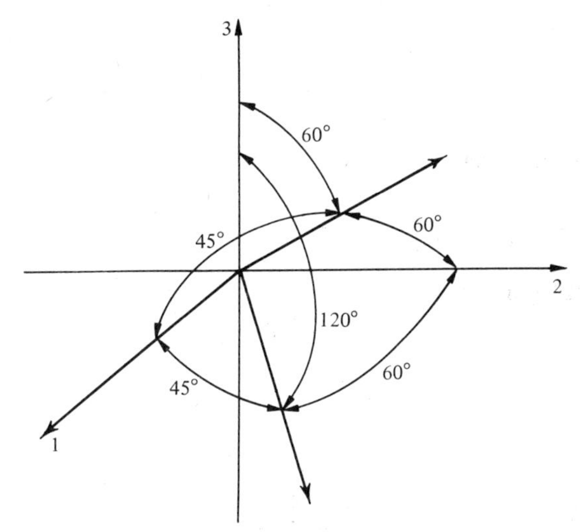

Figure 3-9. *Related Sets of Direction Angles.*

The differences $y_1 - x_1$, $y_2 - x_2$, and $y_3 - x_3$ appearing in (3.4.6) are *ordered* differences: (coordinate of *second* point) minus (coordinate of *first* point) and, by what precedes, the length, direction, and sense of a directed line segment are uniquely determined by these ordered differences. Two directed line segments are defined to be equal if and only if they have the same length, direction, and sense. *The results of this section show that two directed line segments are equal if and only if their ordered differences are respectively equal.*

A given set of ordered differences leads to infinitely many equal directed segments $\overrightarrow{P_1P_2}$, any one of which is determined as soon as its initial point P_1 or its terminal point P_2 is given. For example, if the ordered differences are 3, -1, 2 and $P_1 = (2, 1, -1)$, we have $y_1 - 2 = 3$, $y_2 - 1 = -1$, and $y_3 + 1 = 2$, so $P_2 = (5, 0, 1)$.

The ordered differences of a pair of points are the components of an important vector, namely of $X_2 - X_1$, where X_2 is the vector $\overrightarrow{OP_2}$ and X_1 is the vector $\overrightarrow{OP_1}$ (Figure 3-10). The same difference is determined by infinitely many pairs of vectors.

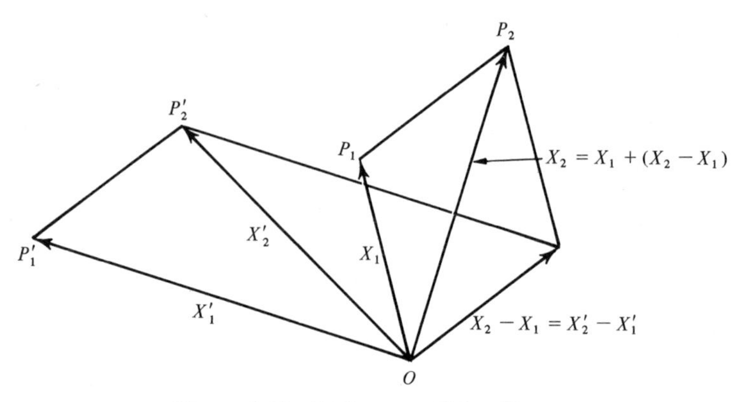

Figure 3-10. *Difference of Two Vectors.*

The length of the vector $X_2 - X_1$ determined by $\overrightarrow{P_1P_2}$ is the same as that of P_1P_2:

(3.4.8) $$|X_2 - X_1| = \sqrt{(y_1 - x_1)^2 + (y_2 - x_2)^2 + (y_3 - x_3)^2}.$$

The specialization of these results to $\mathscr{E}^2$ is not hard. Let us regard $\mathscr{E}^2$ as the 1,2-plane in $\mathscr{E}^3$ (Figure 3-11). For all points in the 1,2-plane, $\alpha_3 = 90°$, so

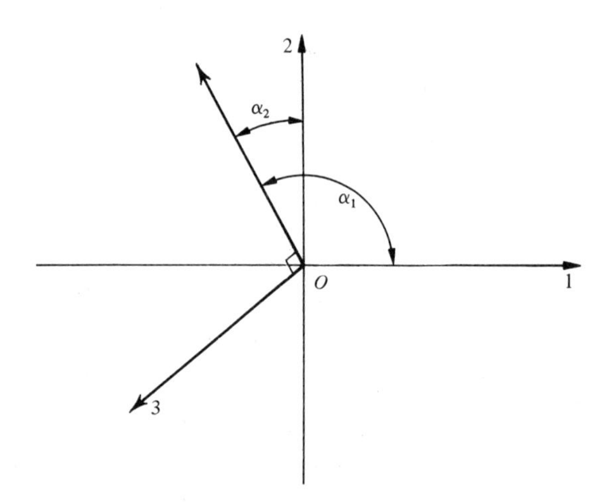

Figure 3-11. *Direction Angles in* $\mathscr{E}^2$.

$\cos \alpha_3 = 0$. If we now drop the third coordinate, which is always 0, and the third direction cosine, which is always 0, we have the usual representation in two dimensions. Formula (3.4.7) now reduces to

(3.4.9) $$\cos^2 \alpha_1 + \cos^2 \alpha_2 = 1 \qquad (\mathscr{E}^2).$$

Similarly, formula (3.4.8) reduces to

(3.4.10) $$|X_2 - X_1| = \sqrt{(y_1 - x_1)^2 + (y_2 - x_2)^2} \qquad (\mathscr{E}^2).$$

3.5 EXERCISES

1. Write the vector $X = X_2 - X_1$ determined by the directed segment $\overrightarrow{P_1P_2}$, where $P_1 = (-1, 0, 2)$ and $P_2 = (2, -1, 1)$. Then find the length and the direction cosines of this vector. Illustrate with a figure.

2. Find the vector which is the sum of the vectors determined by $\overrightarrow{P_1P_2}$ and $\overrightarrow{Q_1Q_2}$, where

$$P_1 = (1, 0, -1), \quad P_2 = (4, 4, -2), \quad Q_1 = (2, -1, 0), \quad Q_2 = (3, 0, 0).$$

Illustrate with a figure.

3. If $\cos \alpha_1 = \frac{1}{3}$, $\cos \alpha_2 = \frac{2}{3}$, what is $\cos \alpha_3$? Illustrate with a figure (two cases).

4. Find a scalar multiple of the vector

$$\begin{bmatrix} 1 \\ 1 \\ 1 \end{bmatrix}$$

which has length 1. Is there more than one such scalar multiple?

5. Show that there exist scalars t_1, t_2, and t_3, not all zero, such that

$$t_1\begin{bmatrix} 1 \\ 2 \\ -3 \end{bmatrix} + t_2\begin{bmatrix} -2 \\ -1 \\ 3 \end{bmatrix} + t_3\begin{bmatrix} 4 \\ -1 \\ -3 \end{bmatrix} = \begin{bmatrix} 0 \\ 0 \\ 0 \end{bmatrix}.$$

Then illustrate the equation with a figure.

6. Find all sets of values of t_1, t_2, and t_3 such that

$$t_1\begin{bmatrix} 1 \\ 2 \\ -3 \end{bmatrix} + t_2\begin{bmatrix} -2 \\ -1 \\ 3 \end{bmatrix} + t_3\begin{bmatrix} 4 \\ -1 \\ -3 \end{bmatrix} = \begin{bmatrix} 3 \\ 0 \\ -3 \end{bmatrix}.$$

***7.** Show that, for a given set of direction cosines,

$$\begin{bmatrix} \cos \alpha_1 \\ \cos \alpha_2 \\ \cos \alpha_3 \end{bmatrix}$$

is a unit vector and illustrate with a figure.

***8.** Show that if

$$X = \begin{bmatrix} x_1 \\ x_2 \\ x_3 \end{bmatrix},$$

then $X/|X|$, that is,

$$\begin{bmatrix} \frac{x_1}{|X|} \\ \frac{x_2}{|X|} \\ \frac{x_3}{|X|} \end{bmatrix},$$

is a unit vector with the same direction and sense as X. (Show that, in fact,

$$\frac{X}{|X|} = \begin{bmatrix} \cos \alpha_1 \\ \cos \alpha_2 \\ \cos \alpha_3 \end{bmatrix},$$

where α_1, α_2, and α_3 are direction angles of X.)

9. If the distance between the points $(-1, a, 2a)$ and $(2, -a, a)$ is $\sqrt{29}$, what are the possible values for a?

10. Illustrate the equation $U + V + W = 0$ with a figure. Do the same for $U + V + W + X = 0$.

***11.** What is the equation of the locus of the endpoints of all vectors $\overrightarrow{OP}$ of length r in $\mathscr{E}^2$? In $\mathscr{E}^3$?

***12.** In $\mathscr{E}^3$, show that for all real numbers α, β, the vector $\alpha X + \beta Y$ is coplanar with X and Y.

3.6 LINES IN SPACE

Consider the line determined by two distinct points, P_0 and P_1. Let the vectors determined by P_0 and P_1 be denoted by X_0 and X_1, respectively. Let X denote the vector determined by P where P is any point on the line. Then Figure 3-12 shows that X is X_0 plus some scalar multiple of $X_1 - X_0$, that is,

(3.6.1) $$X = X_0 + t(X_1 - X_0) \qquad (t \text{ real}).$$

From Figure 3-12 one can see that the values of t are associated with points on the line as follows:

$t < 0$	points of the half-line starting at P_0 and not containing P_1
$t = 0$	point P_0
$0 < t < 1$	points between P_0 and P_1
$t = 1$	point P_1
$t > 1$	points of the half-line starting at P_1 and not containing P_0.

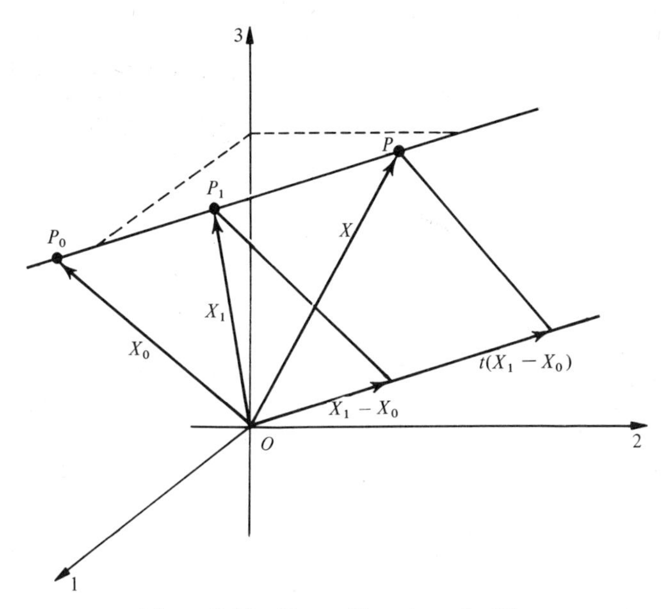

Figure 3-12. *Vector Equation of a Line.*

If P_0 has coordinates (x_{10}, x_{20}, x_{30}) and P_1 has coordinates (x_{11}, x_{21}, x_{31}), equation (3.6.1) is equivalent to the three scalar equations

$$\begin{aligned} x_1 &= x_{10} + t(x_{11} - x_{10}) \\ x_2 &= x_{20} + t(x_{21} - x_{20}) \\ x_3 &= x_{30} + t(x_{31} - x_{30}), \end{aligned} \tag{3.6.2}$$

which give the coordinates of the point P in terms of those of P_0 and P_1 since the coordinates of P are the same as the components of the vector X. These are called **parametric equations** for the line because the variable point P of the line is expressed in terms of the parameter t. In $\mathscr{E}^2$, one employs just the first two of these equations.

If P_0 is the origin and if P_1 is the point (a_1, a_2, a_3), the equations of the line reduce to

$$\begin{aligned} x_1 &= ta_1 \\ x_2 &= ta_2 \\ x_3 &= ta_3 \end{aligned}$$

or, in vector form, if

$$A = \begin{bmatrix} a_1 \\ a_2 \\ a_3 \end{bmatrix},$$

$$X = tA. \tag{3.6.3}$$

This equation shows that a line L lying on the origin carries all scalar multiples of a fixed vector A.

The set of all these vectors tA is **closed** with respect to addition and with respect to multiplication by a scalar, that is, for all real numbers t_1 and t_2,

$$(t_1A) + (t_2A) = (t_1 + t_2)A,$$

so the sum of any two vectors of the set is in the set and

$$t_1(t_2A) = (t_1t_2)A,$$

so any scalar multiple of any member of the set is in the set.

This illustrates an important mathematical concept. A set of vectors in $\mathscr{E}^2$ or $\mathscr{E}^3$ is called a **vector space** if and only if the sum of any two vectors in the set is also in the set and every real scalar multiple of every vector in the set is in the set. Examples of vector spaces are the set consisting of the zero vector alone and also the sets of all vectors in $\mathscr{E}^2$ or in $\mathscr{E}^3$. By what precedes, *the set of all vectors on the line whose equation is $X = tA$ is also a vector space.* Other examples will appear later.

The **direction of a line** is defined to be that of any segment P_0P_1 on it. If we wish to give the line the same *sense* as that of $\overrightarrow{P_0P_1}$, then direction cosines of the line are

$$\cos \alpha_1 = \frac{x_{11} - x_{10}}{|P_0P_1|}$$

$$\cos \alpha_2 = \frac{x_{21} - x_{20}}{|P_0P_1|}$$

$$\cos \alpha_3 = \frac{x_{31} - x_{30}}{|P_0P_1|}.$$

From similar triangles, one can see that any other segment $\overrightarrow{Q_0Q_1}$ on the line and with the same sense as $\overrightarrow{P_0P_1}$ would determine the same direction cosines.

For every pair of points $P{:}(x_1, x_2, x_3)$, $Q{:}(y_1, y_2, y_3)$ on the line, the ordered differences $y_1 - x_1$, $y_2 - x_2$, and $y_3 - x_3$ are proportional to the direction cosines and are called **direction numbers** of the line.

3.7 ORTHOGONALITY OF VECTORS

Let

$$V = \begin{bmatrix} v_1 \\ v_2 \\ v_3 \end{bmatrix} \quad \text{and} \quad W = \begin{bmatrix} w_1 \\ w_2 \\ w_3 \end{bmatrix}$$

be two nonzero vectors in $\mathscr{E}^3$. Then the angle θ between V and W (Figure 3-13) is defined to be the undirected **angle** θ, $0 \leq \theta \leq \pi$, between the directed seg-

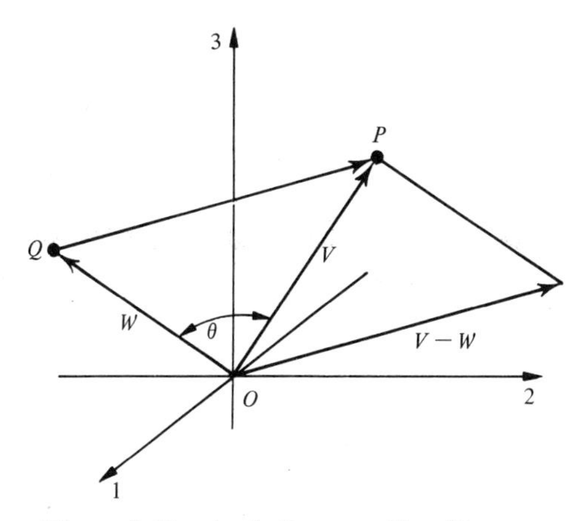

Figure 3-13. *Angle Between Two Vectors.*

ments $\overrightarrow{OP}$ and $\overrightarrow{OQ}$ which represent these vectors. The directed segment $\overrightarrow{QP}$ has the same length as the vector $V - W$. Hence, by the law of cosines,

$$|V - W|^2 = |V|^2 + |W|^2 - 2|V|\,|W| \cos \theta$$

or

$$(v_1 - w_1)^2 + (v_2 - w_2)^2 + (v_3 - w_3)^2$$
$$= (v_1^2 + v_2^2 + v_3^2) + (w_1^2 + w_2^2 + w_3^2) - 2|V|\,|W| \cos \theta.$$

From this, after expanding and simplifying, we have

$$\cos \theta = \frac{v_1 w_1 + v_2 w_2 + v_3 w_3}{|V| \cdot |W|}.$$

The expression in the numerator is the **scalar product** of V and W that we introduced in Section 2.7:

$$V^{\mathsf{T}} W = v_1 w_1 + v_2 w_2 + v_3 w_3.$$

Hence

(3.7.1) $$\cos \theta = \frac{V^{\mathsf{T}} W}{|V| \cdot |W|}, \qquad 0 \le \theta \le \pi.$$

Now

$$\frac{|V^{\mathsf{T}} W|}{|V| \cdot |W|} = |\cos \theta| \le 1,$$

so

(3.7.2) $$|V^{\mathsf{T}} W| \le |V| \cdot |W|.$$

This is one form of the famous **Cauchy–Schwarz inequality**, which is frequently useful. Although $\cos \theta$ was defined only for nonzero V and W, (3.7.2) holds without exception. Why?

Nonzero vectors V and W are **orthogonal** or are at right angles, that is, $\theta = \pi/2$, if and only if $\cos\theta = 0$. This holds if and only if the numerator of the expression in (3.7.1) vanishes, that is, *V and W are orthogonal if and only if*

$$(3.7.3) \qquad V^{\mathsf{T}}W = 0, \quad V \neq 0, \quad W \neq 0.$$

It is often necessary to find a vector V orthogonal to each of two vectors A and B; that is, one has to solve the equations

$$\begin{aligned} A^{\mathsf{T}}V &= 0 \\ B^{\mathsf{T}}V &= 0 \end{aligned} \qquad \text{or} \qquad \begin{aligned} a_1v_1 + a_2v_2 + a_3v_3 &= 0 \\ b_1v_1 + b_2v_2 + b_3v_3 &= 0 \end{aligned}$$

for the v's. For example, let

$$A = \begin{bmatrix} 1 \\ 1 \\ 0 \end{bmatrix}, \qquad B = \begin{bmatrix} 0 \\ 1 \\ 1 \end{bmatrix}.$$

Then

$$\begin{aligned} v_1 + v_2 \phantom{{}+v_3} &= 0 \\ v_2 + v_3 &= 0. \end{aligned}$$

By the methods of Chapter 1, the complete solution is

$$V = \begin{bmatrix} v_1 \\ v_2 \\ v_3 \end{bmatrix} = t\begin{bmatrix} 1 \\ -1 \\ 1 \end{bmatrix},$$

so there are actually infinitely many solutions, all having the same direction since they are all scalar multiples of a common vector, but differing in sense or length or both. The vectors A, B, and V for the case $t = 1$, are shown in Figure 3-14.

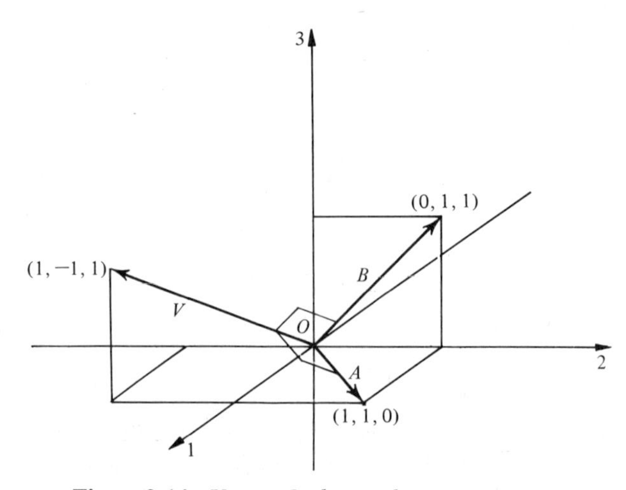

Figure 3-14. *Vector Orthogonal to Two Vectors.*

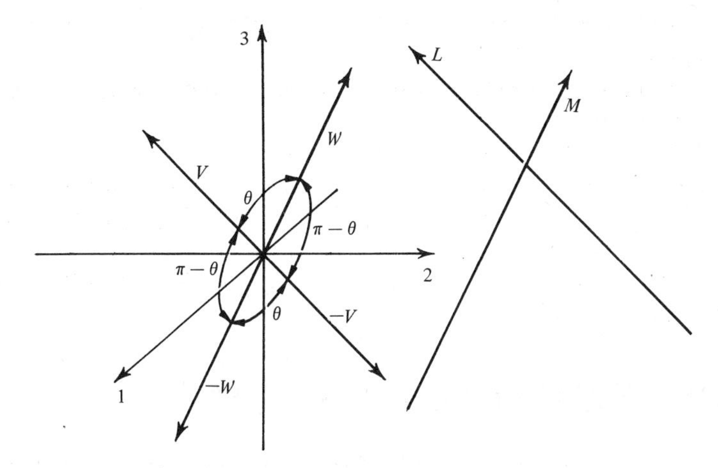

Figure 3-15. *Angle Between Two Lines.*

3.8 THE ANGLE BETWEEN TWO LINES

The concepts of the preceding section extend immediately to lines. Let L and M be any two directed lines in $\mathscr{E}^3$ and let V and W be any two vectors with the same sense and direction as L and M, respectively. We then define the **angle between the directed lines** L and M to be the angle between V and W, as given by (3.7.1). When this angle is a right angle, we say the lines are **orthogonal**. Note that two lines need not intersect for the angle between them to be defined.

For example, if the directed segment $\overrightarrow{P_1P_2}$, where $P_1 = (2, 1, 0)$ and $P_2 = (-1, -1, 5)$, determines L and if the directed segment $\overrightarrow{Q_1Q_2}$, where $Q_1 = (0, 0, 0)$ and $Q_2 = (1, 1, 1)$, determines M, then we may choose

$$V = \begin{bmatrix} -3 \\ -2 \\ 5 \end{bmatrix} \quad \text{and} \quad W = \begin{bmatrix} 1 \\ 1 \\ 1 \end{bmatrix},$$

so

$$\cos \theta = \frac{0}{\sqrt{38}\sqrt{3}} = 0, \qquad \theta = \frac{\pi}{2},$$

and the lines are orthogonal.

If we reverse the sense of *one* of the lines, we replace θ by $\pi - \theta$, as may be seen from Figure 3-15, in which θ is the angle between V and W. For example, if we reverse the sense of M and replace W by $-W$, then the cosine of the angle between the resulting directed lines is given by

$$\frac{V^{\mathsf{T}}(-W)}{|V|\,|-W|} = -\frac{V^{\mathsf{T}}W}{|V|\,|W|} = -\cos\theta = \cos(\pi - \theta).$$

The **angle between two undirected lines** is defined to be the smaller of the angles θ and $\pi - \theta$.

Two undirected lines are defined to be parallel if and only if $\theta = 0$ (0 or π if they are directed).

If one only needs to determine whether or not two lines are orthogonal, one finds V and W (that is, one finds a set of direction numbers for each line) and then determines whether or not $V^{\mathsf{T}}W = 0$. The preceding numerical example illustrates the point.

Direction numbers for a line orthogonal to each of two given lines may be found as in the numerical example of Section 3.7.

3.9 THE EQUATION OF A PLANE

A plane in $\mathscr{E}^3$ is determined by any three noncollinear points on it. Putting it another way, there is a unique plane on three given noncollinear points in $\mathscr{E}^3$. Let three given noncollinear points be $P_0:(x_{10}, x_{20}, x_{30})$, $P_1:(x_{11}, x_{21}, x_{31})$, and $P_2:(x_{12}, x_{22}, x_{32})$ and let the corresponding vectors be X_0, X_1, and X_2, respectively (Figure 3-16). Let $P:(x_1, x_2, x_3)$, with corresponding vector X denote an

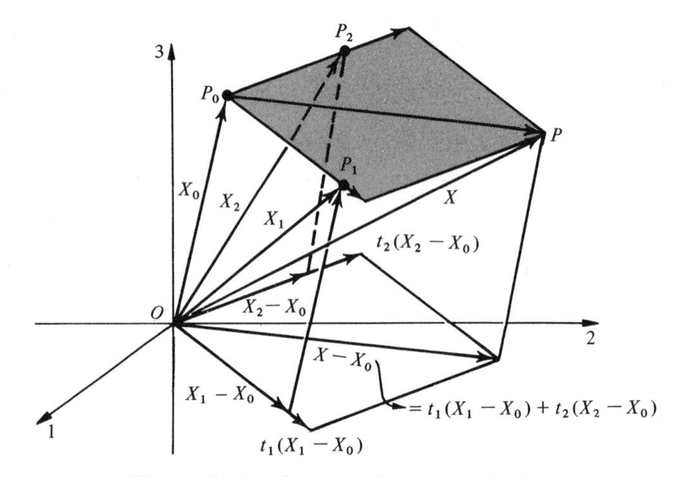

Figure 3-16. *Plane on Three Given Points.*

arbitrary point of the plane determined by P_0, P_1, and P_2. Then, since P_0, P_1, P_2, and P are coplanar, the three vectors $X_1 - X_0$, $X_2 - X_0$, and $X - X_0$ are coplanar. Moreover, since $X_2 - X_0$ and $X_1 - X_0$ necessarily have distinct directions (Why?), $X - X_0$ may be written as the sum of appropriate multiples of these two vectors. That is, there exist scalars t_1 and t_2 such that

$$X - X_0 = t_1(X_1 - X_0) + t_2(X_2 - X_0)$$

or

$$X = X_0 + t_1(X_1 - X_0) + t_2(X_2 - X_0). \tag{3.9.1}$$

This is the **parametric vector equation of a plane** on three given points. It is analogous to the parametric vector equation (3.6.1) of a line on two given points.

For example, consider the plane on the points $P_0:(0, 1, 1)$, $P_1:(1, 0, 1)$, $P_2:(1, 1, 0)$. Its parametric vector equation is

$$\begin{bmatrix} x_1 \\ x_2 \\ x_3 \end{bmatrix} = \begin{bmatrix} 0 \\ 1 \\ 1 \end{bmatrix} + t_1\left(\begin{bmatrix} 1 \\ 0 \\ 1 \end{bmatrix} - \begin{bmatrix} 0 \\ 1 \\ 1 \end{bmatrix}\right) + t_2\left(\begin{bmatrix} 1 \\ 1 \\ 0 \end{bmatrix} - \begin{bmatrix} 0 \\ 1 \\ 1 \end{bmatrix}\right)$$

or

$$\begin{bmatrix} x_1 \\ x_2 \\ x_3 \end{bmatrix} = \begin{bmatrix} 0 \\ 1 \\ 1 \end{bmatrix} + t_1\begin{bmatrix} 1 \\ -1 \\ 0 \end{bmatrix} + t_2\begin{bmatrix} 1 \\ 0 \\ -1 \end{bmatrix}.$$

In scalar form, we have

$$\begin{aligned} x_1 &= t_1 + t_2 \\ x_2 &= 1 - t_1 \\ x_3 &= 1 - t_2. \end{aligned}$$

If we solve the second and third equations for t_1 and t_2, then substitute into the first equation and rearrange terms, we get

$$x_1 + x_2 + x_3 = 2.$$

That is, the plane may be represented by a single, nonparametric equation.

In the particular case when $X_0 = 0$, that is, when the origin is one of the three points determining the plane, (3.9.1) reduces to

$$X = t_1X_1 + t_2X_2. \tag{3.9.2}$$

For example, the plane on O, (1, 1, 0), and (0, 1, 1) has the equation

$$\begin{bmatrix} x_1 \\ x_2 \\ x_3 \end{bmatrix} = t_1\begin{bmatrix} 1 \\ 1 \\ 0 \end{bmatrix} + t_2\begin{bmatrix} 0 \\ 1 \\ 1 \end{bmatrix},$$

which leads to the equivalent scalar system

$$\begin{aligned} x_1 &= t_1 \\ x_2 &= t_1 + t_2 \\ x_3 &= \phantom{t_1 + {}} t_2. \end{aligned} \tag{3.9.3}$$

Note that these three equations imply

$$x_1 - x_2 + x_3 = 0, \tag{3.9.4}$$

an equation that is satisfied by the coordinates of every point of the plane. Moreover, (3.9.4), considered as a system of equations, has the parametric

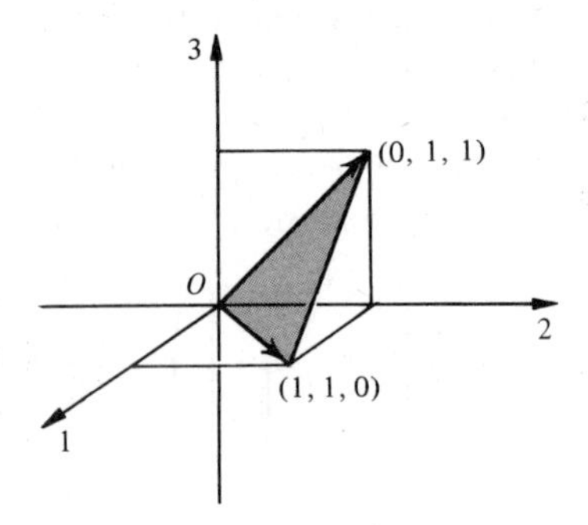

Figure 3-17. *Plane on O Determined by Two Vectors.*

solution (3.9.3), so all points whose coordinates satisfy (3.9.3) are on the plane.

It is easy to sketch a triangular portion of the plane by sketching the vectors that determine it (Figure 3-17).

Now let X and Y be any two vectors of the plane (3.9.2), where

$$X = t_1X_1 + t_2X_2, \qquad Y = s_1X_1 + s_2X_2.$$

Then

$$cX = (ct_1)X_1 + (ct_2)X_2$$

and

$$X + Y = (t_1 + s_1)X_1 + (t_2 + s_2)X_2,$$

so cX and $X + Y$ are also on the plane. Thus *the vectors of a plane on the origin constitute a vector space* (Section 3.6). We have now seen that the zero

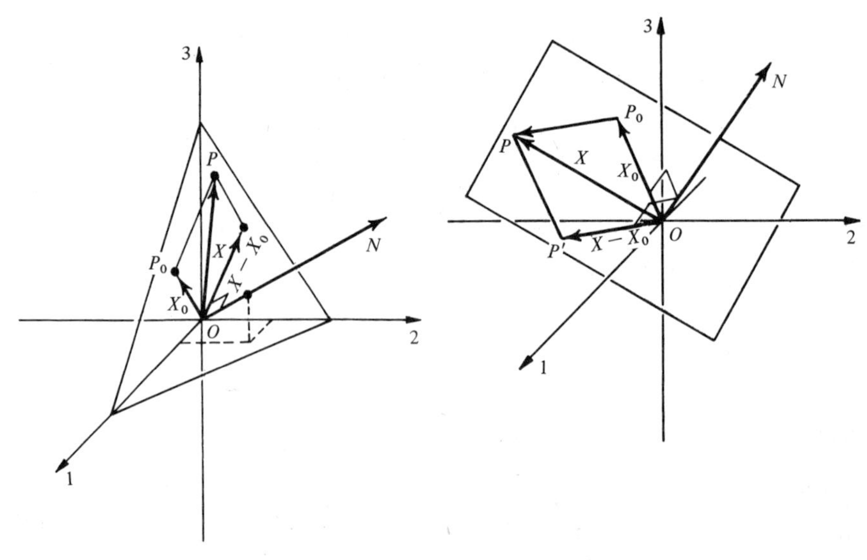

Figure 3-18. *The Normal to a Plane.*

vector alone, the vectors on a line on the origin, the vectors on a plane on the origin, and $\mathscr{E}^3$ itself, all constitute vector spaces. There are no other kinds of vector spaces in $\mathscr{E}^3$. Can you prove this?

We now show how the single scalar equation representing a given plane may be written directly rather than being obtained by the elimination of parameters.

Given a plane in $\mathscr{E}^3$, we can erect a line orthogonal (perpendicular) to it from (or at) the origin (see Figure 3-18). Such a line is called a **normal** to the plane. From elementary Euclidean geometry, a line normal to a plane is orthogonal to every line in the plane. Also, all lines orthogonal to the normal and having one point in the plane lie entirely in the plane.

If we have given both the direction of the normal line, N, to the plane and a point P_0 on the plane, the plane is completely determined since, again from elementary Euclidean geometry, there is a unique plane perpendicular to a given line N and containing a given point P_0. We should therefore be able to write a vector equation for the plane, using this information. Let $P_0{:}(c_1, c_2, c_3)$ be the given point on the plane, let $P{:}(x_1, x_2, x_3)$ denote an arbitrary point on the plane, and let the vector

$$A = \begin{bmatrix} a_1 \\ a_2 \\ a_3 \end{bmatrix}$$

define the direction of the normal. Let X_0 and X be the vectors determined by $\overrightarrow{OP_0}$ and $\overrightarrow{OP}$, respectively. Then the vector A and the vector $X - X_0$ determined by $\overrightarrow{P_0P}$ are orthogonal, and the plane is the locus of all points P for which this is true. That is, for every point P in the plane, and for no points not in the plane, we have

$$A^{\mathsf{T}}(X - X_0) = 0, \tag{3.9.5}$$

or, in scalar form,

$$a_1(x_1 - c_1) + a_2(x_2 - c_2) + a_3(x_3 - c_3) = 0. \tag{3.9.6}$$

Thus a plane in $\mathscr{E}^3$ is represented by a single linear equation. Expanding and transposing the constant terms, we get

$$a_1x_1 + a_2x_2 + a_3x_3 = b, \tag{3.9.7}$$

where

$$b = a_1c_1 + a_2c_2 + a_3c_3.$$

For example, the plane having a normal line with direction numbers 2, -1, 3 and containing the point $(-1, 1, 0)$ is represented by the equation

$$2(x_1 + 1) - (x_2 - 1) + 3(x_3 - 0) = 0$$

or

$$2x_1 - x_2 + 3x_3 = -3.$$

The form (3.9.7) is the commonest way of writing the equation of a plane, and every equation of the form (3.9.7) represents a plane, for by finding a point whose coordinates satisfy (3.9.7), we can rearrange this equation in the form (3.9.6), which identifies the plane completely. For example, the equation

$$2x_1 - 3x_2 + x_3 = 5$$

is satisfied by $(1, -1, 0)$. (To find a point on the plane, substitute two convenient values in the equation, then solve for the third.) Hence

$$2 \cdot 1 - 3(-1) + 1 \cdot 0 = 5.$$

Subtracting this equation from the preceding equation, we get

$$2(x_1 - 1) - 3(x_2 + 1) + 1(x_3 - 0) = 0.$$

The equation therefore represents a plane orthogonal to the vector

$$\begin{bmatrix} 2 \\ -3 \\ 1 \end{bmatrix}$$

and on the point $(1, -1, 0)$. There are infinitely many similar ways of describing this same plane, depending on which point P_0 we choose on it.

We can also solve problems about planes and lines. For example, in what point does the line through $(1, 1, 1)$ and $(0, -1, 2)$ intersect the plane whose equation was just given? By (3.6.1), the line has parametric equations

$$\begin{aligned} x_1 &= 1 + t(0 - 1) = 1 - t \\ x_2 &= 1 + t(-1 - 1) = 1 - 2t \\ x_3 &= 1 + t(2 - 1) = 1 + t. \end{aligned}$$

If the point (x_1, x_2, x_3) is to lie on the plane, we must therefore have

$$2(1 - t) - 3(1 - 2t) + (1 + t) = 5,$$

from which

$$t = 1,$$

so the required point is $(0, -1, 2)$.

Other problems about planes lead to systems of linear equations. Recall again that three points not on the same straight line determine a plane. Instead of using a vector approach in order to find an equation for the plane, we may solve three linear equations simultaneously. To illustrate, let the points be $(1, -1, 0)$, $(1, 1, 2)$, and $(1, 1, -2)$. If (3.9.7) is the equation of the plane, substitution of these sets of coordinates into the equation yields the system of equations:

$$\begin{aligned} a_1 - a_2 \qquad\quad &= b \\ a_1 + a_2 + 2a_3 &= b \\ a_1 + a_2 - 2a_3 &= b. \end{aligned}$$

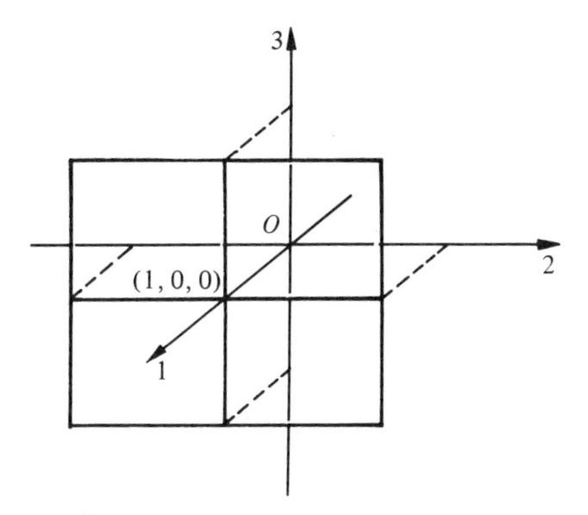

Figure 3-19. *The Plane* $x_1 = 1$.

Inspection shows that $a_2 = a_3 = 0$ and $a_1 = b$, so the equation of the plane is just $bx_1 = b$, with b arbitrary or, if we put $b = 1$,

$$x_1 = 1,$$

which represents a plane perpendicular to the 1-axis (Figure 3-19).

It is not hard to show, in the manner of the preceding example, that the equation of the plane on the three points $(c_1, 0, 0)$, $(0, c_2, 0)$, and $(0, 0, c_3)$, where $c_1c_2c_3 \neq 0$, can be written in the form

(3.9.8) $$\frac{x_1}{c_1} + \frac{x_2}{c_2} + \frac{x_3}{c_3} = 1.$$

This is the **intercept form** of the equation of the plane. The coordinates c_1, c_2, and c_3 are called the **intercepts** of the plane. When three nonzero intercepts c_1, c_2, and c_3 are known, it is easy to sketch a triangular portion of the plane. For example, if we divide the equation

$$2x_1 + 3x_2 - 4x_3 = 5$$

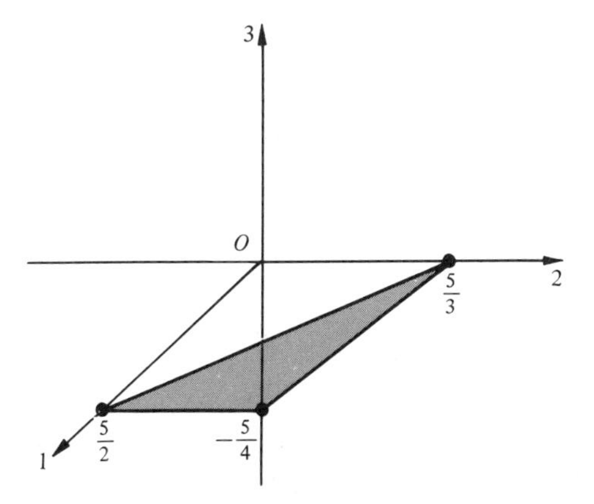

Figure 3-20. *Intercepts of a Plane.*

by 5 and rearrange the coefficients, we have

$$\frac{x_1}{\frac{5}{2}} + \frac{x_2}{\frac{5}{3}} + \frac{x_3}{-\frac{5}{4}} = 1.$$

The intercepts are $\frac{5}{2}$, $\frac{5}{3}$, and $-\frac{5}{4}$, respectively (Figure 3-20).

Planes parallel to an axis have no intercept on that axis and hence cannot be sketched in this way.

3.10 LINEAR COMBINATIONS OF VECTORS IN $\mathscr{E}^3$

An expression of the form

$$t_1V_1 + t_2V_2 + \cdots + t_kV_k,$$

where the t's are scalars, is called a **linear combination** of the V's.

Consider the set of all linear combinations W of two fixed vectors V_1 and V_2 in $\mathscr{E}^3$:

(3.10.1) $$W = t_1V_1 + t_2V_2.$$

The parallelogram law of addition shows that, for all values of t_1 and t_2, the vector W is in the same plane with V_1 and V_2. Whenever three or more vectors lie in the same plane, we call them **coplanar**. Thus all linear combinations of two given vectors are coplanar.

On the other hand, if V_1 and V_2 and any vector W in the same plane are given, we can use W as the diagonal of a parallelogram which is completed by drawing parallels to V_1 and V_2, through the terminal point of W, to intersect the lines

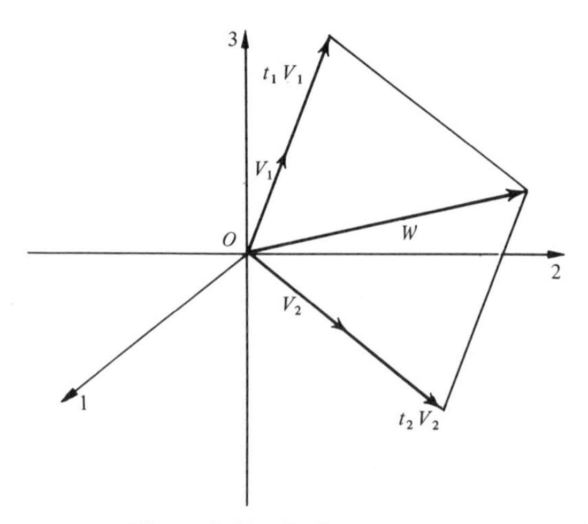

Figure 3-21. *Coplanar Vectors.*

corresponding to V_2 and V_1, respectively (see Figure 3-21). This determines scalar multiples t_1V_1 and t_2V_2 such that $W = t_1V_1 + t_2V_2$. That is, we have

Theorem 3.10.1: *The vector W is coplanar with noncollinear vectors V_1 and V_2 if and only if W can be written as a linear combination of V_1 and V_2.*

The t's are readily determined algebraically when V_1, V_2, and W are given. For example, if

$$V_1 = \begin{bmatrix} 1 \\ 2 \\ -1 \end{bmatrix}, \qquad V_2 = \begin{bmatrix} 3 \\ 0 \\ 2 \end{bmatrix}, \qquad W = \begin{bmatrix} 5 \\ 4 \\ 0 \end{bmatrix},$$

we have to solve

$$\begin{bmatrix} 5 \\ 4 \\ 0 \end{bmatrix} = t_1 \begin{bmatrix} 1 \\ 2 \\ -1 \end{bmatrix} + t_2 \begin{bmatrix} 3 \\ 0 \\ 2 \end{bmatrix},$$

for the t's. This equation is equivalent to the system

$$\begin{aligned} t_1 + 3t_2 &= 5 \\ 2t_1 \qquad &= 4 \\ -t_1 + 2t_2 &= 0. \end{aligned}$$

Here we see at once that $t_1 = 2$, after which the first and third equations both imply that $t_2 = 1$, so the system is consistent even though there are more equations than unknowns. We have then

$$W = 2V_1 + V_2.$$

If we do not know whether or not W is coplanar with V_1 and V_2, we simply attempt to solve the equation (3.10.1) for the t's. If the corresponding system is consistent, W is indeed coplanar with V_1 and V_2. If it is not consistent, then W cannot be written as a linear combination of V_1 and V_2 and hence is not coplanar with them.

Consider now any linear combination

$$W = t_1V_1 + t_2V_2 + t_3V_3, \tag{3.10.2}$$

where the V's are not coplanar. Then W is the diagonal of the parallelepiped determined by t_1V_1, t_2V_2, and t_3V_3.

On the other hand, if W, V_1, V_2, and V_3 are given (V_1, V_2, V_3 not coplanar), we can pass planes parallel to those determined by pairs of V_1, V_2, and V_3 through the terminal point of W (see Figure 3-22), thus determining a parallelepiped and hence scalar multiples of the V's such that (3.10.2) holds. That is, we have

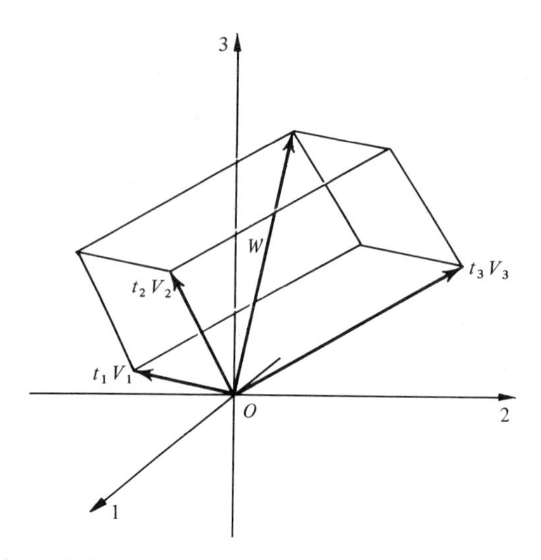

Figure 3-22. *A Linear Combination of Three Vectors.*

Theorem 3.10.2: *Every vector W in $\mathscr{E}^3$ is a linear combination of any three noncoplanar vectors, V_1, V_2, and V_3, in $\mathscr{E}^3$.*

A purely algebraic proof will be given in the next section (Theorem 3.11.4).

Again, for given V_1, V_2, V_3, and W, with V_1, V_2, and V_3 not coplanar, we can determine t_1, t_2, and t_3 algebraically. For example, if

$$V_1 = \begin{bmatrix} 1 \\ 0 \\ 0 \end{bmatrix}, \qquad V_2 = \begin{bmatrix} 1 \\ 1 \\ 0 \end{bmatrix}, \qquad V_3 = \begin{bmatrix} 1 \\ 1 \\ 1 \end{bmatrix},$$

then for any W, the equation

$$t_1 \begin{bmatrix} 1 \\ 0 \\ 0 \end{bmatrix} + t_2 \begin{bmatrix} 1 \\ 1 \\ 0 \end{bmatrix} + t_3 \begin{bmatrix} 1 \\ 1 \\ 1 \end{bmatrix} = \begin{bmatrix} w_1 \\ w_2 \\ w_3 \end{bmatrix}$$

is equivalent to the system of equations

$$\begin{aligned} t_1 + t_2 + t_3 &= w_1 \\ t_2 + t_3 &= w_2 \\ t_3 &= w_3, \end{aligned}$$

so that, since this is already in echelon form, we have readily

$$\begin{aligned} t_1 &= w_1 - w_2 \\ t_2 &= \qquad w_2 - w_3 \\ t_3 &= \qquad\qquad w_3. \end{aligned}$$

That is, *any vector* W in $\mathscr{E}^3$ can be expressed as a linear combination of the three given vectors. The coefficients of the vectors in this combination are given by the preceding equations in terms of the components of W, so

$$W = (w_1 - w_2)\begin{bmatrix}1\\0\\0\end{bmatrix} + (w_2 - w_3)\begin{bmatrix}1\\1\\0\end{bmatrix} + w_3\begin{bmatrix}1\\1\\1\end{bmatrix}.$$

3.11 LINEAR DEPENDENCE OF VECTORS; BASES

The vectors $V_1, V_2, \ldots, V_k$ are said to be **linearly dependent** if and only if there exist scalars $t_1, t_2, \ldots, t_k$, not all of which are zero, such that

(3.11.1) $$t_1V_1 + t_2V_2 + \cdots + t_kV_k = 0.$$

For example,

$$0\begin{bmatrix}1\\-1\\5\end{bmatrix} + 2\begin{bmatrix}1\\2\\-1\end{bmatrix} + (-1)\begin{bmatrix}3\\-2\\2\end{bmatrix} + (-1)\begin{bmatrix}-1\\6\\-4\end{bmatrix} = \begin{bmatrix}0\\0\\0\end{bmatrix},$$

so the four vectors on the left are linearly dependent.

If $V_1, V_2, \ldots, V_k$ are linearly dependent and if $t_i \neq 0$, then we can solve (3.11.1) for V_i:

(3.11.2) $$V_i = \left(-\frac{t_1}{t_i}\right)V_1 + \cdots + \left(-\frac{t_{i-1}}{t_i}\right)V_{i-1} + \left(-\frac{t_{i+1}}{t_i}\right)V_{i+1} + \cdots + \left(-\frac{t_k}{t_i}\right)V_k,$$

that is, at least one of the vectors is dependent on the others in the usual sense of the term. On the other hand, if V_i is a linear combination of the other vectors, thus:

(3.11.3) $$V_i = s_1V_1 + \cdots + s_{i-1}V_{i-1} + s_{i+1}V_{i+1} + \cdots + s_kV_k,$$

then

$$s_1V_1 + s_2V_2 + \cdots + s_{i-1}V_{i-1} + (-1)V_i + s_{i+1}V_{i+1} + \cdots + s_kV_k = 0.$$

Here we take $t_i = -1$ and $t_j = s_j$ for all j except i. Since $t_i = -1$, not all the t's are 0. Hence if one of the vectors is linearly dependent on the others, the set $V_1, V_2, \ldots, V_k$ is a linearly dependent set. The advantage of the definition of linear dependence with which we began is that it does not single out any one vector which must be dependent on the others and hence lends itself more readily to a variety of linear computations.

If equation (3.11.1) holds only when all the t's are 0, then we say that the V's are **linearly independent**.

Suppose that the vectors V_1 and V_2 are linearly dependent. Then there exist t_1 and t_2, not both zero, such that

$$t_1V_1 + t_2V_2 = 0.$$

If $t_1 \neq 0$, then

$$V_1 = \frac{-t_2}{t_1} V_2,$$

so V_1 is a scalar multiple of V_2; that is, V_1 and V_2 are collinear. Similarly if $t_2 \neq 0$. Conversely, collinear (proportional) vectors are linearly dependent, for if

$$V_1 = kV_2,$$

then

$$1 \cdot V_1 + (-k)V_2 = 0.$$

Thus we have

Theorem 3.11.1: *Two vectors in $\mathscr{E}^3$ are linearly dependent if and only if they are collinear.*

If V_1, V_2, and V_3 are linearly dependent, then the equation

$$t_1V_1 + t_2V_2 + t_3V_3 = 0$$

with not all the t's 0 permits us to write at least one of the V's as a linear combination of the other two. For example, if $t_1 \neq 0$,

$$V_1 = \left(-\frac{t_2}{t_1}\right)V_2 + \left(-\frac{t_3}{t_1}\right)V_3,$$

so V_1 is coplanar with V_2 and V_3. Conversely, if

$$V_1 = k_2V_2 + k_3V_3$$

so that V_1 is coplanar with V_2 and V_3, then

$$1 \cdot V_1 + (-k_2)V_2 + (-k_3)V_3 = 0,$$

so the three vectors are linearly dependent. Similarly if $t_2 \neq 0$ or $t_3 \neq 0$. This proves

Theorem 3.11.2: *Three vectors in $\mathscr{E}^3$ are linearly dependent if and only if they are coplanar.*

Finally, consider four vectors V_1, V_2, V_3, and V_4 in $\mathscr{E}^3$. The equation

$$t_1V_1 + t_2V_2 + t_3V_3 + t_4V_4 = 0$$

is equivalent to three scalar equations in four unknowns:

$$\begin{aligned} v_{11}t_1 + v_{12}t_2 + v_{13}t_3 + v_{14}t_4 &= 0 \\ v_{21}t_1 + v_{22}t_2 + v_{23}t_3 + v_{24}t_4 &= 0 \\ v_{31}t_1 + v_{32}t_2 + v_{33}t_3 + v_{34}t_4 &= 0, \end{aligned}$$

where

$$V_j = \begin{bmatrix} v_{1j} \\ v_{2j} \\ v_{3j} \end{bmatrix}.$$

Now, if we reduce this system of equations to echelon form, we can express one, two, or three of the variables, as the case may be, in terms of the remaining variables, of which there is at least one. Hence to at least one variable we can assign a nonzero value, so that nontrivial solutions for the t's certainly exist. That is, we have

Theorem 3.11.3: *Any four vectors in $\mathscr{E}^3$ are necessarily linearly dependent.*

An important special case is this: Let V_1, V_2, and V_3 be any three linearly independent (noncoplanar) vectors in $\mathscr{E}^3$ and let W be any vector in $\mathscr{E}^3$. Then these four vectors are linearly dependent by the preceding theorem. Hence there exist scalars t, t_1, t_2, and t_3 not all zero such that

$$tW + t_1V_1 + t_2V_2 + t_3V_3 = 0.$$

In this equation we must have $t \neq 0$, since otherwise V_1, V_2, and V_3 would be linearly dependent because we would have $t_1V_1 + t_2V_2 + t_3V_3 = 0$ with not all of t_1, t_2, and t_3 zero. Hence we can solve for W:

$$W = \left(-\frac{t_1}{t}\right)V_1 + \left(-\frac{t_2}{t}\right)V_2 + \left(-\frac{t_3}{t}\right)V_3.$$

Thus we have proved

Theorem 3.11.4: *Every vector W of $\mathscr{E}^3$ may be expressed as a linear combination of any three linearly independent (noncoplanar) vectors in $\mathscr{E}^3$.*

A set of three noncoplanar vectors, in terms of which every vector of $\mathscr{E}^3$ can necessarily be expressed as a linear combination, is called a **basis** for $\mathscr{E}^3$. The most obvious basis is the **standard basis** (Figure 3-23) consisting of the set of **elementary vectors,**

$$E_1 = \begin{bmatrix} 1 \\ 0 \\ 0 \end{bmatrix}, \quad E_2 = \begin{bmatrix} 0 \\ 1 \\ 0 \end{bmatrix}, \quad E_3 = \begin{bmatrix} 0 \\ 0 \\ 1 \end{bmatrix},$$

for any vector X can be written thus:

$$\begin{aligned} X = \begin{bmatrix} x_1 \\ x_2 \\ x_3 \end{bmatrix} &= x_1\begin{bmatrix} 1 \\ 0 \\ 0 \end{bmatrix} + x_2\begin{bmatrix} 0 \\ 1 \\ 0 \end{bmatrix} + x_3\begin{bmatrix} 0 \\ 0 \\ 1 \end{bmatrix} \\ &= x_1E_1 + x_2E_2 + x_3E_3. \end{aligned}$$

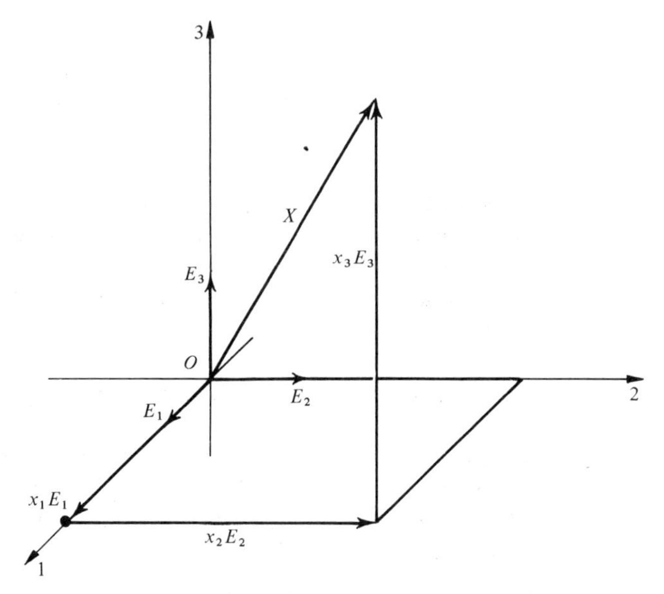

Figure 3-23. *The Standard Basis for* $\mathscr{E}^3$.

Since $E_i^T E_j = 0$ when $i \neq j$, these three vectors are mutually orthogonal. Each also has length 1. Any basis consisting of three mutually orthogonal unit vectors is called an **orthonormal basis** for $\mathscr{E}^3$. (The vectors E_1, E_2, and E_3 are the same as the unit vectors **i, j,** and **k** of vector analysis.)

3.12 HALF-SPACES IN $\mathscr{E}^3$

A particularly useful type of locus in $\mathscr{E}^3$ is that represented by an inequality of the form

$$a_1x_1 + a_2x_2 + a_3x_3 + c > 0.$$

A plane with equation

$$a_1x_1 + a_2x_2 + a_3x_3 + c = 0$$

separates $\mathscr{E}^3$ into what are called **half-spaces**. An important result is

Theorem 3.12.1: *All points such that*

$$a_1x_1 + a_2x_2 + a_3x_3 + c > 0 \tag{3.12.1}$$

lie in one of the half-spaces determined by the plane

$$a_1x_1 + a_2x_2 + a_3x_3 + c = 0, \tag{3.12.2}$$

and all points such that

$$a_1x_1 + a_2x_2 + a_3x_3 + c < 0 \tag{3.12.3}$$

lie in the other half-space.

We call these regions, respectively, the **positive and negative half-spaces** associated with the function $f(X) = a_1x_1 + a_2x_2 + a_3x_3 + c$.

To prove the theorem, let $P:(y_1, y_2, y_3)$ be any point such that

$$a_1y_1 + a_2y_2 + a_3y_3 + c > 0.$$

Then P is not on the plane in question. (Prove that such a point P does, in fact, exist.) Let $B:(b_1, b_2, b_3)$ be the foot of the perpendicular from P to the plane. The directed segment $\overrightarrow{BP}$ (Figure 3-24) determines the vector

$$\begin{bmatrix} y_1 - b_1 \\ y_2 - b_2 \\ y_3 - b_3 \end{bmatrix}.$$

The vector

$$A = \begin{bmatrix} a_1 \\ a_2 \\ a_3 \end{bmatrix}$$

is normal to the plane. Because all lines orthogonal to a fixed plane have the same direction, the vector A and the vector determined by $\overrightarrow{BP}$ are collinear. Hence there must exist a scalar k such that

$$\begin{bmatrix} a_1 \\ a_2 \\ a_3 \end{bmatrix} = k \begin{bmatrix} y_1 - b_1 \\ y_2 - b_2 \\ y_3 - b_3 \end{bmatrix}. \tag{3.12.4}$$

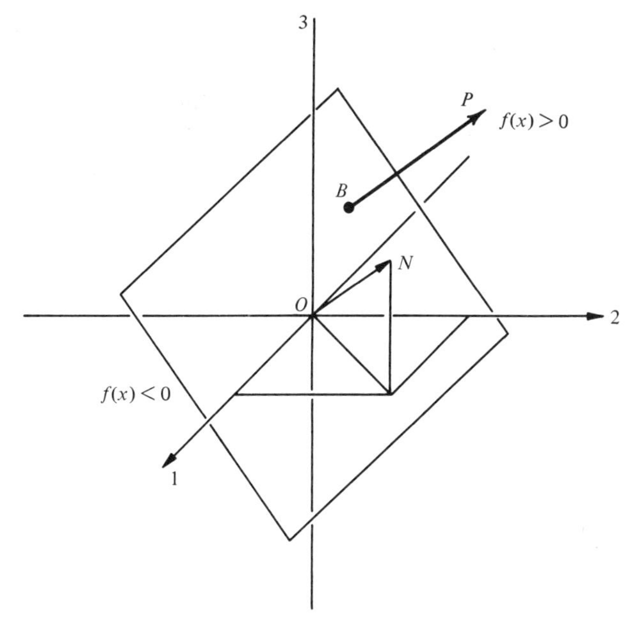

Figure 3-24. *Half-Spaces of* $\mathscr{E}^3$.

Now

$$(3.12.5) \qquad a_1y_1 + a_2y_2 + a_3y_3 + c > 0$$

and, since B is on the plane,

$$(3.12.6) \qquad a_1b_1 + a_2b_2 + a_3b_3 + c = 0.$$

Hence, subtracting (3.12.6) from (3.12.5), we have

$$(3.12.7) \qquad a_1(y_1 - b_1) + a_2(y_2 - b_2) + a_3(y_3 - b_3) > 0.$$

Substituting for the a's from (3.12.4) and simplifying, we get

$$(3.12.8) \qquad k[(y_1 - b_1)^2 + (y_2 - b_2)^2 + (y_3 - b_3)^2] > 0,$$

so that, since the expression in parentheses is positive, k must be positive also. Therefore, for *every* point $P:(y_1, y_2, y_3)$ such that $a_1y_1 + a_2y_2 + a_3y_3 + c > 0$, the corresponding directed segment $\overrightarrow{BP}$ has the same sense as

$$\begin{bmatrix} a_1 \\ a_2 \\ a_3 \end{bmatrix}$$

and hence all such points P must lie in the same half-space. Moreover, all points P in that half-space determine parallel directed segments $\overrightarrow{BP}$ all of which have the same sense, so that for all such points P the parameter k in (3.12.4) must be positive, since it is positive for one of them. Then, beginning with (3.12.8) and reversing the steps until (3.12.5) is obtained, we see that for every point P in that half-space (3.12.5) holds.

Since all those points and only those points such that (3.12.5) holds constitute one of the half-spaces determined by the plane, all those points and only those points such that (3.12.3) holds must constitute the other half-space.

An immediate consequence of all this is

Corollary 3.12.2: *The sign of the function value* $f(X) = a_1x_1 + a_2x_2 + a_3x_3 + c$ *in one of the half-spaces associated with the corresponding plane is determined by any one point of the half-space in question.*

These half-spaces are often called **open half-spaces** because they do not include any points of the plane which is their common boundary. The inequalities

$$a_1x_1 + a_2x_2 + a_3x_3 + c \geq 0$$

and

$$a_1x_1 + a_2x_2 + a_3x_3 + c \leq 0$$

are satisfied by the same points as before and also by all points on the plane. These sets, which now include the boundary plane, are called **closed half-spaces**.

For example, let us identify the open half-space such that

$$f(X) = 6x_1 + 10x_2 - 5x_3 + 30 > 0.$$

The simplest test point to use is the origin. We have

$$f(0) = 30,$$

so the origin is in the positive half-space associated with this function (Figure 3-25). The figure shows that the positive half-space associated with a given

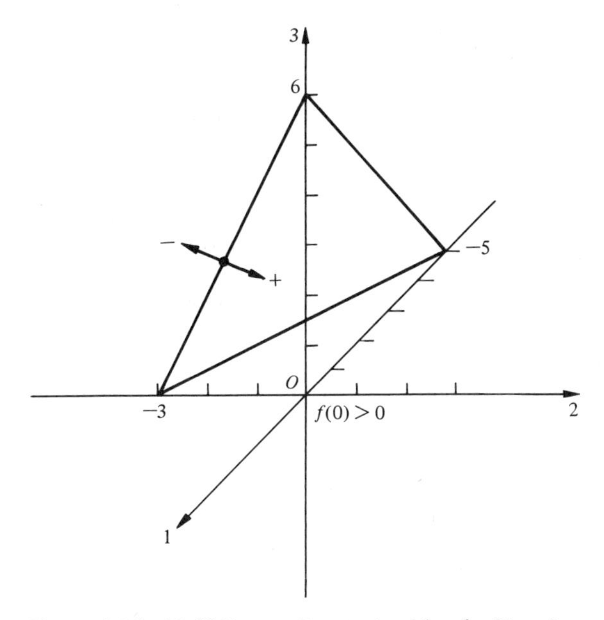

Figure 3-25. *Half-Spaces Determined by the Function* $f(X) = 6x_1 + 10x_2 - 5x_3 + 30.$

nonhomogeneous linear function may well be "below" the corresponding plane.

Frequently one is concerned with the region determined by more than one inequality. Such a region is then the intersection of certain half-spaces. For example, what region in $\mathscr{E}^3$ is the intersection of the positive half-spaces associated with the two functions

$$f(X) = x_1 + x_2 - x_3 + 2$$

and

$$g(X) = 2x_1 + 2x_2 + x_3 + 4?$$

We determine the points where the planes cross the coordinate axes, sketch the two planes, and find that the origin is on the positive side of each. (Figure 3-26). The region in question is an infinite wedge formed by the two planes and opening toward the first octant.

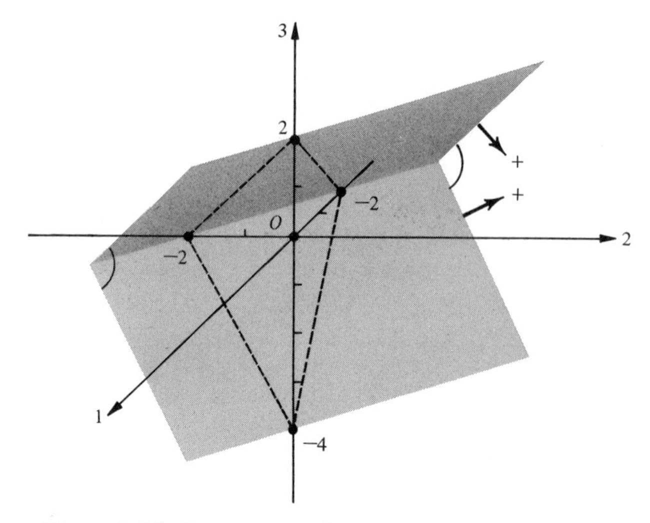

Figure 3-26. *Intersection of Two Positive Half-Spaces.*

In the same way, we can determine the intersection of the positive half-spaces associated with the functions

$$h(X) = -x_1 - x_2 - x_3 + 4,$$

and

$$k(X) = x_1 + x_2 + x_3 + 4.$$

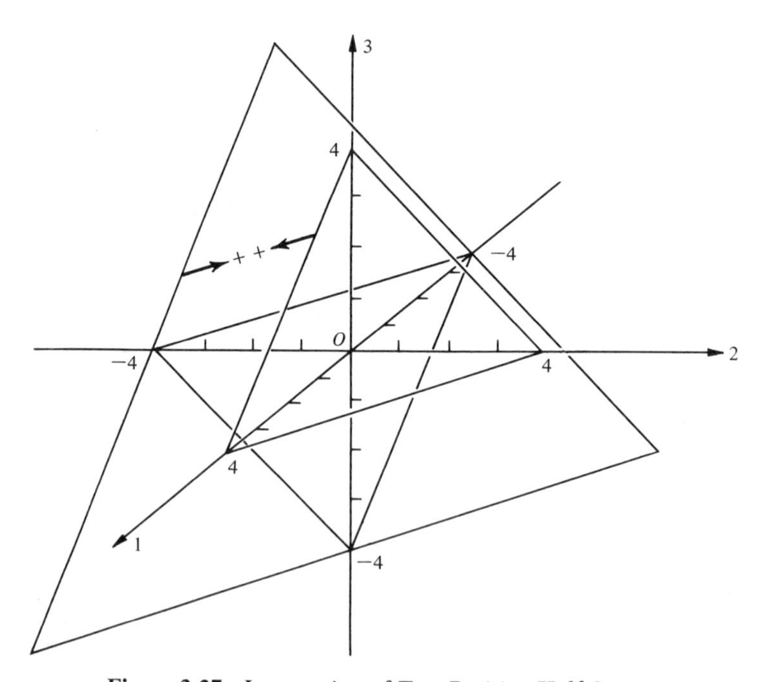

Figure 3-27. *Intersection of Two Positive Half-Spaces.*

As Figure 3-27 shows, the required region is the infinite slab lying between the two planes $h(X) = 0$, $k(X) = 0$.

The intersection of the closed half-spaces determined by the three inequalities $x_1 \geq 0$, $x_2 \geq 0$, and $x_3 \geq 0$ is simply the first octant, inclusive of those portions of the coordinate planes that bound it.

The intersection of the half-spaces determined by the four inequalities

$$\begin{aligned} x_1 &> 0 \\ x_2 &> 0 \\ x_3 &> 0 \\ -x_1 - x_2 - x_3 + 2 &> 0 \end{aligned}$$

is the interior of a tetrahedron with vertex at the origin and three edges along the coordinate axes (Figure 3-28).

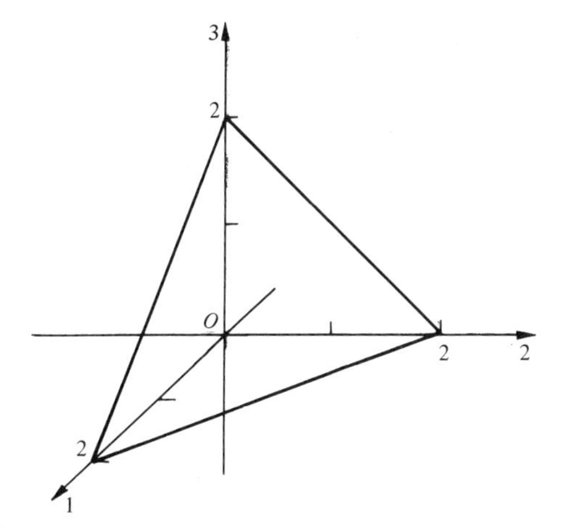

Figure 3-28. *The Tetrahedron Defined by*
$x_1 > 0,\ x_2 > 0,\ x_3 > 0,\ -x_1 - x_2 - x_3 + 2 > 0.$

One cannot enclose in $\mathscr{E}^3$ a region whose maximum diameter is finite with fewer than four linear inequalities. There is no upper limit on the number of inequalities one may use for the determination of either a finite or an infinite region in $\mathscr{E}^3$.

3.13 EXERCISES

1. Find parametric scalar equations for the line on
(a) P_1:$(-1, 4, 3)$ and P_2:$(2, 0, 3)$,
(b) Q_1:$(-4, 4, 3)$ and Q_2:$(2, 4, 3)$.
What is special about each of these lines? Illustrate with figures.

2. Prove that the triangle with vertices P_1:$(-1, 4, 3)$, P_2:$(2, 0, 3)$, and P_3:$(-1, 4, 5)$ is a right triangle.

3. (a) Find the point where the line on P_1:$(1, 1, 1)$ and P_2:$(-1, 0, 3)$ meets the plane $2x_1 - 3x_2 + 7x_3 + 5 = 0$.

(b) Show that the line on Q_1:$(2, -1, 4)$ and Q_2:$(-3, 5, 0)$ does not intersect, that is, is parallel to, the plane

$$2x_1 + x_2 - x_3 = 8.$$

4. For what value of x_3 will the line on $(2, 1, 4)$ and $(3, -1, 2)$ be parallel to the line on $(4, -2, 1)$ and $(2, 2, x_3)$?

5. Let the *unit* vector

$$A = \begin{bmatrix} a_1 \\ a_2 \\ a_3 \end{bmatrix}$$

define the direction of a line and let P_1:(b_1, b_2, b_3) be a point on it. Show that parametric equations for the line are

$$\begin{aligned} z_1 &= b_1 + sa_1 \\ z_2 &= b_2 + sa_2 \\ z_3 &= b_3 + sa_3. \end{aligned}$$

Then show that the parameter s may be used as the directed distance from point P_1 to point P:(z_1, z_2, z_3).

6. The two planes

$$\begin{aligned} x_1 + 2x_2 - x_3 + 4 &= 0 \\ 2x_1 - 3x_2 + x_3 - 3 &= 0 \end{aligned}$$

intersect in a line. Find parametric scalar equations for it.

7. (a) Find the equation of the plane on the three points $(1, 1, 2)$, $(1, 2, 1)$, and $(2, 1, 1)$. Illustrate with a figure.

(b) Find the equation of the plane parallel to the x_3-axis and on the two points $(2, 0, 0)$ and $(0, 2, 0)$. (*Hint*: What are the direction numbers for the normal to the plane?)

***8.** Prove in detail that the equation of the plane through the points $(c_1, 0, 0)$, $(0, c_2, 0)$, and $(0, 0, c_3)$, where $c_1c_2c_3 \neq 0$, can be written in the form

$$\frac{x_1}{c_1} + \frac{x_2}{c_2} + \frac{x_3}{c_3} = 1.$$

This is the intercept form of the equation of the plane introduced in Section 3.13.

9. Reduce the following equations of planes to intercept form. Then sketch the planes. In each case, find the point where the normal from the origin to the plane meets the plane. Show the normal on your sketch.

$$\text{(a)}\ \ 3x_1 - 2x_2 + 6x_3 = 12,$$

$$\text{(b)}\ \ 12x_1 + 3x_2 + 5x_3 = 4.$$

10. Identify the positive half-space determined by the function $f(X) = x_1 + x_2 + 3x_3 + 3$ and illustrate with a sketch.

11. (a) Identify and sketch the intersection of the positive half-spaces determined by the functions

$$\text{(a)} \begin{cases} f(X) = x_1 + x_2 + x_3 + 3 \\ g(X) = x_1 + x_2 + x_3 - 3, \end{cases}$$

$$\text{(b)} \begin{cases} f(X) = x_1 + x_2 + x_3 + 3 \\ g(X) = x_1 - x_2 - x_3 - 3, \end{cases}$$

$$\text{(c)} \begin{cases} f(X) = x_1 + x_2 - x_3 + 2 \\ g(X) = 2x_1 + 2x_2 + x_3 + 4, \end{cases} \qquad \text{(d)} \begin{cases} f(X) = x_1 + x_2 + x_3 - 2 \\ g(X) = -x_1 - x_2 - x_3 + 4, \end{cases}$$

$$\text{(e)} \begin{cases} -x_1 + 1 > 0 & -x_2 + 1 > 0 & -x_3 + 1 \geq 0 \\ x_1 + 1 > 0, & x_2 + 1 > 0, & x_3 + 1 \geq 0. \end{cases}$$

12. (a) Express the vector

$$\begin{bmatrix} a_1 \\ a_2 \\ a_3 \end{bmatrix}$$

as a linear combination of the vectors

$$\begin{bmatrix} 0 \\ 1 \\ 1 \end{bmatrix}, \quad \begin{bmatrix} 1 \\ 0 \\ 1 \end{bmatrix}, \quad \begin{bmatrix} 1 \\ 1 \\ 0 \end{bmatrix}.$$

(b) Show that the vectors

$$\begin{bmatrix} 0 \\ 2 \\ -3 \end{bmatrix}, \quad \begin{bmatrix} 1 \\ 4 \\ 0 \end{bmatrix}, \quad \begin{bmatrix} 3 \\ 8 \\ 6 \end{bmatrix}$$

are linearly dependent.

***13.** Show that if $V_1, V_2, \ldots, V_k$ are vectors in $\mathscr{E}^3$ with $k \geq 4$, then the V's are necessarily linearly dependent.

***14.** Show that the zero vector and any other vector V are linearly dependent.

***15.** Show that if the vectors V_1, V_2, and V_3 are mutually orthogonal, then they are linearly independent.

16. Show that the vectors

$$\begin{bmatrix} 0 \\ \frac{1}{\sqrt{2}} \\ \frac{1}{\sqrt{2}} \end{bmatrix}, \quad \begin{bmatrix} 1 \\ 0 \\ 0 \end{bmatrix}, \quad \begin{bmatrix} 0 \\ \frac{-1}{\sqrt{2}} \\ \frac{1}{\sqrt{2}} \end{bmatrix}$$

constitute an orthonormal basis for $\mathscr{E}^3$.

17. Show that the three vectors E_1, $E_1 + E_2$, $E_1 + E_2 + E_3$ are linearly independent.

18. Are $E_1 - E_2$, $E_2 - E_3$, and $E_3 - E_1$ linearly independent? Why?

19. A *unit* vector is orthogonal to each of

$$\begin{bmatrix} 2 \\ -1 \\ 3 \end{bmatrix}, \quad \begin{bmatrix} -3 \\ 0 \\ 2 \end{bmatrix}.$$

Find it. Is there more than one solution?

20. (a) If the vectors

$$\begin{bmatrix} -1 \\ a \\ 2a \end{bmatrix} \quad \text{and} \quad \begin{bmatrix} 2 \\ -a \\ a \end{bmatrix}$$

are orthogonal, what are the possible values for a?

(b) Interpret geometrically the equation

$$[x_1, x_2, x_3] \begin{bmatrix} x_1 \\ x_2 \\ -x_3 \end{bmatrix} = 0.$$

***21.** Show that an equation of a sphere with center O and radius r is

$$x_1^2 + x_2^2 + x_3^2 = r^2.$$

22. Show that an equation of a sphere with center (h_1, h_2, h_3) and radius r is

$$(x_1 - h_1)^2 + (x_2 - h_2)^2 + (x_3 - h_3)^2 = r^2.$$

23. Prove that

$$\begin{bmatrix} 1 \\ 2 \\ 2 \end{bmatrix}, \quad \begin{bmatrix} 0 \\ 1 \\ 2 \end{bmatrix}, \quad \begin{bmatrix} 0 \\ 0 \\ 3 \end{bmatrix}$$

constitute a basis for $\mathscr{E}^3$ by showing that the equation

$$\begin{bmatrix} w_1 \\ w_2 \\ w_3 \end{bmatrix} = t_1 \begin{bmatrix} 1 \\ 2 \\ 2 \end{bmatrix} + t_2 \begin{bmatrix} 0 \\ 1 \\ 2 \end{bmatrix} + t_3 \begin{bmatrix} 0 \\ 0 \\ 3 \end{bmatrix}$$

can be solved for the t's regardless of the values given to w_1, w_2, and w_3.

***24.** Show that the set of all vectors X in $\mathscr{E}^3$ which satisfy an equation $AX = 0$, where A is an $m \times 3$ matrix, is a vector space.

***25.** Prove that the set of all vectors X in $\mathscr{E}^3$ such that $X^{\mathsf{T}}A = 0$, where A is a fixed vector in $\mathscr{E}^3$, is a vector space. Represent this space by an appropriate figure.

***26.** Prove that the set of all linear combinations of fixed vectors $V_1, V_2, \ldots, V_k$ in $\mathscr{E}^3$ is a vector space.

***27.** Prove that the set of all vectors X in $\mathscr{E}^3$ such that $AX = \lambda X$, where λ is a fixed scalar, is a vector space.

28. Given that X_1, X_2, and X_3 are linearly independent vectors in $\mathscr{E}^3$, prove that X_1, $X_1 + X_2$, and $X_1 + X_2 + X_3$ are linearly independent also.

29. Find a linear combination of

$$\begin{bmatrix} 0 \\ 1 \\ 1 \end{bmatrix} \quad \text{and} \quad \begin{bmatrix} 1 \\ 1 \\ 0 \end{bmatrix}$$

which is orthogonal to

$$\begin{bmatrix} 0 \\ 1 \\ 1 \end{bmatrix}.$$

Illustrate with a figure.

30. Two systems of inequalities are **equivalent** if and only if every solution of either system is also a solution of the other. Are the systems of inequalities

$$\begin{aligned} x_1 \qquad\qquad &\geq 0 \\ x_1 + x_2 \qquad &\geq 0 \\ x_1 + x_2 + x_3 &\geq 0 \end{aligned} \qquad \text{and} \qquad \begin{aligned} x_1 &\geq 0 \\ x_2 &\geq 0 \\ x_3 &\geq 0 \end{aligned}$$

equivalent? Why? Illustrate with figures in $\mathscr{E}^3$. What operations on a system of inequalities transform it into an equivalent system of inequalities?

31. For what sets of points do these systems of inequalities hold?

(a) $x_1^2 + x_2^2 + x_3^2 > 1$,

(b) $x_1^2 + x_2^2 + x_3^2 \leq 1$,

(c) $x_1^2 + x_2^2 + x_3^2 < 1$, $x_3 > 0$,

(d) $x_1^2 + x_2^2 + x_3^2 < 1$, $x_3 > 0$, $x_2 - x_3 > 0$.

CHAPTER 4

Vector Geometry in n-Dimensional Space

4.1 THE REAL n-SPACE $\mathscr{R}^n$

The geometry of spaces of more than three dimensions is a purely mathematical creation since there are no corresponding, simple physical spaces to serve as guides to intuition. One way to develop such a geometry is to generalize, in the most natural way, various concepts and results of the analytic geometry of two and three dimensions. The reason for doing this is that many useful algebraic formulas and processes involving n variables are precisely analogous to formulas and processes involving two or three variables with which we have associated familiar geometric names. The development of n-dimensional geometry involves the association of a set of geometric names, similar to those used in two and three dimensions, with analogous arithmetic and algebraic objects and processes involving n variables. What happens in two and three dimensions helps us to know what to expect algebraically in higher-dimensional spaces, but we do not try to imagine, for example, four or more mutually perpendicular axes. When $n > 3$, "n-space," as we develop it, is nothing more or less than an arithmetic and algebraic construction. A synthetic development analogous to that of classical Euclidean geometry is also possible but is much less useful and is much more difficult.

We first define **real, arithmetic n-dimensional space**, $\mathscr{R}^n$, to be the set of all ordered n-tuples $(a_1, a_2, \ldots, a_n)$ of real numbers. An n-tuple $(a_1, a_2, \ldots, a_n)$ is called a **point** of $\mathscr{R}^n$. The numbers $a_1, a_2, \ldots, a_n$ are called the **coordinates** of the point. The point $O{:}(0, 0, \ldots, 0)$ is called the **origin** of $\mathscr{R}^n$. The set of all points $(0, \ldots, 0, a_i, 0, \ldots, 0)$ whose ith coordinate is an arbitrary real number is called the ***i*-axis of** $\mathscr{R}^n$.

A little later we shall introduce the concept of distance and then define what we mean by Euclidean n-space, $\mathscr{E}^n$, but first we examine some of the linear geometry that is independent of the definition of distance.

4.2 VECTORS IN $\mathscr{R}^n$

Let $P_0{:}(a_1, a_2, \ldots, a_n)$ and $P_1{:}(b_1, b_2, \ldots, b_n)$ be two points of $\mathscr{R}^n$. Then, as in $\mathscr{E}^3$, the set of points given by

$$(4.2.1) \quad (a_1 + t(b_1 - a_1), a_2 + t(b_2 - a_2), \ldots, a_n + t(b_n - a_n)), \quad 0 \le t \le 1,$$

is called the **line segment** P_0P_1. [Recall (3.6.2) and the paragraph preceding those equations.] If we give to the points of this segment the order determined by increasing t continuously from 0 to 1 inclusive, we call the resulting ordered set the **directed line segment** $\overrightarrow{P_0P_1}$ of which P_0 is the **initial point** and P_1 is the **terminal point**.

With each real n-vector $X = [x_1, x_2, \ldots, x_n]^{\mathsf{T}}$, we associate a directed line segment $\overrightarrow{OP}$, where O is the origin and P is the point $(x_1, x_2, \ldots, x_n)$. As in $\mathscr{E}^3$, this directed segment will be called a **geometric vector**. In this chapter, we shall not distinguish verbally between X and $\overrightarrow{OP}$ and for brevity we shall call P "the terminal point of X." Since there is a one-to-one correspondence between points $P{:}(x_1, x_2, \ldots, x_n)$ and geometric vectors $\overrightarrow{OP}$, $\mathscr{R}^n$ may be considered to be either a space of points or a space of vectors.

In $\mathscr{E}^3$ there are familiar, purely geometric rules for adding vectors (the parallelogram law) and for multiplying them by scalars (with the aid of similar triangles). We saw in Chapter 3 that by describing geometric vectors by column matrices, we could replace the geometric operations by corresponding algebraic operations on the matrices representing the vectors. By contrast, in $\mathscr{R}^n$ there are no purely geometric rules for these same operations. Hence, in $\mathscr{R}^n$, we define the addition of vectors and the multiplication of vectors by scalars a priori in terms of these operations on the column matrices which represent the vectors. As a result, these operations have the same algebraic properties as do the corresponding operations on vectors in $\mathscr{E}^3$ or on matrices in general.

With each directed segment $\overrightarrow{P_0P_1}$ as defined above, we associate the n-vector of ordered differences

$$X_1 - X_0 = \begin{bmatrix} b_1 - a_1 \\ b_2 - a_2 \\ \vdots \\ b_n - a_n \end{bmatrix},$$

where X_1 and X_0 are the vectors with terminal points P_1 and P_0, respectively. Thus $\overrightarrow{P_0P_1}$ determines the geometric vector $\overrightarrow{OP}$, where P is the point $(b_1 - a_1, b_2 - a_2, \ldots, b_n - a_n)$. (Recall here Figure 3-10.)

Two directed segments are defined to be **equal** if and only if their associated n-vectors of ordered differences are equal. For example, the ordered pairs of points in $\mathscr{R}^4$, $P_0:(2, 1, -3, 0)$ and $P_1:(1, -1, 2, 1)$, $Q_0:(1, 3, -1, 2)$ and $Q_1:(0, 1, 4, 3)$, determine the equal directed segments $\overrightarrow{P_0P_1}$, $\overrightarrow{Q_0Q_1}$ and the associated vector

$$X = \begin{bmatrix} -1 \\ -2 \\ 5 \\ 1 \end{bmatrix}.$$

Recall here Figure 3-10.

Each given n-vector X in turn determines a class of infinitely many directed segments whose columns of ordered differences all equal X. Given the initial point of such a segment, the terminal point is easily found since the ordered differences are known. Thus, for the vector X in our example, if $(1, 1, 1, 1)$ is the initial point of one such segment, the terminal point must be $(0, -1, 6, 2)$.

4.3 LINES AND PLANES IN $\mathscr{R}^n$

Given two distinct points $P_0:(x_{10}, x_{20}, \ldots, x_{n0})$ and $P_1:(x_{11}, x_{21}, \ldots, x_{n1})$, the **line** P_0P_1 in $\mathscr{R}^n$ is defined to be the set of all points $P:(x_1, x_2, \ldots, x_n)$ determined by

$$X = X_0 + t(X_1 - X_0) \qquad (t \text{ arbitrary}), \tag{4.3.1}$$

where the vectors X, X_0, and X_1 have terminal points P, P_1, and P_2, respectively. [This is the extension to $\mathscr{R}^n$ of the equation (3.6.1) of a line in 3-space.] It can be shown that the same set of points is determined by any two distinct points of the line.

Note that, as in $\mathscr{E}^3$, P_1 and P_2 are on the same line on O if and only if all points of the directed segments $\overrightarrow{OP_1}$ and $\overrightarrow{OP_2}$ are on such a line. (Prove this.)

Three points are defined to be **collinear** if and only if they are not all distinct or two of them are distinct and the vector X corresponding to the third satisfies the equation (4.3.1) determined by the other two for some value of t. (Compare with Figure 3-12.)

For example, are the points $(1, 1, 2, -1)$, $(3, 0, 1, 4)$, and $(-5, 4, 5, -16)$ collinear? The problem is, in effect, to find a value for t, if one exists, such that

$$\begin{aligned} -5 &= 1 + (3 - 1)t &&= 1 + 2t \\ 4 &= 1 + (0 - 1)t &&= 1 - t \\ 5 &= 2 + (1 - 2)t &&= 2 - t \\ -16 &= -1 + [4 - (-1)]t &&= -1 + 5t. \end{aligned}$$

The solution of the first equation is $t = -3$ and this also satisfies the other three, so the three points are indeed collinear. This problem provides a good illustration of the usefulness of systems of linear equations having more equations than unknowns.

Given three noncollinear points P_0, P_1, and P_2, where $P_i = (x_{1i}, x_{2i}, \ldots, x_{ni})$, $i = 0, 1, 2$, the **plane** on P_0, P_1, P_2 is defined to be the set of all terminal points $(x_1, x_2, \ldots, x_n)$ of vectors X given by

$$X = X_0 + t_1(X_1 - X_0) + t_2(X_2 - X_0), \tag{4.3.2}$$

where P_i is the terminal point of X_i, $i = 1, 2, 3$. (Compare Figure 3-16.) It can be shown that the same set of points is determined by any three noncollinear points of the plane.

For example, in $\mathscr{R}^4$, the plane on the three noncollinear points (1, 1, 1, 1), (1, 1, 0, 0), and (0, 0, 2, 2) has the vector equation

$$X = \begin{bmatrix} 1 \\ 1 \\ 1 \\ 1 \end{bmatrix} + t_1 \begin{bmatrix} 0 \\ 0 \\ -1 \\ -1 \end{bmatrix} + t_2 \begin{bmatrix} -1 \\ -1 \\ 1 \\ 1 \end{bmatrix}. \tag{4.3.3}$$

These ideas are readily generalized. A line in $\mathscr{R}^n$ is called a **1-flat**; a plane in $\mathscr{R}^n$ is called a **2-flat**. If we assume that a $(k - 1)$-flat has been defined, then a ***k*-flat** is the set of terminal points P of the vectors X defined by

$$X = X_0 + t_1(X_1 - X_0) + t_2(X_2 - X_0) + \cdots + t_k(X_k - X_0), \tag{4.3.4}$$

where $X_0, X_1, \ldots, X_k$ correspond to $k + 1$ points $P_0, P_1, \ldots, P_k$ which are not all in the same $(k - 1)$-flat. (As in the case of lines and planes, the k-flat is determined by *any* $k + 1$ of it points, provided that they are not all in the same $(k - 1)$-flat. This we can prove rigorously a little later.) The maximum value of k is n, in which case the n-flat is simply $\mathscr{R}^n$ itself, as our study of linear dependence in Chapter 5 will reveal.

A particularly important case for applications is the case $k = n - 1$. An $(n - 1)$-**flat**, or **hyperplane** as it is usually called, may also be defined by a single linear equation (as may a line in $\mathscr{E}^2$ and a plane in $\mathscr{E}^3$):

$$a_1x_1 + a_2x_2 + \cdots + a_nx_n = b. \tag{4.3.5}$$

We shall be able to prove, using the results of Chapter 5, the equivalence of the representations (4.3.4) and (4.3.5).

In $\mathscr{E}^2$, a line ordinarily meets another line in a point, and in $\mathscr{E}^3$ a line ordinarily meets a plane in a point. Similarly, in $\mathscr{R}^n$ a line ordinarily meets a hyperplane in a point. To see this, rewrite (4.3.5) in matrix form:

$$A^{\mathsf{T}}X = b \tag{4.3.6}$$

and substitute from (4.3.1) into (4.3.6):

$$A^{\mathsf{T}}(X_0 + t(X_1 - X_0)) = b,$$

so

$$tA^{\mathsf{T}}(X_1 - X_0) = b - A^{\mathsf{T}}X_0$$

and hence, when $A^{\mathsf{T}}(X_1 - X_0) \neq 0$,

$$t = \frac{b - A^{\mathsf{T}}X_0}{A^{\mathsf{T}}(X_1 - X_0)}. \tag{4.3.7}$$

Since there is, when $A^{\mathsf{T}}(X_1 - X_0) \neq 0$, a unique value of t, in this case the line meets the hyperplane in precisely one point. When $A^{\mathsf{T}}(X_1 - X_0) = 0$, the line may lie entirely on the hyperplane or may not meet it at all, as following examples show. In the latter case, the line and the hyperplane are said to be **parallel**.

For example, in $\mathscr{R}^4$, the line on P_0:(1, 0, 0, 0) and P_1:(0, 1, 0, 0) and the hyperplane

$$x_1 + x_2 + x_3 + x_4 = 4$$

may be represented, respectively, by the matrix equations

$$X = \begin{bmatrix} 1 \\ 0 \\ 0 \\ 0 \end{bmatrix} + t\begin{bmatrix} -1 \\ 1 \\ 0 \\ 0 \end{bmatrix} \qquad \text{and} \qquad [1, 1, 1, 1]X = 4.$$

From these, by substitution from the first equation into the second, we obtain

$$[1, 1, 1, 1]\left(\begin{bmatrix} 1 \\ 0 \\ 0 \\ 0 \end{bmatrix} + t\begin{bmatrix} -1 \\ 1 \\ 0 \\ 0 \end{bmatrix}\right) = 4$$

or

$$1 + 0 \cdot t = 4 \qquad \text{(inconsistent)}.$$

Since no point of the line P_0P_1 is on the given hyperplane, the line is parallel to it.

On the other hand, the line on Q_0:(1, 1, 1, 1) and Q_1:(0, 2, 0, 2) has the equation

$$X = \begin{bmatrix} 1 \\ 1 \\ 1 \\ 1 \end{bmatrix} + t\begin{bmatrix} -1 \\ 1 \\ -1 \\ 1 \end{bmatrix}.$$

Its intersection with the same hyperplane is defined by the equation

$$4 + 0 \cdot t = 4,$$

which is satisfied by all values of t, so the entire line is on the given hyperplane. This is what one would expect since Q_0 and Q_1 are both on the hyperplane.

4.4 LINEAR DEPENDENCE AND INDEPENDENCE IN $\mathscr{R}^n$

Let $X_1, X_2, \ldots, X_k$ be vectors in $\mathscr{R}^n$. The expression $t_1X_1 + t_2X_2 + \cdots + t_kX_k$, where the t's are scalars, is called a **linear combination** of $X_1, X_2, \ldots, X_k$. If, for given vectors $X_1, X_2, \ldots, X_k$, there exist scalars $t_1, t_2, \ldots, t_k$, *not all zero*, such that

$$t_1X_1 + t_2X_2 + \cdots + t_kX_k = 0,$$

then we say that $X_1, X_2, \ldots, X_k$ are **linearly dependent**. If this equation holds true only when all the t's are zero, we say that $X_1, X_2, \ldots, X_k$ are **linearly independent**.

As in $\mathscr{E}^3$, linear dependence in $\mathscr{R}^n$ has a geometrical interpretation. For example, two vectors X_1 and X_2 in $\mathscr{R}^n$ are said to be **collinear vectors** if and only if their terminal points P_1 and P_2 lie on the same line on the origin. The zero vector and any vector X are collinear. If $X_1 \neq 0$ and $X_2 \neq 0$ are collinear and if $X = tA$ is the equation of the corresponding line on the origin, there exist scalars $t_1 \neq 0$ and $t_2 \neq 0$ such that $X_1 = t_1A$ and $X_2 = t_2A$. Then $t_2X_1 + (-t_1)X_2 = t_2t_1A - t_1t_2A = 0$, so that X_1 and X_2 are linearly dependent.

Suppose now that X_1 and X_2 are linearly dependent so that there exist t_1 and t_2, not both zero, such that $t_1X_1 + t_2X_2 = 0$. If $t_2 \neq 0$, then $X_2 = -(t_1/t_2)X_1$, so the terminal point P_2 of X_2 is on the line on the origin determined by P_1. If $t_1 \neq 0$, then $X_1 = -(t_2/t_1)X_2$, so P_1 is on the line on the origin determined by P_2. Since at least one of t_1, t_2 is not 0, we have proved

Theorem 4.4.1: *Two vectors in $\mathscr{R}^n$ are collinear if and only if they are linearly dependent.*

In general, suppose that $X_1, X_2, \ldots, X_k$ are linearly dependent. Then if $t_j \neq 0$ in

$$t_1X_1 + t_2X_2 + \cdots + t_jX_j + \cdots + t_kX_k = 0,$$

we can write

$$X_j = 0 + \left(-\frac{t_1}{t_j}\right)(X_1 - 0) + \cdots + \left(-\frac{t_{j-1}}{t_j}\right)(X_j - 0)$$
$$+ \left(-\frac{t_{j+1}}{t_j}\right)(X_{j+1} - 0) + \cdots + \left(-\frac{t_k}{t_j}\right)(X_k - 0),$$

so that all the vectors $X_1, X_2, \ldots, X_k$ belong to the same $(k-1)$-flat on the origin, or to some lower-dimensional flat space on O if $X_1, \ldots, X_{j-1}, X_{j+1}, \ldots,$

X_k are linearly dependent. The problem of dimension and its computation will be treated in detail in Chapter 5.

4.5 VECTOR SPACES IN $\mathscr{R}^n$

A set $\mathscr{V}$ of vectors in $\mathscr{R}^n$ is called a **vector space** if and only if, whenever X and Y belong to $\mathscr{V}$, $X + Y$ and tX also belong to $\mathscr{V}$, that is, if and only if $\mathscr{V}$ is **closed** with respect to addition and with respect to multiplication by a scalar. The simplest vector spaces are the set consisting of the zero vector alone and the set consisting of all vectors in $\mathscr{R}^n$.

The set of all linear combinations of k given vectors $X_1, X_2, \ldots, X_k$ in $\mathscr{R}^n$ is a basic type of vector space. Indeed, if two such linear combinations are

$$X = t_1X_1 + t_2X_2 + \cdots + t_kX_k$$

and

$$Y = s_1X_1 + s_2X_2 + \cdots + s_kX_k,$$

then $X + Y$ and tX are also such linear combinations since

$$X + Y = (t_1 + s_1)X_1 + (t_2 + s_2)X_2 + \cdots + (t_k + s_k)X_k$$

and

$$tX = (tt_1)X_1 + (tt_2)X_2 + \cdots + (tt_k)X_k,$$

so the required closure properties hold.

In the case $k = 1$, if $X_1 \neq 0$, such a vector space is the set of all vectors X given by

$$X = tX_1 = 0 + t(X_1 - 0).$$

Hence this vector space is represented geometrically by a line on the origin. As in $\mathscr{E}^3$, the vector X corresponding to any point P on this line belongs to this vector space and every vector X of the space has its terminal point P on this line.

More generally, the vector space which is the set of all vectors X given by

$$\begin{aligned} X &= t_1X_1 + t_2X_2 + \cdots + t_kX_k \\ &= 0 + t_1(X_1 - 0) + t_2(X_2 - 0) + \cdots + t_k(X_k - 0) \end{aligned}$$

is represented geometrically by a k-flat on the origin provided that $X_1, X_2, \ldots, X_k$ are not linearly dependent. These flat spaces will be treated more adequately in Chapter 5.

4.6 EXERCISES

1. Find the vector equation of the line on the points (0, 1, 1, 0) and (1, 0, 0, 1) in $\mathscr{R}^4$. Then eliminate the parameter from the corresponding scalar system of equations. How many hyperplanes must one intersect in order to define a line in $\mathscr{R}^4$? In $\mathscr{R}^n$?

2. Prove that the points (0, 1, 1, 0), (1, 0, 0, 1), and (1, 1, 1, 1) are not collinear.

3. Write the vector equation of the line in $\mathscr{R}^n$ on O and P, where $P = (0, \ldots, 0, 1, 0, \ldots, 0)$, the 1 being the ith coordinate. Thus show that the i-axis is indeed a line.

4. Two points P_0 and P_1 of $\mathscr{R}^n$ are always collinear. Does it follow that the associated vectors X_0 and X_1 are always collinear? Illustrate with a figure in $\mathscr{E}^3$.

***5.** Show that two nonzero vectors in $\mathscr{R}^n$ are collinear if and only if they are proportional.

***6.** Show that two nonzero vectors in $\mathscr{R}^n$ are linearly dependent if and only if they are proportional.

7. Show that in $\mathscr{R}^n$ *any* two points of the line $X = X_0 + t(X_1 - X_0)$, where $X_1 \neq X_0$, determine the same line, that is, determine the same total set of points.

8. Show that

$$X = \begin{bmatrix} 2 \\ 1 \\ -1 \\ 4 \end{bmatrix} + t\begin{bmatrix} 3 \\ 0 \\ 1 \\ -5 \end{bmatrix} \quad \text{and} \quad X = \begin{bmatrix} 5 \\ 1 \\ 0 \\ -1 \end{bmatrix} + t\begin{bmatrix} -6 \\ 0 \\ -2 \\ 10 \end{bmatrix}$$

represent the same line in $\mathscr{R}^4$.

9. Write the vector equation of the plane in $\mathscr{R}^4$ on the three points given in Exercise 2. Show that this plane lies on the origin.

***10.** Three vectors in $\mathscr{R}^n$ are said to be **coplanar** if and only if their terminal points all lie on the same plane on O. Show that three vectors in $\mathscr{R}^n$ are coplanar if and only if they are linearly dependent.

11. Show that the four vectors in $\mathscr{R}^4$,

$$\begin{bmatrix} 1 \\ -1 \\ 2 \\ 3 \end{bmatrix}, \quad \begin{bmatrix} 2 \\ 1 \\ -1 \\ 4 \end{bmatrix}, \quad \begin{bmatrix} -2 \\ 3 \\ 0 \\ 5 \end{bmatrix}, \quad \begin{bmatrix} -5 \\ 4 \\ 3 \\ 9 \end{bmatrix},$$

are linearly dependent, whereas the first three are linearly independent. Show that, in consequence, these four vectors lie in a 3-flat on the origin and write its vector equation.

12. Use (4.3.4) to write the vector equations of the hyperplane (3-flat) in $\mathscr{R}^4$ determined by the points (4, 0, 0, 0), (0, 4, 0, 0), (0, 0, 4, 0), and (0, 0, 0, 4). Then eliminate the parameters from the corresponding set of scalar equations and show in this way that an equation for the 3-flat is $x_1 + x_2 + x_3 + x_4 = 4$.

***13.** Generalize the intercept form (3.9.8) for the equation of a plane to an intercept form for the equation of a hyperplane in $\mathscr{R}^4$ and then in $\mathscr{R}^n$. What happens to this equation if the hyperplane fails to intersect a given axis? Can a hyperplane in $\mathscr{R}^n$ fail to intersect all n of the axes?

14. Find the point where the line OP, where $P = (1, 1, 1, 1)$, meets the hyperplane of Exercise 12.

***15.** Show that if two distinct points of a line lie on a k-flat, then every point of the line lies on the k-flat. This property is the reason for calling a flat space *flat*. In $\mathscr{E}^3$, only a line and a plane contain all the points of *every* line which intersects them in two distinct points. A curved surface does not do this.

16. Show that every k-flat on the origin can be represented by an equation of the form

$$X = t_1X_1 + t_2X_2 + \cdots + t_kX_k,$$

where $X_1, X_2, \ldots, X_k$ are linearly independent.

17. Under what conditions will the equations $X = \sum_1^k t_iX_i$ and $X = \sum_1^k s_iY_i$ represent the same vector space in $\mathscr{R}^n$?

18. Show that the vectors with terminal points on any k-flat *not* on the origin do *not* constitute a vector space. Illustrate with figures in $\mathscr{E}^2$ and $\mathscr{E}^3$.

19. Let a k-flat in $\mathscr{R}^n$ be defined by

$$X = X_0 + t_1(X_1 - X_0) + t_2(X_2 - X_0) + \cdots + t_k(X_k - X_0).$$

Show that the set of all vectors Y where

$$Y = X - X_0 = t_1(X_1 - X_0) + t_2(X_2 - X_0) + \cdots + t_k(X_k - X_0)$$

is a vector space. Interpret geometrically and sketch figures in $\mathscr{E}^2$ and $\mathscr{E}^3$.

20. Given a plane in $\mathscr{E}^3$, not on the origin, in how many ways can one derive a vector space in the manner of Exercise 19? What would you guess is the answer for a k-flat in $\mathscr{R}^n$? For a given k-flat, are these vector spaces distinct? Why?

4.7 LENGTH AND THE CAUCHY–SCHWARZ INEQUALITY

The length of a vector in $\mathscr{R}^n$ can be defined in various sensible ways, each suitable to a particular purpose or application. On the basis of previous experience, it is natural to define the **length** $|X|$ of the vector X by

$$|X| = (X^{\mathsf{T}}X)^{1/2} = \sqrt{x_1^2 + x_2^2 + \cdots + x_n^2}. \tag{4.7.1}$$

This function is also called the **norm** of X. (In other contexts, other functions of the components of X are also called norms.) Since a sum of squares of real numbers is zero if and only if each of the real numbers is zero, that is,

$$(4.7.2) \quad x_1^2 + x_2^2 + \cdots + x_n^2 = 0 \qquad \text{if and only if } x_1 = x_2 = \cdots = x_n = 0,$$

it follows that *a vector X in $\mathcal{R}^n$ has length zero if and only if it is the zero vector.*

A vector is called a **unit vector** if and only if its length is 1. Some examples are

$$(4.7.3) \qquad E_1 = \begin{bmatrix} 1 \\ 0 \\ 0 \\ \vdots \\ 0 \end{bmatrix}, E_2 = \begin{bmatrix} 0 \\ 1 \\ 0 \\ \vdots \\ 0 \end{bmatrix}, \ldots, E_n = \begin{bmatrix} 0 \\ 0 \\ \vdots \\ 0 \\ 1 \end{bmatrix}, \quad \text{and} \quad \begin{bmatrix} \frac{1}{\sqrt{n}} \\ \frac{1}{\sqrt{n}} \\ \vdots \\ \frac{1}{\sqrt{n}} \end{bmatrix}.$$

The first n of these will be called the **elementary unit vectors.**

If X is the vector associated with the directed segment $\overrightarrow{AB}$, then the **distance between A and B** is defined to be the length of X, which in this case is given by

$$(4.7.4) \qquad d(A, B) = \sqrt{(b_1 - a_1)^2 + (b_2 - a_2)^2 + \cdots + (b_n - a_n)^2}.$$

In view of this last definition, the length of a vector X is the distance between the origin O and the terminal point P of the vector X.

The space $\mathcal{R}^n$, with this definition of distance imposed on it, is called **Euclidean space of n dimensions** and is denoted by $\mathcal{E}^n$.

There are four basic properties that should be satisfied by any useful definition of distance. For all points A, B, and C of $\mathcal{R}^n$, the distance function $d(A, B)$ should be such that

$$(4.7.5) \qquad \begin{aligned} &\text{(a) } d(A, B) \geq 0 \\ &\text{(b) } d(A, B) = 0 \qquad \text{if and only if } A = B \\ &\text{(c) } d(A, B) = d(B, A) \\ &\text{(d) } d(A, B) + d(B, C) \geq d(A, C) \qquad \text{(the triangle inequality).} \end{aligned}$$

In the case of (4.7.4), properties (a) and (c) follow at once from the definition. Property (c) implies that distance is *not directed.* In the case of (b), one employs the result of (4.7.2). Property (d) is somewhat more difficult to prove. It rests on the following form of the famous Cauchy–Schwarz inequality:

$$(4.7.6) \qquad |X| \cdot |Y| \geq |X^{\mathsf{T}} Y|$$

or, in scalar form,

$$(4.7.7) \qquad \sqrt{\sum x_i^2} \cdot \sqrt{\sum y_i^2} \geq \left|\sum x_i y_i\right|,$$

where the summations run from 1 to n on i.

This last formula is proved by the following computation, where again summations on i and j run from 1 to n so far as allowed by any indicated restrictions:

$$\begin{aligned}
\sum x_i^2 \cdot \sum y_i^2 - \left(\sum x_i y_i\right)^2 &= \sum_{i \neq j} x_i^2 y_j^2 + \sum x_i^2 y_i^2 - \sum x_i^2 y_i^2 - 2 \sum_{i<j} x_i y_i x_j y_j \\
&= \sum_{i \neq j} x_i^2 y_j^2 - 2 \sum_{i<j} x_i y_i x_j y_j \\
&= \sum_{i<j} x_i^2 y_j^2 - 2 \sum_{i<j} x_i y_i x_j y_j + \sum_{i<j} x_j^2 y_i^2 \\
&= \sum_{i<j} (x_i^2 y_j^2 - 2x_i y_i x_j y_j + x_j^2 y_i^2) \\
&= \sum_{i<j} (x_i y_j - x_j y_i)^2.
\end{aligned}$$

(If this is a little difficult to follow, try writing it out in full for $n = 3$.) Since the last member is necessarily nonnegative, we have

$$\sum x_i^2 \cdot \sum y_i^2 \geq \left(\sum x_i y_i\right)^2.$$

The inequalities (4.7.7) and (4.7.6) now follow when we take the positive square root of each side.

From (4.7.7) we have

$$\sqrt{\sum x_i^2} \cdot \sqrt{\sum y_i^2} \geq \left|\sum x_i y_i\right| \geq \sum x_i y_i.$$

Multiplying by 2 and adding $\sum x_i^2$ and $\sum y_i^2$ to the extreme members, we get

$$\begin{aligned}
\sum x_i^2 + 2\sqrt{\sum x_i^2}\sqrt{\sum y_i^2} + \sum y_i^2 &\geq \sum x_i^2 + 2\sum x_i y_i + \sum y_i^2 \\
&= \sum (x_i^2 + 2x_i y_i + y_i^2);
\end{aligned}$$

that is,

$$\left(\sqrt{\sum x_i^2} + \sqrt{\sum y_i^2}\right)^2 \geq \sum (x_i + y_i)^2.$$

Taking the positive square root on both sides, we have

$$\sqrt{\sum x_i^2} + \sqrt{\sum y_i^2} \geq \sqrt{\sum (x_i + y_i)^2},$$

which is the **vector form of the triangle inequality**:

(4.7.8) $$|X| + |Y| \geq |X + Y|.$$

(See Figure 4-1).

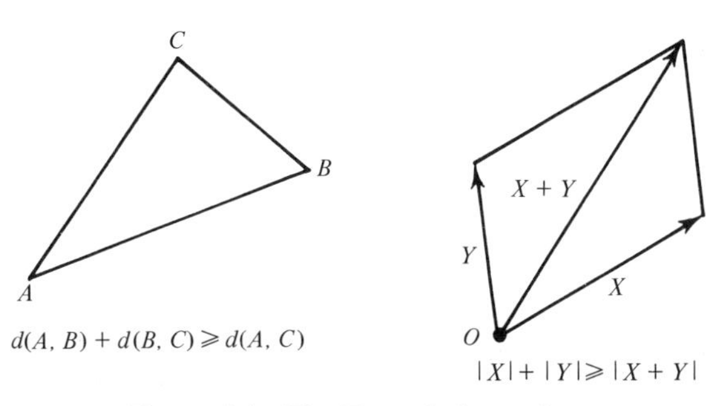

Figure 4-1. *The Triangle Inequality.*

Now let

$$X = \begin{bmatrix} b_1 - a_1 \\ b_2 - a_2 \\ \vdots \\ b_n - a_n \end{bmatrix}, \qquad Y = \begin{bmatrix} c_1 - b_1 \\ c_2 - b_2 \\ \vdots \\ c_n - b_n \end{bmatrix},$$

so

$$X + Y = \begin{bmatrix} c_1 - a_1 \\ c_2 - a_2 \\ \vdots \\ c_n - a_n \end{bmatrix}.$$

Then $|X| = d(A, B)$, $|Y| = d(B, C)$, and $|X + Y| = d(A, C)$, so (4.7.8) becomes (4.7.5d) and the proof is complete. The inequalities (4.7.6) and (4.7.8) are frequently useful.

4.8 ANGLES AND ORTHOGONALITY IN $\mathscr{E}^n$

The angle between two vectors X and Y in $\mathscr{E}^n$ is defined to be the angle θ, $0 \leq \theta \leq \pi$, which is determined by

(4.8.1) $$\cos\theta = \frac{X^{\mathsf{T}}Y}{|X| \cdot |Y|}, \qquad 0 \leq \theta \leq \pi.$$

Since, by the Cauchy–Schwarz inequality,

$$|X^{\mathsf{T}}Y| \leq |X| \cdot |Y|,$$

we have from (4.8.1)

$$-1 \leq \cos\theta \leq 1,$$

so this is a reasonable definition.

The angles $\alpha_1, \alpha_2, \ldots, \alpha_n$ between a vector X and the elementary unit vectors $E_1, E_2, \ldots, E_n$ are called the **direction angles** of X and, by (4.8.1), are determined by

(4.8.2) $$\cos\alpha_i = \frac{X^{\mathsf{T}}E_i}{|X|} = \frac{x_i}{|X|}, \qquad i = 1, 2, \ldots, n.$$

These expressions are called the **direction cosines** of X. The direction angles determine both the sense and the direction of X. The vector $-X$ has direction cosines which are the negatives of those of X, so the direction angles of $-X$ are the supplements of those of X. We describe this by saying that X and $-X$ have **opposite senses** but the same direction. Squaring the direction cosines and adding, we find that

(4.8.3) $$\sum \cos^2\alpha_i = 1.$$

The vector

$$(4.8.4) \qquad \frac{X}{|X|} = \begin{bmatrix} \dfrac{x_1}{|X|} \\ \dfrac{x_2}{|X|} \\ \vdots \\ \dfrac{x_n}{|X|} \end{bmatrix}$$

has the same direction cosines as X itself, so that it has the same direction and sense as X, but it has length 1. This unit vector is called the **normalized vector** associated with X. The process of dividing X by $|X|$ is called **normalization.**

The endpoints $(x_1, x_2, \ldots, x_n)$ of the members of the set of all *unit* vectors satisfy the equation

$$(4.8.5) \qquad x_1^2 + x_2^2 + \cdots + x_n^2 = 1.$$

The locus of these endpoints is called the **unit n-sphere with center at the origin.** For a unit vector $(x_1, x_2, \ldots, x_n)$, we have

$$(4.8.6) \qquad x_i = \cos \alpha_i, \qquad i = 1, 2, \ldots, n.$$

The direction angles of a directed segment $\overrightarrow{AB}$ are defined to be those of the corresponding vector and are therefore given by

$$(4.8.7) \qquad \cos \alpha_i = \frac{b_i - a_i}{d(A, B)}, \qquad i = 1, 2, \ldots, n.$$

We say that the vectors X and Y are orthogonal if and only if the angle between them is $\pi/2$. From (4.8.1) it follows that $\theta = \pi/2$ if and only if $X^\mathsf{T}Y = 0$; that is, X and Y are orthogonal if and only if $X^\mathsf{T} Y = 0$. For example, the vectors

$$X = \begin{bmatrix} 2 \\ -1 \\ 4 \\ 2 \end{bmatrix} \quad \text{and} \quad Y = \begin{bmatrix} -1 \\ 4 \\ 5 \\ -7 \end{bmatrix}$$

in $\mathscr{E}^4$ are orthogonal because $X^\mathsf{T} Y = 0$. Of particular importance is the fact that the elementary unit vectors are mutually orthogonal:

$$(4.8.8) \qquad E_i^\mathsf{T} E_j = 0, \qquad i \neq j.$$

Since in $\mathscr{E}^n$, just as in $\mathscr{E}^3$, we can write any vector X as a linear combination of the E_j's, thus:

$$\begin{bmatrix} x_1 \\ x_2 \\ \vdots \\ x_n \end{bmatrix} = x_1 \begin{bmatrix} 1 \\ 0 \\ 0 \\ \vdots \\ 0 \end{bmatrix} + x_2 \begin{bmatrix} 0 \\ 1 \\ 0 \\ \vdots \\ 0 \end{bmatrix} + \cdots + x_n \begin{bmatrix} 0 \\ 0 \\ \vdots \\ 0 \\ 1 \end{bmatrix},$$

and since all the E_j's are necessary here, we say that the E_j's form a **basis** for the vectors of $\mathscr{E}^n$. Since the E_j's are mutually orthogonal unit vectors, we call the basis **orthonormal**. We shall investigate such bases more extensively later.

4.9 HALF-LINES AND DIRECTED DISTANCES

A point A on a line L divides the line into two **half-lines**. Let B be a point on L distinct from A and denote the corresponding vectors by X_B and X_A, respectively, so the line has the equation

$$X = X_A + t(X_B - X_A). \tag{4.9.1}$$

Then the **positive half** of L, as determined by $\overrightarrow{AB}$, is defined to be all those points of L such that $t > 0$ and the **negative half** of L consists of all those points of L such that $t < 0$. These are **open half-lines**. If the conditions on t are replaced by $t \geq 0$ and $t \leq 0$, respectively, we get **closed half-lines**.

A simple but important example is the j-axis, for which we choose A as the origin and B as the **unit point** $(0, \ldots, 0, 1, 0, \ldots, 0)$ with a 1 as the jth coordinate, all other coordinates being zero. The equation of the j-axis is then

$$X = tE_j,$$

so the positive j-axis consists of those points $(0, \ldots, 0, t, 0, \ldots, 0)$ for which $t > 0$, as one would wish.

Two directed segments $\overrightarrow{AB}$ and $\overrightarrow{PQ}$ in $\mathscr{E}^n$, with associated nonzero vectors V and W, respectively, are said to have the **same direction** or to be **parallel** if and only if there exists a scalar k such that $W = kV$. They are further said to have the **same sense** if $k > 0$ and **opposite senses** if $k < 0$. The nonzero segment $\overrightarrow{PQ}$ on a line L whose positive half is determined by $\overrightarrow{AB}$ in the manner described above, will be said to be in the **positive sense** if its sense agrees with that of $\overrightarrow{AB}$ and in the **negative sense** if its sense is opposite to that of $\overrightarrow{AB}$.

Let P and Q on L have associated vectors X_P and X_Q, respectively. Then, from (4.9.1),

$$X_P - X_Q = (t_P - t_Q)(X_B - X_A),$$

so that the sense of $\overrightarrow{PQ}$ is the same as or opposite to that of $\overrightarrow{AB}$ according as $t_P - t_Q$ is positive or negative. We may therefore define the **positive sense on the line** L to be the sense of increasing t. A line with a positive sense assigned to it is called a **directed line**.

We may now define the **directed distance** on L, from P to Q, denoted by $\vec{d}(P, Q)$, to be $d(P, Q)$ or $-d(P, Q)$ according as the sense of $\overrightarrow{PQ}$ is positive or negative.

For example, on the line in $\mathscr{E}^4$ for which $A = (0, 0, 0, 0)$ and $B = (1, 1, 1, 1)$, the *undirected* distance between the point $P{:}(t_1, t_1, t_1, t_1)$ and the point $Q{:}(t_2, t_2, t_2, t_2)$ is given by $\sqrt{4(t_2 - t_1)^2} = 2|t_2 - t_1|$. The *directed* distance from P to Q is then given by $2(t_2 - t_1)$, since this is positive when $t_2 > t_1$ and negative when $t_2 < t_1$.

4.10 HALF-SPACES

In $\mathscr{E}^n$, a linear function $f(X) = A^\mathsf{T}X - b$ determines two nonintersecting regions called **open half-spaces**: (1) a **positive half-space** consisting of all those points for which $A^\mathsf{T}X - b > 0$ and (2) a **negative half-space** consisting of all those points for which $A^\mathsf{T}X - b < 0$. The hyperplane with equation $A^\mathsf{T}X - b = 0$, that is, the set of all points such that $A^\mathsf{T}X - b = 0$, is said to constitute the **boundary** of each of these spaces.

The function $f(X)$ also determines two **closed half-spaces**: the region for which $A^\mathsf{T}X - b \geq 0$ and the region for which $A^\mathsf{T}X - b \leq 0$. Each of these half-spaces is the union of an open half-space and its boundary.

We say that two points are **on the same side of the hyperplane** with equation $A^\mathsf{T}X - b = 0$ if and only if they are in the same one of the two open half-spaces just defined.

Consider now a k-flat represented by the parametric vector equation

$$X = X_0 + t_1(X_1 - X_0) + t_2(X_2 - X_0) + \cdots + t_k(X_k - X_0).$$

If X is the vector determined by a point P of the k-flat and if Y is determined by the point Q of the k-flat where

$$Y = X_0 + s_1(X_1 - X_0) + s_2(X_2 - X_0) + \cdots + s_k(X_k - X_0),$$

then, for all α, the terminal point of the vector $X + \alpha(Y - X)$ lies on the k-flat, since

$$X + \alpha(Y - X) = X_0 + (\alpha s_1 + (1 - \alpha)t_1)(X_1 - X_0) + \cdots + (\alpha s_k + (1 - \alpha)t_k)(X_k - X_0).$$

Thus we have proved

Theorem 4.10.1: *If the points P and Q lie on a k-flat, so does every point of the line PQ.*

Using this, we can easily prove a basic result showing how hyperplanes separate $\mathscr{E}^n$:

Theorem 4.10.2: *If a k-flat has no points in common with a hyperplane, then all its points are on the same side of the hyperplane.*

Let P and Q, with corresponding vectors X and Y, be on the k-flat. Let the hyperplane have the equation $A^\mathsf{T}X - b = 0$ and suppose, contrary to what the theorem asserts, that $A^\mathsf{T}X - b < 0$ and $A^\mathsf{T}Y - b > 0$, so that P and Q

lie on opposite sides of the hyperplane. Then some point of the line PQ will lie *on* the hyperplane if we can choose α so that

$$A^{\mathsf{T}}(X + \alpha(Y - X)) - b = 0,$$

that is, so that

$$(A^{\mathsf{T}}X - b) + \alpha((A^{\mathsf{T}}Y - b) - (A^{\mathsf{T}}X - b)) = 0.$$

Since $A^{\mathsf{T}}Y - b > 0$ and $A^{\mathsf{T}}X - b < 0$, the coefficient of α is a positive number. Hence we can solve for α:

$$\alpha = -\frac{A^{\mathsf{T}}X - b}{A^{\mathsf{T}}(Y - X)}.$$

Thus some point of PQ, which by Theorem 4.10.1 lies on the k-flat, also lies on the hyperplane, which contradicts the hypothesis of the theorem. Hence the k-flat cannot have points P and Q on opposite sides of the hyperplane and the proof is complete.

In linear programming problems, one has to study intersections of sets of half-spaces of the types described above. In particular, the set of points in $\mathscr{E}^n$ such that $x_1 > 0, x_2 > 0, \ldots, x_n > 0$ corresponds to the first quadrant in $\mathscr{E}^2$ and to the first octant in $\mathscr{E}^3$. It is called the **positive orthant** of $\mathscr{E}^n$; the **nonnegative orthant** of $\mathscr{E}^n$ is the intersection of the closed half-spaces defined by $x_1 \geq 0$, $x_2 \geq 0, \ldots, x_n \geq 0$.

4.11 UNITARY n-SPACE

For certain physical applications, vectors with components from the complex field are useful, and definitions of length and orthogonality are needed. If X is such a vector and if we use $\sqrt{X^{\mathsf{T}}X}$ for the length of X, peculiarities arise. For example, if

$$X = \begin{bmatrix} 3 \\ 4 \\ 5i \end{bmatrix},$$

then $X^{\mathsf{T}}X = 3^2 + 4^2 + (5i)^2 = 0$, so a nonzero vector can have zero length, which is not ordinarily desirable. All such peculiarities are eliminated if we make this definition of length:

$$|X| = \sqrt{X^*X}, \tag{4.11.1}$$

where X^* denotes the tranjugate of X (see Section 2.13). Then length is a non-negative real number, for $\bar{x}_i x_i = |x_i|^2$, so that

$$|X| = \sqrt{\bar{x}_1 x_1 + \bar{x}_2 x_2 + \cdots + \bar{x}_n x_n} = \sqrt{\sum |x_i|^2} \geq 0.$$

In the case of our example, the length is now $\sqrt{3 \cdot 3 + 4 \cdot 4 + (-5i)(5i)} = \sqrt{50}$ instead of 0. Since $\sum |x_i|^2 = 0$ if and only if each $x_i = 0$, the zero vector is the only vector whose length is zero according to this definition of length.

Similarly, if we were to use $X^{\mathsf{T}} Y = 0$ as a test for orthogonality, peculiarities would again appear. For example, the vector used in the example above would be orthogonal to itself since $X^{\mathsf{T}} X = 0$. Once again, we replace the transpose by the tranjugate in order to eliminate the peculiarity: We now define two vectors X and Y to be orthogonal if and only if neither is zero and

$$X^* Y = 0 \qquad \text{(orthogonality condition).} \tag{4.11.2}$$

The set of all n-vectors over the field of complex numbers, with the definitions of length and orthogonality given in (4.11.1) and (4.11.2), is called **unitary n-space** and we denote it by $\mathscr{U}^n$.

If X and Y are both real, both (4.11.1) and (4.11.2) reduce to the same rules we used before since then X^* and X^{T} are identical. Thus the geometry of $\mathscr{E}^n$ is a specialization of the geometry of $\mathscr{U}^n$. In $\mathscr{U}^n$, just as in $\mathscr{E}^n$, the elementary unit vectors $E_1, E_2, \ldots, E_n$ constitute an orthonormal basis since, for all X,

$$X = x_1 E_1 + x_2 E_2 + \cdots + x_n E_n.$$

In $\mathscr{U}^n$, the **Cauchy–Schwarz inequality** takes the form

$$|X| \cdot |Y| \geq |X^* Y|. \tag{4.11.3}$$

We give a different type of proof for the unitary case to illustrate another technique. If $X = 0$, the inequality is satisfied. Hence assume that $X \neq 0$ and consider the quadratic polynomial $p(t)$ defined as follows:

$$\begin{aligned} p(t) &= \sum (|x_j| t - |y_j|)^2 \\ &= \left(\sum |x_j|^2\right) t^2 - 2\left(\sum |x_j| \cdot |y_j|\right) t + \sum |y_j|^2. \end{aligned} \tag{4.11.4}$$

If $p(t_0) = 0$, where t_0 is real, then $\sum (|x_j| t_0 - |y_j|)^2$ is a vanishing sum of squares of real numbers, so that

$$|x_j| t_0 - |y_j| = 0, \qquad j = 1, 2, \ldots, n.$$

Solving this equation for $|y_j|$ and substituting in (4.11.4), we obtain

$$p(t) = \sum (|x_j| t - |x_j| t_0)^2 = (t - t_0)^2 \sum |x_j|^2 = (t - t_0)^2 |X|^2.$$

By hypothesis, $X \neq 0$, so that $|X| \neq 0$ also and therefore $p(t) = 0$ *only* when $t = t_0$. That is, if $p(t) = 0$ has a real root, that root is a *double* root. Hence the equation $p(t) = 0$ has either real and equal roots or conjugate complex roots, so that the discriminant of the quadratic polynomial $p(t)$ must be zero or negative:

$$4\left(\sum |x_j|\,|y_j|\right)^2 - 4 \sum |x_j|^2 \cdot \sum |y_j|^2 \leq 0.$$

From this, since $|x_j| = |\bar{x}_j|$ and $|\bar{x}_j| \cdot |y_j| = |\bar{x}_j \cdot y_j|$, we obtain

$$\sum |x_j|^2 \cdot \sum |y_j|^2 \geq \left(\sum |x_j|\,|y_j|\right)^2 = \left(\sum |\bar{x}_j y_j|\right)^2. \tag{4.11.5}$$

Using now the fact that the absolute value of a sum of terms is equal to or less than the sum of the absolute values of the separate terms, we have

$$\sum |x_j|^2 \cdot \sum |y_j|^2 \geq \left|\sum \bar{x}_j y_j\right|^2.$$

Taking square roots on both sides, we have (4.11.3), and the proof is complete.

Starting with the inequality of (4.11.5), we can obtain the triangle inequality for $\mathscr{U}^n$,

$$|X| + |Y| \geq |X + Y|, \tag{4.11.6}$$

in much the same manner as we obtained (4.7.8) from (4.7.7). The details are left as an exercise.

4.12 EXERCISES

1. The distance between the points $(a, 2a, -a, 3, 1)$ and $(3a, 2a, 2a, 1, 4)$ in $\mathscr{E}^5$ is $\sqrt{26}$. What are the possible values for a?

2. The vectors

$$\begin{bmatrix} a \\ 2a \\ 2a \\ 4 \end{bmatrix} \quad \text{and} \quad \begin{bmatrix} 4 \\ 2a \\ 2a \\ a \end{bmatrix}$$

in $\mathscr{E}^4$ are orthogonal. What are the possible values for a?

3. The vector $[\frac{1}{2}, \frac{1}{2}, a, \sqrt{2}/2, b]^{\mathsf{T}}$ in $\mathscr{E}^5$ is a unit vector. What can you say about a and b?

4. Verify the Cauchy–Schwarz inequality for the vectors $[1, -2, 3, 4]^{\mathsf{T}}$ and $[4, 1, -2, -3]^{\mathsf{T}}$.

5. Find $\cos\theta$, where θ is the angle between the two vectors of Exercise 4.

6. Show that the points $(0, 0, 0, 0)$, $(1, -2, 3, 4)$, and $(4, -1, 2, -3)$ are the vertices of a right triangle in $\mathscr{E}^4$.

7. Show that the four vectors

$$V_1 = \begin{bmatrix} \frac{\sqrt{2}}{2} \\ \frac{\sqrt{2}}{2} \\ 0 \\ 0 \end{bmatrix}, \quad V_2 = \begin{bmatrix} \frac{\sqrt{2}}{2} \\ \frac{-\sqrt{2}}{2} \\ 0 \\ 0 \end{bmatrix}, \quad V_3 = \begin{bmatrix} 0 \\ 0 \\ \frac{\sqrt{2}}{2} \\ \frac{\sqrt{2}}{2} \end{bmatrix}, \quad V_4 = \begin{bmatrix} 0 \\ 0 \\ \frac{\sqrt{2}}{2} \\ \frac{-\sqrt{2}}{2} \end{bmatrix}$$

constitute an orthonormal set. Show that they are a basis for $\mathscr{E}^4$ by solving the equation

$$X = t_1V_1 + t_2V_2 + t_3V_3 + t_4V_4$$

for the t's in terms of the components of an arbitrary vector X.

8. Find a vector orthogonal to each of the vectors

$$\begin{bmatrix} 1 \\ -1 \\ 2 \end{bmatrix} \quad \text{and} \quad \begin{bmatrix} 2 \\ 0 \\ -1 \end{bmatrix}$$

and then normalize all three. Is the result an orthonormal basis for the set of all real 3-vectors?

***9.** Show that every pair of points P, Q of a hyperplane $A^{\mathsf{T}}X = b$ determines a vector orthogonal to the coefficient vector A of the equation. We describe this by saying that A is **orthogonal** or **normal** to the hyperplane.

***10.** Show that the line defined by $X = X_0 + tA$ is normal to the hyperplane $A^{\mathsf{T}}X = b$.

***11.** Given that A is a unit vector, determine the point where the line $X = tA$ meets the hyperplane $A^{\mathsf{T}}X = b$. Show that the distance between the origin and this point is $|b|$. Illustrate with figures in $\mathscr{E}^2$ and $\mathscr{E}^3$.

***12.** Show that (a) the line with equation $X = X_0 + tB$ meets the hyperplane with equation $A^{\mathsf{T}}X = b$ in a unique point unless $A^{\mathsf{T}}B = 0$; (b) if $A^{\mathsf{T}}B = 0$ but $A^{\mathsf{T}}X_0 \neq b$, the line does not intersect the hyperplane, that is, is parallel to it; (c) if $A^{\mathsf{T}}B = 0$ and $A^{\mathsf{T}}X_0 = b$, all points of the line lie on the hyperplane.

***13.** How does one define and compute the midpoint of a line segment AB in $\mathscr{E}^n$?

14. Find the length of the vector

$$X = \begin{bmatrix} 1 - i \\ 1 + i \\ 1 \\ 0 \end{bmatrix}$$

in $\mathscr{U}^4$, then normalize X.

15. Determine whether or not the vectors

$$\begin{bmatrix} i \\ -i \\ i \\ -i \\ 1 \end{bmatrix} \quad \text{and} \quad \begin{bmatrix} i \\ i \\ i \\ i \\ 1 \end{bmatrix}$$

in $\mathscr{U}^5$ are orthogonal.

***16.** If X is a unit vector in $\mathscr{U}^n$, for what scalars α will αX still be a unit vector? (The set of scalars is now the field of complex numbers.)

17. The vectors

$$\begin{bmatrix} i \\ 1 \\ 0 \end{bmatrix} \quad \text{and} \quad \begin{bmatrix} 0 \\ 1 \\ i \end{bmatrix}$$

are each orthogonal to a *unit* vector

$$X = \begin{bmatrix} x_1 \\ x_2 \\ x_3 \end{bmatrix}$$

in $\mathscr{U}^3$. Find such a vector X. (How many solutions?)

***18.** Prove that for all scalars α and all vectors X in $\mathscr{U}^n$,

$$|\alpha X| = |\alpha|\,|X|.$$

***19.** Prove that $X^*Y = \overline{Y^*X}$.

***20.** Show that if $A_1, A_2, \ldots, A_p$ are all orthogonal in $\mathscr{U}^n$ to a vector X, so is any linear combination $t_1A_1 + t_2A_2 + \cdots + t_pA_p$ of these vectors.

***21.** Prove that if X and Y are vectors in $\mathscr{U}^n$ such that $|X| \leq 1$ and $|Y| \leq 1$, then $|X^*Y| \leq 1$ also.

22. Prove that in $\mathscr{E}^n$, the vectors X_P and X_Q with terminal points P and Q, respectively, are orthogonal if and only if $d^2(P, Q) = d^2(O, P) + d^2(O, Q)$. Does the same theorem hold in $\mathscr{U}^n$?

***23.** Prove that mutually orthogonal vectors in $\mathscr{E}^n$, and in $\mathscr{U}^n$, are linearly independent.

24. Show that in $\mathscr{U}^n$ the set of all vectors of the form

$$X = t_1X_1 + t_2X_2 + \cdots + t_kX_k,$$

where the X_j's are given and the t_j's are arbitrary complex numbers, constitutes a vector space.

25. Show that in $\mathscr{E}^n$ the parameter s can be used as a directed distance from the terminal point of X_0 to the terminal point of X on the line with equation

$$X = X_0 + s\begin{bmatrix} \cos\alpha_1 \\ \cos\alpha_2 \\ \vdots \\ \cos\alpha_n \end{bmatrix}.$$

26. Let the positive sense on a line AB be that of $\overrightarrow{AB}$. Let P be a point on the line. Find formulas for

(a) $\dfrac{\vec{d}(A, P)}{\vec{d}(A, B)}$, (b) $\dfrac{\vec{d}(A, P)}{\vec{d}(P, B)}$.

Illustrate with figures in $\mathscr{E}^2$.

27. Let the line PQ intersect the hyperplane with equation $A^{\mathsf{T}}X - b = 0$ in the point P and in no other point. Show that the line has points on both sides of the hyperplane.

28. Show that the set of all points in $\mathscr{E}^n$ which are equidistant from $P:(a_1, a_2, \ldots, a_n)$ and $Q:(-a_1, -a_2, \ldots, -a_n)$ is a hyperplane on the origin and find its equation. This hyperplane is the **perpendicular bisector** of the line segment PQ.

CHAPTER 5

Vector Spaces

5.1 THE GENERAL DEFINITION OF A VECTOR SPACE

In the preceding chapters we studied special sets of vectors in $\mathscr{E}^3$ and $\mathscr{E}^n$, sets we called "vector spaces." This geometrical concept may be generalized as follows. Assume first that we have a number field $\mathscr{F}$ of "scalars" over which to work and a collection $\mathscr{V}$ of mathematical objects which we call "vectors" (even though they may not resemble geometric vectors). Equality of two vectors means they are identical. We call the set $\mathscr{V}$ a **vector space over** $\mathscr{F}$ if and only if its members satisfy the following requirements:

(A) There is an operation called "addition" such that, corresponding to any two members V_1 and V_2 of $\mathscr{V}$, there exists a unique "sum" $V_1 + V_2$. Moreover, addition of vectors from $\mathscr{V}$ obeys these rules:

(A_1) $V_1 + V_2$ is always a member of $\mathscr{V}$ (closure of $\mathscr{V}$ with respect to addition).

(A_2) For all V_1, V_2, V_3 in $\mathscr{V}$, $V_1 + (V_2 + V_3) = (V_1 + V_2) + V_3$ (the associative law of addition).

(A_3) For all V_1 and V_2 in $\mathscr{V}$, $V_1 + V_2 = V_2 + V_1$ (the commutative law of addition).

(A_4) There is a special member of $\mathscr{V}$, called the **zero vector** and denoted by 0, such that for each V of $\mathscr{V}$

$$V + 0 = V.$$

(A_5) To every member V of $\mathscr{V}$ there corresponds a **negative**, $-V$, which is also a member of $\mathscr{V}$, such that

$$V + (-V) = 0.$$

(S) There is an operation of multiplication by a scalar such that if α is any element of $\mathscr{F}$ and if V is any member of $\mathscr{V}$, there exists a uniquely defined product αV. Scalar multiplication is assumed to obey these rules, for all α, β in $\mathscr{F}$ and for all V, V_1, V_2 in $\mathscr{V}$:

(S_1) αV belongs to $\mathscr{V}$ (closure of $\mathscr{V}$ with respect to multiplication by scalars).

(S_2) $V\alpha = \alpha V$ (commutative law of multiplication by scalars).

(S_3) $\alpha(V_1 + V_2) = \alpha V_1 + \alpha V_2$ } distributive laws of

(S_4) $(\alpha + \beta)V = \alpha V + \beta V$ } multiplication by scalars.

(S_5) $\alpha(\beta V) = (\alpha\beta)V$ (associative law of multiplication by scalars).

(S_6) $0 \cdot V = 0.$

(S_7) $1 \cdot V = V.$

Note that in (S_6), the left 0 denotes the *scalar* zero, whereas the right zero denotes the *vector* zero.

The set consisting of the zero vector alone is a vector space, the **zero space** over $\mathscr{F}$. This space will be denoted by $\mathscr{Z}$.

The reader will recall from Chapter 4 that the vector spaces of $\mathscr{E}^n$, over the field of real numbers, exhibit all these properties. They provide the most familiar illustrations of this general definition. There are other simple examples, however. To illustrate, consider the set $\mathscr{Q}$ of all quadratic polynomials

$$q = \alpha_1 x^2 + \alpha_2 xy + \alpha_3 y^2$$

where α_1, α_2, and α_3 are arbitrary real numbers. Addition of two of these yields a third, addition is associative and commutative, the zero element is the zero polynomial $0x^2 + 0xy + 0y^2$ and the negative of $\alpha_1 x^2 + \alpha_2 xy + \alpha_3 y^2$ is $-\alpha_1 x^2 - \alpha_2 xy - \alpha_3 y^2$. If α is a real number, αq is $\alpha\alpha_1 x^2 + \alpha\alpha_2 xy + \alpha\alpha_3 y^2$, which is again a quadratic polynomial in the set. The required properties for multiplication by a scalar are easily verified. That is, $\mathscr{Q}$ is a vector space over the field of real numbers.

When we add these quadratic polynomials or multiply them by scalars, we operate only on the coefficients, not on the symbols x^2, xy, and y^2. In fact, if we establish a one-to-one correspondence

$$\alpha_1 x^2 + \alpha_2 xy + \alpha_3 y^2 \leftrightarrow \begin{bmatrix} \alpha_1 \\ \alpha_2 \\ \alpha_3 \end{bmatrix},$$

which assigns to each quadratic polynomial $\alpha_1 x^2 + \alpha_2 xy + \alpha_3 y^2$ the unique 3-vector $\begin{bmatrix} \alpha_1 \\ \alpha_2 \\ \alpha_3 \end{bmatrix}$ and to each 3-vector $\begin{bmatrix} \alpha_1 \\ \alpha_2 \\ \alpha_3 \end{bmatrix}$ the unique quadratic polynomial $\alpha_1 x^2 + \alpha_2 xy + \alpha_3 y^2$, then to the sum of polynomials

$$(\alpha_1 x^2 + \alpha_2 xy + \alpha_3 y^2) + (\beta_1 x^2 + \beta_2 xy + \beta_3 y^2) = (\alpha_1 + \beta_1)x^2 + (\alpha_2 + \beta_2)xy + (\alpha_3 + \beta_3)y^2,$$

there corresponds the sum of 3-vectors

$$\begin{bmatrix} \alpha_1 \\ \alpha_2 \\ \alpha_3 \end{bmatrix} + \begin{bmatrix} \beta_1 \\ \beta_2 \\ \beta_3 \end{bmatrix} = \begin{bmatrix} \alpha_1 + \beta_1 \\ \alpha_2 + \beta_2 \\ \alpha_3 + \beta_3 \end{bmatrix}$$

and to the scalar multiple

$$\alpha(\alpha_1 x^2 + \alpha_2 xy + \alpha_3 y^2) = \alpha\alpha_1 x^2 + \alpha\alpha_2 xy + \alpha\alpha_3 y^2$$

there corresponds the scalar multiple

$$\alpha \begin{bmatrix} \alpha_1 \\ \alpha_2 \\ \alpha_3 \end{bmatrix} = \begin{bmatrix} \alpha\alpha_1 \\ \alpha\alpha_2 \\ \alpha\alpha_3 \end{bmatrix}.$$

That is, instead of adding quadratic polynomials and multiplying them by scalars, we could just as well perform the same operations on the corresponding 3-vectors. (This is similar to omitting the variables in synthetic elimination.) Thus, the distinction between the set of real quadratic polynomials and the set of real 3-vectors is one of notation only, as far as vector-space operations are concerned.

Similarly, the set of all real linear polynomials

$$\alpha_1 x + \alpha_2 y + \alpha_3$$

is a vector space over the field of real numbers that differs only in notation from the space of all real 3-vectors. The concept of vector spaces that are fundamentally alike is a very important one, as we shall see.

An example of a vector space over the complex number field is the set of all polynomials

$$\alpha_n x^n + \alpha_{n-1} x^{n-1} + \cdots + \alpha_1 x + \alpha_0, \qquad n = 0, 1, 2, \ldots,$$

with complex numbers as coefficients. It is simple to verify that the properties of the definition all hold, for they are familiar properties of the algebra of polynomials.

Another kind of example is given by the set of all real functions $f(x)$ which are continuous on the closed interval $0 \leq x \leq 1$. Indeed, the sum of two such functions is also a real function continuous on $0 \leq x \leq 1$ and a scalar multiple of such a function is again a member of the set, as is the function which is 0 on $0 \leq x \leq 1$. The other properties are all familiar properties of algebra. Thus these functions constitute a vector space over the field of real numbers.

A most important kind of vector space for our purposes is given in

Theorem 5.1.1: *Over a given field $\mathscr{F}$, the set of all solutions of a system of homogeneous equations in n unknowns is a vector space.*

The set of homogeneous equations can be represented in matrix form by the equation

$$AX = 0,$$

where the coefficient matrix A is over $\mathscr{F}$. Now let Y_1 and Y_2 be vectors over $\mathscr{F}$ such that

$$AY_1 = 0 \qquad \text{and} \qquad AY_2 = 0.$$

Then

$$AY_1 + AY_2 = 0,$$

so

$$A(Y_1 + Y_2) = 0;$$

that is, $Y_1 + Y_2$ belongs to the set if Y_1 and Y_2 do (property A_1).

Also, for any scalar α,

$$\alpha AY_1 = A(\alpha Y_1) = 0;$$

that is, αY_1 belongs to the set if Y_1 does (property S_1).

Moreover, the vector 0 is a solution (the trivial solution) and

$$-AY_1 = A(-Y_1) = 0,$$

so (A_4) and (A_5) hold. The other properties of a vector space hold in this case because our vectors are matrices, and the proof is complete.

The preceding examples illustrate the fact that the vector-space structure is one of the basic structures of mathematics.

5.2 LINEAR COMBINATIONS AND LINEAR DEPENDENCE

Let $V_1, V_2, \ldots, V_k$ be members of a vector space $\mathscr{V}$ over $\mathscr{F}$ and let $\alpha_1, \alpha_2, \ldots, \alpha_k$ be arbitrary scalars of $\mathscr{F}$. Then, by the properties (A_1), (A_2), and (S_1), repeatedly applied, the expression

$$\alpha_1 V_1 + \alpha_2 V_2 + \cdots + \alpha_k V_k \tag{5.2.1}$$

represents a vector of $\mathscr{V}$. This vector is called a **linear combination** of $V_1, V_2, \cdots, V_k$.

An important result is

Theorem 5.2.1: *A subset $\mathscr{S}$ of a vector space $\mathscr{V}$ is itself a vector space if and only if $\mathscr{S}$ is closed with respect to addition and with respect to multiplication by scalars.*

Such a vector space $\mathscr{S}$ is called a **subspace** of $\mathscr{V}$.

First assume that $\mathscr{S}$ is a subset of $\mathscr{V}$ that is closed with respect to addition and with respect to multiplication by a scalar. Let V belong to $\mathscr{S}$. Since $\mathscr{S}$ is closed with respect to multiplication by scalars, the vectors $0 \cdot V = 0$ and $(-1)V = -V$ also belong to $\mathscr{S}$. Then since $\mathscr{S}$ is closed with respect to addition and since $\mathscr{V}$ is a vector space, all the remaining properties hold at once.

The converse is immediate.

This theorem shows that the definitions given earlier of vector spaces in $\mathscr{E}^2$, $\mathscr{E}^3$, and $\mathscr{E}^n$ are consistent with the general definition given in the preceding section.

An application of this result is

Theorem 5.2.2: *The set of all linear combinations over $\mathscr{F}$ of the vectors $V_1, V_2, \cdots, V_k$ of a vector space $\mathscr{V}$ is a subspace of $\mathscr{V}$.*

By the preceding theorem, we need only show that (A_1) and (S_1) hold.

If

$$\alpha_1 V_1 + \alpha_2 V_2 + \cdots + \alpha_k V_k$$

and

$$\beta_1 V_1 + \beta_2 V_2 + \cdots + \beta_k V_k$$

are arbitrary vectors of the set, then their sum is the linear combination

$$(5.2.2) \qquad (\alpha_1 + \beta_1)V_1 + (\alpha_2 + \beta_2)V_2 + \cdots + (\alpha_k + \beta_k)V_k,$$

which is also in the set, so (A_1) holds. Moreover,

$$(5.2.3) \qquad (\alpha\alpha_1)V_1 + (\alpha\alpha_2)V_2 + \cdots + (\alpha\alpha_k)V_k$$

is in the set, so (S_1) holds. Hence we have a vector space.

For example, the set of all real 2×2 matrices is a vector space over the set of real numbers. (Verify that all the properties hold.) Hence, by Theorem 5.2.2, the set of all matrices

$$\alpha_1\begin{bmatrix}1 & 0\\0 & 1\end{bmatrix} + \alpha_2\begin{bmatrix}0 & 1\\-1 & 0\end{bmatrix} \qquad (\alpha_1, \alpha_2 \text{ real})$$

is also a vector space over the set of real numbers.

We say that the vectors $V_1, V_2, \ldots, V_k$ of a vector space $\mathscr{V}$ are **linearly**

dependent over $\mathscr{F}$ if and only if there exist scalars $\alpha_1, \alpha_2, \ldots, \alpha_k$, *not all zero*, in $\mathscr{F}$ such that

$$\alpha_1 V_1 + \alpha_2 V_2 + \cdots + \alpha_k V_k = 0. \tag{5.2.4}$$

In particular, *when* $k = 1$, *the zero vector is dependent* since $\alpha \cdot 0 = 0$ for all scalars α. *Again, when* $k = 1$, *a nonzero vector* V_1 *is independent* since $\alpha V_1 = 0$ if and only if $\alpha = 0$. Calling a single vector dependent or independent avoids the necessity of stating exceptional cases of many theorems.

For example, since

$$2\begin{bmatrix}1\\0\\0\end{bmatrix} + 3\begin{bmatrix}0\\1\\0\end{bmatrix} + (-5)\begin{bmatrix}0\\0\\1\end{bmatrix} + (-1)\begin{bmatrix}2\\3\\-5\end{bmatrix} = \begin{bmatrix}0\\0\\0\end{bmatrix},$$

the vectors

$$\begin{bmatrix}1\\0\\0\end{bmatrix}, \quad \begin{bmatrix}0\\1\\0\end{bmatrix}, \quad \begin{bmatrix}0\\0\\1\end{bmatrix}, \quad \begin{bmatrix}2\\3\\-5\end{bmatrix}$$

are linearly dependent. Similarly, the linear polynomials

$$2x_1 - 3x_2, \quad x_1 + 2x_2, \quad 2x_1 + 4x_2$$

are linearly dependent since

$$0 \cdot (2x_1 - 3x_2) + (-2)(x_1 + 2x_2) + 1 \cdot (2x_1 + 4x_2) \equiv 0,$$

that is, since this linear combination reduces identically to the zero polynomial.

This last example illustrates the fact that *some*, but *not all*, of the α's may be 0, as the definition says.

We say that $V_1, V_2, \ldots, V_k$ are **linearly independent** if and only if equation (5.2.4) holds only when (implies that) all the α's are 0. For example, we show that the three expressions

$$\begin{aligned} 2x_1 + 3x_2 - x_3 \\ x_1 - 2x_2 + x_3 \\ 3x_1 + x_2 - 4x_3 \end{aligned}$$

are linearly independent. We have

$$\alpha_1(2x_1 + 3x_2 - x_3) + \alpha_2(x_1 - 2x_2 + x_3) + \alpha_3(3x_1 + x_2 - 4x_3) \equiv 0$$

if and only if

$$(2\alpha_1 + \alpha_2 + 3\alpha_3)x_1 + (3\alpha_1 - 2\alpha_2 + \alpha_3)x_2 + (-\alpha_1 + \alpha_2 - 4\alpha_3)x_3 \equiv 0.$$

This identity is true if and only if the coefficients of the x's all vanish, that is, if and only if

$$\begin{aligned} 2\alpha_1 + \alpha_2 + 3\alpha_3 &= 0 \\ 3\alpha_1 - 2\alpha_2 + \alpha_3 &= 0 \\ -\alpha_1 + \alpha_2 - 4\alpha_3 &= 0. \end{aligned}$$

The computation

$$\begin{array}{rrrl} 1 & -1 & 4 & \text{(from equation 3)} \\ 0 & 1 & -11 & \text{(from equation 2)} \\ 0 & 3 & -5 & \text{(from equation 1)} \\ \hline 1 & 0 & -7 & \\ 0 & 1 & -11 & \\ 0 & 0 & 28 & \\ \hline \end{array}$$

shows that the only solution of these equations in the α's is the trivial solution $(0, 0, 0)$, so the given expressions are indeed linearly independent.

Questions of linear dependence typically lead to systems of linear equations, as in the preceding example. To illustrate further, we determine whether or not the vectors

$$\begin{bmatrix} 2 \\ -1 \end{bmatrix}, \begin{bmatrix} 1 \\ 2 \end{bmatrix}, \begin{bmatrix} 7 \\ 4 \end{bmatrix}$$

are linearly dependent. The vector equation

$$\alpha_1 \begin{bmatrix} 2 \\ -1 \end{bmatrix} + \alpha_2 \begin{bmatrix} 1 \\ 2 \end{bmatrix} + \alpha_3 \begin{bmatrix} 7 \\ 4 \end{bmatrix} = \begin{bmatrix} 0 \\ 0 \end{bmatrix}$$

is equivalent to the scalar equations

$$\begin{aligned} 2\alpha_1 + \alpha_2 + 7\alpha_3 &= 0 \\ -\alpha_1 + 2\alpha_2 + 4\alpha_3 &= 0, \end{aligned}$$

from which, by the usual methods,

$$\begin{aligned} \alpha_1 &= -2\alpha_3 \\ \alpha_2 &= -3\alpha_3. \end{aligned}$$

A convenient choice for α_3 is the value -1. Then a solution is $(2, 3, -1)$, so

$$2\begin{bmatrix} 2 \\ -1 \end{bmatrix} + 3\begin{bmatrix} 1 \\ 2 \end{bmatrix} - \begin{bmatrix} 7 \\ 4 \end{bmatrix} = \begin{bmatrix} 0 \\ 0 \end{bmatrix}.$$

A particularly useful special case is given in

Theorem 5.2.3: *Every nonempty subset of a set of vectors of the form*

$$\begin{bmatrix} 1 \\ a_{21} \\ a_{31} \\ a_{41} \\ \vdots \\ a_{n1} \end{bmatrix}, \begin{bmatrix} 0 \\ 1 \\ a_{32} \\ a_{42} \\ \vdots \\ a_{n2} \end{bmatrix}, \ldots, \begin{bmatrix} 0 \\ \vdots \\ 0 \\ 1 \\ a_{k+1,k} \\ \vdots \\ a_{nk} \end{bmatrix}, \tag{5.2.5}$$

where the components belong to a field $\mathscr{F}$, is linearly independent over $\mathscr{F}$.

Indeed, if a linear combination of any of these is to be zero, the coefficient of the first vector must be zero to eliminate the leading 1. Thereafter, the coefficient of the second vector must be zero to eliminate *its* leading 1. Continuing in this way, we see that all the coefficients must be zero, so the vectors are indeed linearly independent.

A special case of this is the set of elementary n-vectors $E_1, E_2, \ldots, E_n$, any subset of which is a linearly independent set.

We say that a vector V is a **linear combination** of the vectors $V_1, V_2, \ldots, V_k$ if and only if there exist scalars $\alpha_1, \alpha_2, \ldots, \alpha_k$ such that

$$V = \alpha_1 V_1 + \alpha_2 V_2 + \cdots + \alpha_k V_k. \tag{5.2.6}$$

Thus, since

$$\begin{bmatrix} 7 \\ 4 \end{bmatrix} = 2\begin{bmatrix} 2 \\ -1 \end{bmatrix} + 3\begin{bmatrix} 1 \\ 2 \end{bmatrix},$$

the vector $\begin{bmatrix} 7 \\ 4 \end{bmatrix}$ is a linear combination of $\begin{bmatrix} 2 \\ -1 \end{bmatrix}$ and $\begin{bmatrix} 1 \\ 2 \end{bmatrix}$. Indeed, we can write *any* 2-vector as a linear combination of these last two vectors, for the vector equation

$$\begin{bmatrix} x_1 \\ x_2 \end{bmatrix} = \alpha_1\begin{bmatrix} 2 \\ -1 \end{bmatrix} + \alpha_2\begin{bmatrix} 1 \\ 2 \end{bmatrix}$$

is equivalent to the pair of scalar equations

$$\begin{aligned} 2\alpha_1 + \alpha_2 &= x_1 \\ -\alpha_1 + 2\alpha_2 &= x_2, \end{aligned}$$

the solution of which is

$$\begin{aligned} \alpha_1 &= \tfrac{1}{5}(2x_1 - x_2) \\ \alpha_2 &= \tfrac{1}{5}(x_1 + 2x_2), \end{aligned}$$

so that as soon as $\begin{bmatrix} x_1 \\ x_2 \end{bmatrix}$ is given, we can compute the corresponding values of α_1 and α_2.

As another illustration, consider the complete solution of the system of equations

$$\begin{aligned} 5x_1 - 8x_2 + 2x_3 - x_4 &= 0 \\ -7x_1 + 3x_2 + x_3 + x_4 &= 0, \end{aligned}$$

which we find by the methods of Chapter 1 to be, in parametric form,

$$\begin{aligned} x_1 &= 3u \\ x_2 &= 3v \\ x_3 &= 2u + 5v \\ x_4 &= 19u - 14v. \end{aligned}$$

This can be written in vector form as

$$\begin{bmatrix} x_1 \\ x_2 \\ x_3 \\ x_4 \end{bmatrix} = u\begin{bmatrix} 3 \\ 0 \\ 2 \\ 19 \end{bmatrix} + v\begin{bmatrix} 0 \\ 3 \\ 5 \\ -14 \end{bmatrix}.$$

The two vectors on the right are linearly independent solutions of the given system. That is, the complete solution is the set of all linear combinations of a set of linearly independent particular solutions. We shall see in Chapter 6 that this holds true in general for systems of homogeneous linear equations.

We conclude this section with two simple theorems.

Theorem 5.2.4: *If the vectors $V_1, V_2, \ldots, V_k$ are linearly dependent, then at least one of them can be written as a linear combination of the rest.*

By hypothesis, the equation

$$\alpha_1 V_1 + \alpha_2 V_2 + \cdots + \alpha_n V_n = 0$$

holds with at least one of the α's different from 0. Suppose that $\alpha_i \neq 0$. Then we can solve for V_i:

$$V_i = \left(-\frac{\alpha_1}{\alpha_i}\right)V_i + \cdots + \left(-\frac{\alpha_{i-1}}{\alpha_i}\right)V_{i-1} + \left(-\frac{\alpha_{i+1}}{\alpha_i}\right)V_{i+1} + \cdots + \left(-\frac{\alpha_n}{\alpha_i}\right)V_n.$$

This proves the theorem. A converse result is this:

Theorem 5.2.5: *If V can be expressed as a linear combination of $V_1, V_2, \ldots, V_k$, then the vectors $V, V_1, V_2, \ldots, V_k$ are linearly dependent.*

By hypothesis, there exist $\alpha_1, \alpha_2, \ldots, \alpha_k$ such that

$$V = \alpha_1 V_1 + \alpha_2 V_2 + \cdots + \alpha_k V_k.$$

This equation can be written

$$(-1)\,V + \alpha_1 V_1 + \alpha_2 V_2 + \cdots + \alpha_k V_k = 0.$$

Since at least the coefficient of V is not zero, the desired conclusion follows.

5.3 EXERCISES

1. Show that the set of all real n-vectors X which simultaneously satisfy two systems of homogeneous equations, $AX = 0$ and $BX = 0$, with real coefficients, is a vector space over the field of real numbers.

2. Let A and B be fixed $n \times n$ real matrices. Show that the set of all $n \times n$ real matrices W such that $AWB = 0$ is a vector space, the scalars being the real numbers.

3. Show that the set of all real linear polynomials of the form

$$a_1x_1 + a_2x_2 + \cdots + a_nx_n,$$

where n is fixed, is a vector space over the field of real numbers. Then do the same for the set of all real quadratic polynomials of the form

$$a_{11}x_1^2 + a_{12}x_1x_2 + \cdots + a_{1n}x_1x_n + a_{22}x_2^2 + a_{23}x_2x_3 + \cdots + a_{nn}x_n^2,$$

where again n is fixed.

4. Is the set of all Hermitian forms in n complex variables, that is, the set of all expressions of the form

$$\sum a_{ij}\bar{z}_i z_j \qquad (a_{ji} = \bar{a}_{ij})$$

a vector space over the field of complex numbers? Why? Is it a vector space over the field of real numbers? Why?

5. Show that the set of all real $n \times n$ matrices A such that $AG = GA$, where G is a fixed real $n \times n$ matrix, is a vector space over the field of real numbers.

6. Prove that the set of all real n-vectors X such that $X^{\mathsf{T}}A = 0$, where A is a fixed real n-vector, is a vector space over the field of real numbers. Interpret geometrically in $\mathscr{E}^3$.

7. Prove that the vectors

$$\begin{bmatrix}1\\0\\0\\0\end{bmatrix}, \begin{bmatrix}1\\1\\0\\0\end{bmatrix}, \begin{bmatrix}1\\1\\1\\0\end{bmatrix}, \begin{bmatrix}1\\1\\1\\1\end{bmatrix}$$

are linearly independent.

8. Prove that the vectors of Exercise 7 together with any real vector

$$\begin{bmatrix}a_1\\a_2\\a_3\\a_4\end{bmatrix}$$

are linearly dependent.

***9.** Show that vectors $V_1, V_2, \ldots, V_k$ are linearly dependent if and only if $\alpha_1V_1, \alpha_2V_2, \ldots, \alpha_kV_k$, where the α's are nonzero scalars, are linearly dependent.

10. For each of the following systems of equations, represent in matrix form the vector space of all solutions of the system:

(a) $\begin{aligned} 2x_1 - x_2 + 4x_3 - 3x_4 &= 0 \\ 5x_1 - 2x_2 - x_3 + x_4 &= 0, \end{aligned}$ (b) $x_1 - x_2 + x_3 = 0,$

$$\text{(c)}\quad \begin{aligned} x_1 + x_2 + x_3 + x_4 &= 0 \\ x_1 - x_2 + x_3 + x_4 &= 0 \\ x_1 + x_2 - x_3 + x_4 &= 0 \\ x_1 + x_2 + x_3 - x_4 &= 0. \end{aligned}$$

11. Given that V_1, V_2, and V_3 are linearly independent vectors of an arbitrary vector space $\mathscr{V}$, prove that V_1, $V_1 + V_2$, and $V_1 + V_2 + V_3$ are also linearly independent vectors of $\mathscr{V}$. Generalize.

12. Prove that vectors

$$A = \begin{bmatrix} a_1 \\ a_2 \end{bmatrix} \quad \text{and} \quad B = \begin{bmatrix} b_1 \\ b_2 \end{bmatrix}$$

in $\mathscr{E}^2$ are linearly dependent if and only if $a_1b_2 - a_2b_1 = 0$.

13. Given that $a_1b_2 - a_2b_1 \neq 0$, show how to express an arbitrary vector

$$X = \begin{bmatrix} x_1 \\ x_2 \end{bmatrix}$$

in $\mathscr{E}^2$ as a linear combination of

$$A = \begin{bmatrix} a_1 \\ a_2 \end{bmatrix}, \quad B = \begin{bmatrix} b_1 \\ b_2 \end{bmatrix};$$

that is, find the coefficients α and β in the equation

$$X = \alpha A + \beta B.$$

Then solve the problem for the special case

$$A = \begin{bmatrix} 1 \\ 2 \end{bmatrix}, \quad B = \begin{bmatrix} 2 \\ 1 \end{bmatrix}.$$

14. Prove that the vectors

$$A = \begin{bmatrix} a_1 \\ a_2 \\ \vdots \\ a_n \end{bmatrix} \quad \text{and} \quad B = \begin{bmatrix} b_1 \\ b_2 \\ \vdots \\ b_n \end{bmatrix}$$

in $\mathscr{E}^n$ are linearly dependent if and only if $a_ib_j - a_jb_i = 0$ whenever $1 \leq i < j \leq n$.

15. Given that the vectors $V_1, V_2, \ldots, V_k$ of a vector space $\mathscr{V}$ are linearly independent, under what conditions on the coefficients α_{ij} are the vectors $\alpha_{11}V_1$, $\alpha_{21}V_1 + \alpha_{22}V_2$, $\alpha_{31}V_1 + \alpha_{32}V_2 + \alpha_{33}V_3, \ldots, \alpha_{k1}V_1 + \alpha_{k2}V_2 + \cdots + \alpha_{kk}V_k$ also linearly independent?

16. Which columns of the **upper triangular matrix**

$$\begin{bmatrix} a_{11} & a_{12} & \cdots & a_{1n} \\ 0 & a_{22} & \cdots & a_{2n} \\ \vdots & & & \\ 0 & 0 & \cdots & a_{nn} \end{bmatrix}$$

are linearly independent if $\prod_{i=1}^{n} a_{ii} \neq 0$? If precisely one of the a_{ii}'s is zero?

***17.** Prove that any p nonzero vectors $V_1, V_2, \ldots, V_p$ of a vector space $\mathscr{V}$ are linearly dependent if and only if for some integer $k \geq 2$, V_k is a linear combination of $V_1, V_2, \ldots, V_{k-1}$.

18. Prove that k given vectors $A_1, A_2, \ldots, A_k$ in $\mathscr{E}^n$ are linearly dependent if and only if the equation

$$[A_1, A_2, \ldots, A_k]\begin{bmatrix} x_1 \\ x_2 \\ \vdots \\ x_k \end{bmatrix} = 0$$

has a nontrivial solution.

19. Are the matrices

$$\begin{bmatrix} 2 & -1 \\ 4 & 6 \end{bmatrix}, \quad \begin{bmatrix} 3 & 2 \\ 8 & 3 \end{bmatrix}, \quad \begin{bmatrix} -5 & -8 \\ -16 & 4 \end{bmatrix}, \quad \begin{bmatrix} 0 & -7 \\ -4 & 13 \end{bmatrix}$$

linearly dependent?

20. Find a linear combination of

$$\begin{bmatrix} 0 \\ 1 \\ 1 \end{bmatrix} \quad \text{and} \quad \begin{bmatrix} 1 \\ -1 \\ 1 \end{bmatrix}$$

that is orthogonal to

$$\begin{bmatrix} 1 \\ 2 \\ 0 \end{bmatrix}.$$

Illustrate with a figure.

5.4 BASIC THEOREMS ON LINEAR DEPENDENCE

The following theorems make it much easier to apply the concept of linear dependence.

Theorem 5.4.1: *If some p ($p > 0$) vectors of the set of k vectors $V_1, V_2, \ldots, V_k$ are linearly dependent, then all k of the vectors are linearly dependent.*

Suppose that, in fact, $V_1, V_2, \ldots, V_p$ are dependent. Then there exist scalars $\alpha_1, \alpha_2, \ldots, \alpha_p$, not all 0, such that

$$\alpha_1 V_1 + \alpha_2 V_2 + \cdots + \alpha_p V_p = 0.$$

Hence

$$\alpha_1 V_1 + \alpha_2 V_2 + \cdots + \alpha_p V_p + 0 \cdot V_{p+1} + \cdots + 0 \cdot V_k = 0.$$

Since the coefficients in this latter equation are not all 0, the k vectors are linearly dependent. Similarly, if any other subset of p vectors is dependent.

We have at once

Corollary 5.4.2: *If the vectors $V_1, V_2, \ldots, V_k$ are linearly independent, then no nonempty subset of these is linearly dependent.*

A special case of Theorem 5.4.1 is

Corollary 5.4.3: *If one of the vectors $V_1, V_2, \ldots, V_k$ is the zero vector, the set is linearly dependent.*

This is because the zero vector alone is linearly dependent, so we can apply the theorem for the case $p = 1$.

Theorem 5.4.4: *If the vectors $V_1, V_2, \ldots, V_k$ are linearly independent, but the vectors $V_1, V_2, \ldots, V_k, W$ are linearly dependent, then W can be written as a linear combination of the V's.*

There exists, by hypothesis, a relation

$$\alpha W + \alpha_1 V_1 + \alpha_2 V_2 + \cdots + \alpha_k V_k = 0$$

with not all the α's zero. In particular, $\alpha \neq 0$, for if it were zero, the V's would be linearly dependent, contrary to hypothesis. Hence

$$W = \left(-\frac{\alpha_1}{\alpha}\right)V_1 + \left(-\frac{\alpha_2}{\alpha}\right)V_2 + \cdots + \left(-\frac{\alpha_k}{\alpha}\right)V_k.$$

Theorem 5.4.5: *Any $k + 1$ linear combinations of $V_1, V_2, \ldots, V_k$ are linearly dependent.*

Let the linear combinations be

$$\begin{aligned} W_1 &= \alpha_{11}V_1 + \alpha_{12}V_2 + \cdots + \alpha_{1k}V_k \\ W_2 &= \alpha_{21}V_1 + \alpha_{22}V_2 + \cdots + \alpha_{2k}V_k \\ &\vdots \\ W_{k+1} &= \alpha_{k+1,1}V_1 + \alpha_{k+1,2}V_2 + \cdots + \alpha_{k+1,k}V_k. \end{aligned}$$

We must show there exist scalars $\beta_1, \beta_2, \ldots, \beta_{k+1}$, not all zero, such that

$$\beta_1 W_1 + \beta_2 W_2 + \cdots + \beta_{k+1} W_{k+1} = 0.$$

By substituting for the W's, expanding, and collecting, we reduce this condition to the form

$$\sum_{i=1}^{k} (\beta_1\alpha_{1i} + \beta_2\alpha_{2i} + \cdots + \beta_{k+1}\alpha_{k+1,i})V_i = 0.$$

This equation will certainly hold if all the coefficients of the V's are zero:

$$\begin{aligned}
\beta_1\alpha_{11} + \beta_2\alpha_{21} + \cdots + \beta_{k+1}\alpha_{k+1,1} &= 0\\
\beta_1\alpha_{12} + \beta_2\alpha_{22} + \cdots + \beta_{k+1}\alpha_{k+1,2} &= 0\\
&\vdots\\
\beta_1\alpha_{1k} + \beta_2\alpha_{2k} + \cdots + \beta_{k+1}\alpha_{k+1,k} &= 0.
\end{aligned}$$

Here we have k homogeneous equations in $k + 1$ unknowns. By transforming this system to reduced echelon form, we are able to express some subset of p of these variables in terms of the remaining $k + 1 - p$ variables, $p \leq k$. Thus there always exist nontrivial solutions for the β's, so the theorem is proved.

From this result and Theorem 5.4.1, we have

Corollary 5.4.6: *More than k linear combinations of $V_1, V_2, \ldots, V_k$ are always linearly dependent.*

A particularly useful special case is

Corollary 5.4.7: *More than n n-vectors over a field $\mathscr{F}$ are always linearly dependent.*

In fact, if $E_1, E_2, \ldots, E_n$ denote the elementary n-vectors, then any n-vector X over $\mathscr{F}$ can be written

$$X = x_1E_1 + x_2E_2 + \cdots + x_nE_n,$$

as we saw in Chapter 4. Since all n-vectors are linear combinations of the n n-vectors $E_1, E_2, \ldots, E_n$, more than n n-vectors are linearly dependent by the preceding corollary.

For example, we know at once by this corollary that

$$\begin{bmatrix}2\\1\end{bmatrix}, \quad \begin{bmatrix}1\\2\end{bmatrix}, \quad \begin{bmatrix}-4\\5\end{bmatrix}$$

are linearly dependent over the real field.

5.5 DIMENSION AND BASIS

We call a vector space $\mathscr{V}$ **finite dimensional** if and only if there exists a finite subset $\mathscr{S}$ of vectors of $\mathscr{V}$ such that every vector of $\mathscr{V}$ can be written as a linear combination of the vectors of $\mathscr{S}$. Such a finite subset $\mathscr{S}$ is said to **span** or to **generate** the vector space $\mathscr{V}$.

For example, the set of all n-vectors over a field $\mathscr{F}$ is finite dimensional, since for every vector X,

$$X = x_1E_1 + x_2E_2 + \cdots + x_nE_n,$$

where the E_j's are the elementary n-vectors mentioned earlier.

The set of all real 2×2 matrices is finite dimensional, since

$$\begin{bmatrix} a_{11} & a_{12} \\ a_{21} & a_{22} \end{bmatrix} = a_{11}\begin{bmatrix} 1 & 0 \\ 0 & 0 \end{bmatrix} + a_{12}\begin{bmatrix} 0 & 1 \\ 0 & 0 \end{bmatrix} + a_{21}\begin{bmatrix} 0 & 0 \\ 1 & 0 \end{bmatrix} + a_{22}\begin{bmatrix} 0 & 0 \\ 0 & 1 \end{bmatrix}.$$

The **zero space** $\mathscr{Z}$, consisting of the zero vector alone, is spanned by the zero vector and hence is finite dimensional.

The set of all real quadratic polynomials

$$ax^2 + bxy + cy^2$$

is finite dimensional since each is a linear combination of the three quadratic polynomials x^2, xy, y^2.

On the other hand, the set of all real polynomials

$$a_n x^n + a_{n-1}x^{n-1} + \cdots + a_1 x + a_0, \qquad n = 0, 1, \ldots,$$

is not finite dimensional, for a linear combination of the polynomials of a finite set can have degree no higher than that of the highest degree polynomial in the set. Hence we cannot represent all polynomials in this way.

The finite-dimensional vector spaces are the ones most commonly seen in applications, and from here on, we treat them exclusively.

Consider next the equations

$$\begin{bmatrix} x_1 \\ x_2 \end{bmatrix} = x_1\begin{bmatrix} 1 \\ 2 \end{bmatrix} + x_2\begin{bmatrix} -3 \\ 0 \end{bmatrix} + x_2\begin{bmatrix} 3 \\ 3 \end{bmatrix} + (x_1 + x_2)\begin{bmatrix} 0 \\ -2 \end{bmatrix};$$

$$\begin{bmatrix} x_1 \\ x_2 \end{bmatrix} = x_1\begin{bmatrix} 1 \\ 0 \end{bmatrix} + x_2\begin{bmatrix} 0 \\ 1 \end{bmatrix}.$$

They show that each of the sets of vectors

$$\begin{bmatrix} 1 \\ 2 \end{bmatrix}, \begin{bmatrix} -3 \\ 0 \end{bmatrix}, \begin{bmatrix} 3 \\ 3 \end{bmatrix}, \begin{bmatrix} 0 \\ -2 \end{bmatrix} \quad \text{and} \quad \begin{bmatrix} 1 \\ 0 \end{bmatrix}, \begin{bmatrix} 0 \\ 1 \end{bmatrix}$$

spans $\mathscr{E}^2$, but the second set does so more economically. The vectors of the second (economical) set are linearly independent but those of the first (extravagant) set are not.

Moreover, if we omit either vector from the second set, we cannot get all vectors of $\mathscr{E}^2$, for multiples of one of these vectors always have at least one component zero. On the other hand, since

$$\begin{bmatrix} x_1 \\ x_2 \end{bmatrix} = \frac{x_2}{2}\begin{bmatrix} 1 \\ 2 \end{bmatrix} + \frac{x_2 - 2x_1}{6}\begin{bmatrix} -3 \\ 0 \end{bmatrix},$$

we could omit the last two vectors of the first set and still obtain all vectors of $\mathscr{E}^2$.

These two examples illustrate the following definition and theorem.

We define a set of nonzero vectors spanning a vector space $\mathscr{V}$ to be a **basis** for $\mathscr{V}$ if and only if no proper subset of the spanning set also spans $\mathscr{V}$. The simplest example of a basis is the set of elementary vectors $E_1, E_2, \ldots, E_n$ of $\mathscr{E}^n$. The first two vectors of the preceding numerical example,

$$\begin{bmatrix}1\\2\end{bmatrix} \quad \text{and} \quad \begin{bmatrix}-3\\0\end{bmatrix},$$

are a basis for $\mathscr{E}^2$. The reader should show in each case that no proper subset spans the space in question.

Theorem 5.5.1: *Every basis for a finite-dimensional vector space $\mathscr{V}$ contains only linearly independent vectors.*

Indeed, if the vectors of a spanning set are not linearly independent, we can write one of them as a linear combination of the others. Then, substituting the resulting expression for this vector into any linear combination containing it, and rearranging a bit, we can express any vector of $\mathscr{V}$ as a linear combination of the remaining vectors of the spanning set. The spanning set is thus not a basis, since a proper subset of it also spans $\mathscr{V}$. Hence a basis contains only linearly independent vectors.

To illustrate, we return to the numerical example. Expressing the last two vectors of the spanning set of four vectors as linear combinations of the first two, we have

$$\begin{bmatrix}3\\3\end{bmatrix} = \tfrac{3}{2}\begin{bmatrix}1\\2\end{bmatrix} + (-\tfrac{1}{2})\begin{bmatrix}-3\\0\end{bmatrix}$$

and

$$\begin{bmatrix}0\\-2\end{bmatrix} = (-1)\begin{bmatrix}1\\2\end{bmatrix} + (-\tfrac{1}{3})\begin{bmatrix}-3\\0\end{bmatrix}.$$

Hence, substituting and collecting, we find that

$$\alpha_1\begin{bmatrix}1\\2\end{bmatrix} + \alpha_2\begin{bmatrix}-3\\0\end{bmatrix} + \alpha_3\begin{bmatrix}3\\3\end{bmatrix} + \alpha_4\begin{bmatrix}0\\-2\end{bmatrix}$$
$$= (\alpha_1 + \tfrac{3}{2}\alpha_3 - \alpha_4)\begin{bmatrix}1\\2\end{bmatrix} + (\alpha_2 - \tfrac{1}{2}\alpha_3 - \tfrac{1}{3}\alpha_4)\begin{bmatrix}-3\\0\end{bmatrix},$$

so any linear combination of the four vectors can be expressed as a linear combination of the first two. Hence the four are not a basis for $\mathscr{E}^2$. The first two, being linearly independent, *are* a basis for $\mathscr{E}^2$, as the next theorem reveals:

Theorem 5.5.2: *Every set of linearly independent vectors which spans a finite-dimensional vector space $\mathscr{V}$ is a basis for $\mathscr{V}$.*

Suppose that $V_1, V_2, \ldots, V_k$ are linearly independent and span $\mathscr{V}$. Then, in particular, $V_1, V_2, \ldots, V_k$ belong to $\mathscr{V}$. Now, if these vectors are not a basis for $\mathscr{V}$, then some proper subset thereof spans $\mathscr{V}$. Hence there must exist some V_i that is not in this proper subset. Since V_i is in $\mathscr{V}$, it must be a linear combination of the vectors in the spanning subset. This implies that $V_1, V_2, \ldots, V_k$ are linearly dependent, contrary to hypothesis. The contradiction shows that $V_1, V_2, \ldots, V_k$ must indeed constitute a basis for $\mathscr{V}$.

Theorem 5.5.3: *Every finite-dimensional vector space $\mathscr{V} \neq \mathscr{Z}$ has at least one basis.*

Since $\mathscr{V}$ is finite dimensional and does not consist of the zero vector alone, there exists a finite set $V_1, V_2, \ldots, V_k$ of nonzero vectors of $\mathscr{V}$ which span $\mathscr{V}$. If these are linearly independent, they are a basis for $\mathscr{V}$ by the previous theorem. If not, we can write one of them, say V_i, as a linear combination of the others, and replace it by this expression. Then every linear combination of $V_1, V_2, \ldots, V_k$ can be expressed as a linear combination of $V_1, V_2, \ldots, V_{i-1}, V_{i+1}, \ldots, V_k$, in the manner of the last preceding numerical example. If this remaining set of vectors is not linearly independent, we can eliminate another in the same way, and so on, as long as the reduced set is linearly dependent. Since we cannot eliminate *all* of $V_1, V_2, \ldots, V_k$ and still span $\mathscr{V}$, which is nonempty, we must eventually arrive at a linearly independent set. By the preceding theorem, this is a basis for $\mathscr{V}$.

Theorem 5.5.4: *Any two bases of a finite-dimensional vector space $\mathscr{V}$ contain the same number of vectors.*

Let $\mathscr{B}_1$ and $\mathscr{B}_2$ denote any two bases of $\mathscr{V}$. Every vector of $\mathscr{B}_1$ can be written as a linear combination of the vectors of $\mathscr{B}_2$. If there were more vectors in $\mathscr{B}_1$ than in $\mathscr{B}_2$, they would then be linearly dependent, by Corollary 5.4.6. Since they are linearly independent by hypothesis, the number of vectors in $\mathscr{B}_1$ is no greater than the number of vectors in $\mathscr{B}_2$. In the same way, the number of vectors in $\mathscr{B}_2$ is no greater than the number of vectors in $\mathscr{B}_1$. Hence $\mathscr{B}_1$ and $\mathscr{B}_2$ contain the same number of vectors.

The unique number of vectors in a basis of $\mathscr{V}$ is called the **dimension** of $\mathscr{V}$. For example, the dimension of the set of all vectors in $\mathscr{E}^n$ is n, because this set of vectors has the basis $E_1, E_2, \ldots, E_n$. The space consisting of the zero vector alone is defined to have dimension 0.

Theorem 5.5.5: *If $V_1, V_2, \ldots, V_k$ is a basis for a vector space $\mathscr{V}$, if $V = \alpha_1 V_1 + \alpha_2 V_2 + \cdots + \alpha_k V_k$, and if $\alpha_i \neq 0$, then the set $V_1, V_2, \ldots, V_{i-1}, V, V_{i+1}, \ldots, V_k$ is also a basis for $\mathscr{V}$.*

In fact, since $\alpha_i \neq 0$,

$$V_i = \left(-\frac{\alpha_1}{\alpha_i}\right)V_1 + \cdots + \left(-\frac{\alpha_{i-1}}{\alpha_i}\right)V_{i-1} + \frac{1}{\alpha_i}V + \left(-\frac{\alpha_{i+1}}{\alpha_i}\right)V_{i+1} + \cdots + \left(-\frac{\alpha_k}{\alpha_i}\right)V_k.$$

Hence, if W is any vector of $\mathscr{V}$ and if

$$W = \beta_1 V_1 + \beta_2 V_2 + \cdots + \beta_k V_k,$$

by substituting for V_i and rearranging, we have

$$W = \left(\beta_1 - \frac{\alpha_1\beta_i}{\alpha_i}\right)V_1 + \cdots + \left(\beta_{i-1} - \frac{\alpha_{i-1}\beta_i}{\alpha_i}\right)V_{i-1} + \frac{\beta_i}{\alpha_i}V$$
$$+ \left(\beta_{i+1} - \frac{\alpha_{i+1}\beta_i}{\alpha_i}\right)V_{i+1} + \cdots + \left(\beta_k - \frac{\alpha_k\beta_i}{\alpha_i}\right)V_k,$$

so that any vector W can be written as a linear combination of $V_1, \ldots, V_{i-1}, V, V_{i+1}, \ldots, V_k$. Since the dimension of $\mathscr{V}$ is k, no proper subset of these k vectors can span $\mathscr{V}$. Hence this set is a basis for $\mathscr{V}$.

Theorem 5.5.6: *If a vector space $\mathscr{V}$ has dimension k, then any k linearly independent vectors of $\mathscr{V}$ form a basis* for $\mathscr{V}$.

Since $\mathscr{V}$ has dimension k, it has a basis $V_1, V_2, \ldots, V_k$. Let $W_1, W_2, \ldots, W_k$ be any linearly independent set of k vectors. Since $W_1 \neq 0$, there exist scalars α_i not all zero such that

$$W_1 = \alpha_1 V_1 + \alpha_2 V_2 + \cdots + \alpha_k V_k.$$

If $\alpha_i \neq 0$, then, by the preceding theorem,

$$V_1, V_2, \ldots, V_{i-1}, W_1, V_{i+1}, \ldots, V_k$$

is a basis for $\mathscr{V}$. Since $W_2 \neq 0$, there exist scalars β not all zero such that

$$W_2 = \beta_1 V_1 + \beta_2 V_2 + \cdots + \beta_{i-1}V_{i-1} + \beta_i W_1 + \beta_{i+1}V_{i+1} + \cdots + \beta_k V_k.$$

Now some β_j, $j \neq i$, is not zero, since otherwise we would have $W_2 = \beta_i W_1$, which contradicts the fact that the W's are linearly independent. Then, by the preceding theorem, we can replace V_j by W_2, so that

$$V_1, \ldots, V_{i-1}, W_1, V_{i+1}, \ldots, V_{j-1}, W_2, V_j, \ldots, V_k$$

is a basis for $\mathscr{V}$.

We now show in the same way that W_3 can replace one of the remaining V's, and so on, until all the V's have been replaced by the W's, and the proof is complete.

We close this section with

Theorem 5.5.7: *The representation of a vector V of a finite-dimensional vector space $\mathscr{V}$ as a linear combination of the vectors of a given basis for $\mathscr{V}$ is unique.*

Let $V_1, V_2, \ldots, V_k$ be a basis for $\mathscr{V}$. Then there exist scalars $\alpha_1, \alpha_2, \ldots, \alpha_k$ such that

$$V = \alpha_1 V_1 + \alpha_2 V_2 + \cdots + \alpha_k V_k.$$

If also

$$V = \beta_1 V_1 + \beta_2 V_2 + \cdots + \beta_k V_k,$$

then by subtraction we have

$$(\alpha_1 - \beta_1)V_1 + (\alpha_2 - \beta_2)V_2 + \cdots + (\alpha_k - \beta_k)V_k = 0.$$

Since $V_1, V_2, \ldots, V_k$ are linearly independent, the coefficients here are all zero, and hence

$$\alpha_1 = \beta_1, \alpha_2 = \beta_2, \ldots, \alpha_k = \beta_k.$$

This proves the theorem.

5.6 COMPUTATION OF THE DIMENSION OF A VECTOR SPACE

Suppose that a vector space $\mathscr{V}$ is spanned by $V_1, V_2, \ldots, V_k$, so that it is the set of all linear combinations of the form

$$\alpha_1 V_1 + \alpha_2 V_2 + \cdots + \alpha_i V_i + \cdots + \alpha_j V_j + \cdots + \alpha_k V_k. \tag{5.6.1}$$

By what precedes, to determine the dimension of $\mathscr{V}$, we must determine the maximum number of linearly independent vectors that exist in $\mathscr{V}$. To accomplish this, we can use the following three **elementary operations** on the spanning set:

1. Interchange any two of the V's. That is, the vectors $V_1, \ldots, V_j, \ldots, V_i, \ldots, V_k$ span the same space as do $V_1, \ldots, V_i, \ldots, V_j, \ldots, V_k$. Indeed, by the commutative law of addition, we can interchange $\alpha_i V_i$ and $\alpha_j V_j$ in the sum (5.6.1), which shows that the totality of linear combinations is unaffected by the interchange.

Observe that, by repeated interchange of two of the V's, we can rearrange the V's in any order we wish without altering the space they span.

2. Multiply any vector V_i of the set by an arbitrary nonzero scalar. That is, if $\alpha \neq 0$, the vectors $V_1, V_2, \ldots, \alpha V_i, \ldots, V_k$ span the same space as do $V_1, V_2, \ldots, V_i, \ldots, V_k$. In fact, $\alpha_1 V_1 + \cdots + \alpha_i V_i + \cdots + \alpha_k V_k = \alpha_1 V_1 + \ldots + (\alpha_i/\alpha)(\alpha V_i) + \cdots + \alpha_k V_k$, so that any linear combination of the vectors of one set is expressible as a linear combination of those of the other set. Hence from either set we obtain the same totality of linear combinations. By applying this operation repeatedly, we can multiply each of the V's by an arbitrary nonzero scalar and still obtain the same vector space $\mathscr{V}$.

3. Add any scalar multiple of one of the V's, say αV_j, to any other vector, V_i, of the set. Again, this does not alter the totality of linear combinations since

$$\begin{aligned}\alpha_1 V_1 + \cdots + \alpha_i V_i + \cdots + \alpha_j V_j + \cdots + \alpha_k V_k \\ = \alpha_1 V_1 + \cdots + \alpha_i(V_i + \alpha V_j) + \cdots + (\alpha_j - \alpha_i\alpha)V_j + \cdots + \alpha_k V_k.\end{aligned}$$

Hence the space spanned by $V_1, \ldots, V_i + \alpha V_j, \ldots, V_j, \ldots, V_k$ is the same as that spanned by $V_1, \ldots, V_i, \ldots, V_j, \ldots, V_k$.

Repeated use of these three operations often makes it possible to transform a spanning set of vectors to a set with a recognizably independent subset. For example, let $V_1, V_2, \ldots, V_k$ be vectors in $\mathscr{E}^n$. We first arrange them for convenience as the columns of an $n \times k$ matrix:

$$\begin{bmatrix} a_{11} & a_{12} & \cdots & a_{1k} \\ a_{21} & a_{22} & \cdots & a_{2k} \\ \vdots & & & \\ a_{n1} & a_{n2} & \cdots & a_{nk} \end{bmatrix}.$$

We now apply the sweepout procedure to the *columns* of this matrix, using only the three operations outlined above. The object is to obtain at the left a maximal set of columns whose first nonzero entries are 1 and such that if $p > j$, the 1 in column p is in a lower row than is the 1 in column j. All remaining columns are zero vectors. The end result will be said to be in **column echelon form**. At the end of the process, we have a set of vectors of $\mathscr{E}^n$, the independent members of which are easily recognizable by Theorem 5.2.3.

Note that we do *not* work on rows. To do so would not necessarily leave invariant the vector space spanned by the columns of the resulting matrices. For example, the columns of

$$\begin{bmatrix} 0 & 0 \\ 1 & 0 \\ 0 & 1 \end{bmatrix}$$

span a space for all of whose vectors $x_1 = 0$. If we add the second row to the first, this is no longer true, so the operation has changed the column space.

We illustrate the procedure just described by finding the dimension of, and a basis for, the subspace of $\mathscr{E}^4$ spanned by the columns of the matrix

$$\begin{bmatrix} 1 & 2 & -1 & 0 \\ 2 & 4 & -2 & 0 \\ -1 & 4 & -5 & -2 \\ 3 & 3 & 0 & 1 \end{bmatrix},$$

namely for the **column space** of the matrix. We use the upper left 1 to sweep the first row:

$$\begin{bmatrix} 1 & 0 & 0 & 0 \\ 2 & 0 & 0 & 0 \\ -1 & 6 & -6 & -2 \\ 3 & -3 & 3 & 1 \end{bmatrix}.$$

Now we divide the second column by 6 and use the resulting column to sweep the remaining columns. This yields the matrix

$$\begin{bmatrix} 1 & 0 & 0 & 0 \\ 2 & 0 & 0 & 0 \\ -1 & 1 & 0 & 0 \\ 3 & -\frac{1}{2} & 0 & 0 \end{bmatrix}.$$

It is now apparent that the column space of this matrix has dimension 2 and that a basis for it is the pair of linearly independent vectors

$$\begin{bmatrix} 1 \\ 2 \\ -1 \\ 3 \end{bmatrix} \quad \text{and} \quad \begin{bmatrix} 0 \\ 0 \\ 1 \\ -\frac{1}{2} \end{bmatrix}.$$

Often it is not necessary to reduce precisely to column echelon form in order to determine a basis. In the case of this example, one sees by inspection after the first sweep that columns 1 and 4 are independent and that columns 2 and 3 can be swept clean with column 4 so that the dimension is 2 and a basis is the pair of vectors

$$\begin{bmatrix} 1 \\ 2 \\ -1 \\ 3 \end{bmatrix} \quad \text{and} \quad \begin{bmatrix} 0 \\ 0 \\ -2 \\ 1 \end{bmatrix}.$$

A simple but important special case is that of the identity matrix, I_n, whose columns are the elementary vectors, so that its column space is the set of all n-vectors over whatever field $\mathscr{F}$ one may be using. When $\mathscr{F}$ is the real field, the column space of I_n is just the set of all vectors in $\mathscr{E}^n$.

5.7 EXERCISES

1. Find a basis for the subspace of vectors in $\mathscr{E}^3$ spanned by

$$\begin{bmatrix} 2 \\ 1 \\ 3 \end{bmatrix}, \quad \begin{bmatrix} -1 \\ 4 \\ 0 \end{bmatrix}, \quad \begin{bmatrix} 4 \\ 2 \\ 6 \end{bmatrix}, \quad \begin{bmatrix} 2 \\ -8 \\ 0 \end{bmatrix}.$$

2. Prove that the vectors $E_1, E_1 + E_2, E_1 + E_2 + E_3, \ldots, E_1 + E_2 + \cdots + E_n$ constitute a basis for the vectors of $\mathscr{E}^n$.

3. Do the vectors $E_1 + E_2, E_2 + E_3, \ldots, E_{n-1} + E_n, E_n$ constitute a basis for the vectors of $\mathscr{E}^n$?

4. Show that if $A_1, A_2, \ldots, A_k$ are linearly independent vectors of $\mathscr{E}^n$, then so are

$$A_1, A_1 + 2A_2, A_1 + 2A_2 + 3A_3, \ldots, A_1 + 2A_2 + 3A_3 + \cdots + kA_k.$$

5. Determine a basis for the column space of each matrix:

(a) $\begin{bmatrix} 1 & 3 & 0 & 0 \\ 2 & 2 & 4 & 0 \\ -1 & 4 & -7 & 0 \\ 3 & 0 & 9 & 1 \end{bmatrix}$, (b) $\begin{bmatrix} 0 & 0 & 1 & 1 & 1 & 0 \\ 0 & 1 & 1 & 1 & 0 & 0 \\ 1 & 1 & 1 & 0 & 0 & 0 \\ 0 & 1 & 1 & 1 & 0 & 0 \\ 0 & 0 & 1 & 1 & 1 & 0 \end{bmatrix}$.

Then express each of the columns as a linear combination of the vectors of the basis.

6. Show how to express any vector X in $\mathscr{E}^3$ as a linear combination of

$$\begin{bmatrix} 1 \\ 0 \\ 0 \end{bmatrix}, \quad \begin{bmatrix} 1 \\ 2 \\ 0 \end{bmatrix}, \quad \begin{bmatrix} 1 \\ 2 \\ 3 \end{bmatrix}.$$

Then prove without further computation that these three vectors constitute a basis for the vectors of $\mathscr{E}^3$.

7. Prove without any computation that

$$\begin{bmatrix} 1 \\ 2 \\ 2 \end{bmatrix}, \quad \begin{bmatrix} 0 \\ 1 \\ 2 \end{bmatrix}, \quad \begin{bmatrix} 0 \\ 0 \\ 3 \end{bmatrix}$$

constitute a basis for the vectors of $\mathscr{E}^3$. Then express

$$\begin{bmatrix} 3 \\ 6 \\ 9 \end{bmatrix}$$

as a linear combination of the vectors of this basis. Which of the vectors of this basis may be replaced by

$$\begin{bmatrix} 3 \\ 6 \\ 9 \end{bmatrix},$$

the resulting set still being a basis?

8. Which vectors of the basis

$$\begin{bmatrix} 1 \\ 2 \\ -1 \\ 3 \end{bmatrix}, \quad \begin{bmatrix} 2 \\ 1 \\ 1 \\ 1 \end{bmatrix}, \quad \begin{bmatrix} 3 \\ -2 \\ 1 \\ 4 \end{bmatrix}, \quad \begin{bmatrix} 1 \\ 1 \\ 1 \\ 1 \end{bmatrix}$$

for the vectors of $\mathscr{E}^4$ can be replaced by the vector

$$\begin{bmatrix} 2 \\ -3 \\ 0 \\ 3 \end{bmatrix},$$

the result still being a basis for $\mathscr{E}^4$?

9. Do the vectors $E_1 + E_2, E_2 + E_3, \ldots, E_{n-1} + E_n, E_n + E_1$ constitute a basis for the vectors of $\mathscr{E}^n$? (Consider two cases: n even, n odd.)

***10.** Prove that the vectors $X_1, X_2, \ldots, X_n$ in $\mathscr{E}^n$ constitute a basis for $\mathscr{E}^n$ if and only if every vector X of $\mathscr{E}^n$ can be written as a linear combination of these vectors. (This requires *no* computation.)

11. Prove that a finite-dimensional vector space $\mathscr{V} \neq \mathscr{Z}$ over a number field $\mathscr{F}$ has infinitely many bases.

12. Denote an $n \times n$ matrix with a 1 in the ij-position and zeros elsewhere by E_{ij}. Use these matrices to show that the set of all $n \times n$ matrices over a number field $\mathscr{F}$ is a finite-dimensional vector space of dimension n^2.

13. Prove that the set of all real, upper-triangular matrices

$$\begin{bmatrix} a_{11} & a_{12} & \cdots & a_{1n} \\ 0 & a_{22} & \cdots & a_{2n} \\ \vdots & & & \\ 0 & 0 & \cdots & a_{nn} \end{bmatrix}$$

is a vector space over the real number field. Find the dimension of this space of matrices and give a basis for it.

14. Show that if $A_1, A_2, \ldots, A_k$ are n-vectors over a number field $\mathscr{F}$ and if

$$X = \begin{bmatrix} x_1 \\ x_2 \\ \vdots \\ x_n \end{bmatrix},$$

then the linear expressions $A_1^{\mathsf{T}}X, A_2^{\mathsf{T}}X, \ldots, A_k^{\mathsf{T}}X$ are linearly independent if and only if $A_1, A_2, \ldots, A_k$ are linearly independent.

15. Show that m linear expressions

$$a_{i1}x_1 + a_{i2}x_2 + \cdots + a_{in}x_n, \qquad i = 1, 2, \ldots, m$$

are linearly independent if (a) each expression contains, with nonzero coefficient, a variable appearing in no other expression of the set, or (b) each expression contains, with nonzero coefficient, a variable appearing in no previous linear expression of the set.

***16.** Prove that n vectors in $\mathscr{E}^n$ form a basis for the vectors of $\mathscr{E}^n$ if and only if each of the n equations

$$\alpha_{i1}X_1 + \alpha_{i2}X_2 + \cdots + \alpha_{in}X_n = E_i, \qquad i = 1, 2, \ldots, n$$

can be solved for the α's. This requires no computation.

17. Use mathematical induction to prove that $P_0, P_1, \ldots, P_k$ lie on a k-flat and not on some lower-dimensional flat space if and only if $X_1 - X_0$, $X_2 - X_0$, $\ldots$, $X_k - X_0$ are linearly independent, where X_i is the vector determined by P_i.

18. Under what conditions do the real n-vectors $X_1, X_2, \ldots, X_n$ determine a k-flat on the origin?

19. Show that if $P_0, P_1, \ldots, P_k$ determine a k-flat, if the terminal points $Q_0, Q_1, \ldots, Q_k$ of the vectors $Y_0, Y_1, \ldots, Y_k$ are on this k-flat, and if $Y_1 - Y_0$, $Y_2 - Y_0, \ldots, Y_k - Y_0$ are linearly independent, then $Q_0, Q_1, \ldots, Q_k$ determine the same k-flat.

***20.** Prove that if $X_1, X_2, \ldots, X_k$ are linearly independent vectors of a vector space $\mathscr{V}$ of dimension n, then there exist vectors $X_{k+1}, \ldots, X_n$ in $\mathscr{V}$ such that $X_1, X_2, \ldots, X_k, X_{k+1}, \ldots, X_n$ is a basis for $\mathscr{V}$. Moreover, $X_{k+1}, \ldots, X_n$ may be chosen from any given basis of $\mathscr{V}$.

5.8 ORTHONORMAL BASES

In Chapter 4 we pointed out that the basis $E_1, E_2, \ldots, E_n$ for $\mathscr{E}^n$ consists of mutually orthogonal unit vectors. In general, a basis $V_1, V_2, \ldots, V_k$ of a finite-dimensional vector space $\mathscr{V}$ is called an **orthonormal basis** for $\mathscr{V}$ if and only if it consists of mutually orthogonal unit vectors. Such bases are particularly useful.

When a basis $V_1, V_2, \ldots, V_k$ is known for a subspace $\mathscr{V}$ of $\mathscr{E}^n$, an orthonormal basis $U_1, U_2, \ldots, U_k$, is readily determined by what is known as the **Gram–Schmidt process.** In this process, the first vector of the orthonormal basis is chosen as the unit vector

$$U_1 = \frac{V_1}{|V_1|}.$$

To obtain the second vector, we first find a vector W_2 which is a special sort of linear combination of the *next* of the V_j's, namely V_2, and the *previously found* U_j's, namely U_1:

$$W_2 = V_2 - \alpha_{21} U_1.$$

The coefficient α_{21} is chosen so as to make U_1 and W_2 orthogonal:

$$U_1^{\mathsf{T}} W_2 = U_1^{\mathsf{T}} V_2 - \alpha_{21} U_1^{\mathsf{T}} U_1 = U_1^{\mathsf{T}} V_2 - \alpha_{21} = 0.$$

That is,

$$\alpha_{21} = U_1^{\mathsf{T}} V_2,$$

so

$$W_2 = V_2 - (U_1^{\mathsf{T}} V_2) U_1.$$

We now define U_2 as the unit vector

$$U_2 = \frac{W_2}{|W_2|},$$

and observe that $U_1^{\mathsf{T}} U_2 = 0$. Moreover, U_2 belongs to $\mathscr{V}$ since it is actually a linear combination of V_1 and V_2.

Next we construct W_3 as a special linear combination of V_3 and the previously found U_j's:

$$W_3 = V_3 - \alpha_{31} U_1 - \alpha_{32} U_2,$$

which is orthogonal to each of the unit vectors U_1 and U_2:

$$\begin{aligned} U_1^{\mathsf{T}} W_3 &= U_1^{\mathsf{T}} V_3 - \alpha_{31} &&= 0 \\ U_2^{\mathsf{T}} W_3 &= U_2^{\mathsf{T}} V_3 - \alpha_{32} &&= 0. \end{aligned}$$

Here we find

$$\alpha_{31} = U_1^{\mathsf{T}} V_3, \qquad \alpha_{32} = U_2^{\mathsf{T}} V_3,$$

so

$$W_3 = V_3 - (U_1^{\mathsf{T}} V_3) U_1 - (U_2^{\mathsf{T}} V_3) U_2,$$

after which we define

$$U_3 = \frac{W_3}{|W_3|}.$$

Now $U_1^{\mathsf{T}} U_3 = 0$, $U_2^{\mathsf{T}} U_3 = 0$, and U_3 is actually a linear combination of V_1, V_2, and V_3, so that it, too, is in $\mathscr{V}$.

The general pattern is now clear. We define

$$W_1 = V_1,$$

(5.8.1a) $$W_j = V_j - (U_1^{\mathsf{T}} V_j) U_1 - (U_2^{\mathsf{T}} V_j) U_2 - \cdots - (U_{j-1}^{\mathsf{T}} V_j) U_{j-1}, \qquad j = 2, 3, \ldots, k,$$

and

(5.8.1b) $$U_j = \frac{W_j}{|W_j|}, \qquad j = 1, 2, 3, \ldots, k.$$

It is not hard to show that each such U_j is orthogonal to all previous U_i's, so that we have indeed an orthonormal set. Also, each U_j is in $\mathscr{V}$, since it is expressible as a linear combination of $V_1, V_2, \ldots, V_j$. Moreover, as we have seen before, mutually orthogonal unit vectors are linearly independent, so U_1, U_2, $\cdots$, U_k constitute a basis for $\mathscr{V}$.

As an example, we find an orthonormal basis for the vector space in $\mathscr{E}^3$ spanned by

$$V_1 = \begin{bmatrix} 1 \\ 0 \\ 1 \end{bmatrix} \quad \text{and} \quad V_2 = \begin{bmatrix} 1 \\ 1 \\ 0 \end{bmatrix}.$$

We have first

$$U_1 = \begin{bmatrix} \frac{1}{\sqrt{2}} \\ 0 \\ \frac{1}{\sqrt{2}} \end{bmatrix}.$$

Then

$$W_2 = \begin{bmatrix} 1 \\ 1 \\ 0 \end{bmatrix} - \left(\left[\frac{1}{\sqrt{2}}, 0, \frac{1}{\sqrt{2}} \right] \begin{bmatrix} 1 \\ 1 \\ 0 \end{bmatrix} \right) \begin{bmatrix} \frac{1}{\sqrt{2}} \\ 0 \\ \frac{1}{\sqrt{2}} \end{bmatrix}$$

$$= \begin{bmatrix} 1 \\ 1 \\ 0 \end{bmatrix} - \frac{1}{\sqrt{2}} \begin{bmatrix} \frac{1}{\sqrt{2}} \\ 0 \\ \frac{1}{\sqrt{2}} \end{bmatrix} = \begin{bmatrix} 1 \\ 1 \\ 0 \end{bmatrix} - \begin{bmatrix} \frac{1}{2} \\ 0 \\ \frac{1}{2} \end{bmatrix} = \begin{bmatrix} \frac{1}{2} \\ 1 \\ -\frac{1}{2} \end{bmatrix}.$$

We have $|W_2| = \sqrt{\frac{3}{2}}$, so

$$U_2 = \begin{bmatrix} \frac{1}{\sqrt{6}} \\ \frac{2}{\sqrt{6}} \\ -\frac{1}{\sqrt{6}} \end{bmatrix}.$$

The computations begin to get messy in even this simple example. The procedure is, indeed, not adapted to efficient paper-and-pencil work, but it is readily programmed for a computer. For the purpose of deriving other results, it often suffices to know that an orthonormal basis exists.

5.9 EXERCISES

1. For what real values of a is the n-vector

$$\begin{bmatrix} a \\ \vdots \\ a \\ 0 \\ 0 \\ \vdots \\ 0 \end{bmatrix},$$

where there are k entries a and $n - k$ entries 0, a unit vector?

2. Find a linear combination of

$$\begin{bmatrix} -1 \\ 2 \\ 0 \end{bmatrix} \quad \text{and} \quad \begin{bmatrix} 0 \\ 2 \\ 1 \end{bmatrix}$$

that is orthogonal to

$$\begin{bmatrix} -1 \\ 2 \\ 0 \end{bmatrix}.$$

Then find an orthonormal basis for the subspace of vectors of $\mathscr{E}^3$ which is spanned by the first two given vectors. Illustrate with a figure.

3. Use the Gram–Schmidt process to construct orthonormal bases for the subspaces of $\mathscr{E}^4$ having the following bases:

(a) $V_1 = \begin{bmatrix} 1 \\ 1 \\ 1 \\ 0 \end{bmatrix}, \quad V_2 = \begin{bmatrix} 0 \\ 1 \\ -1 \\ 0 \end{bmatrix}, \quad V_3 = \begin{bmatrix} 1 \\ 2 \\ 3 \\ 0 \end{bmatrix},$

(b) $V_1 = \begin{bmatrix} 1 \\ 0 \\ -1 \\ 0 \end{bmatrix}, \quad V_2 = \begin{bmatrix} 1 \\ 0 \\ 1 \\ 1 \end{bmatrix}, \quad V_3 = \begin{bmatrix} -1 \\ 1 \\ 0 \\ 0 \end{bmatrix},$

(c) $V_1 = \begin{bmatrix} 1 \\ 0 \\ 0 \\ 0 \end{bmatrix}, \quad V_2 = \begin{bmatrix} 1 \\ 1 \\ 0 \\ 0 \end{bmatrix}, \quad V_3 = \begin{bmatrix} 1 \\ 1 \\ 1 \\ 0 \end{bmatrix}.$

***4.** Prove that if $U_1, U_2, \ldots, U_k$ are an orthonormal basis for a subspace of $\mathscr{E}^n$ and if

$$X = \alpha_1 U_1 + \alpha_2 U_2 + \cdots + \alpha_k U_k,$$

then

$$|X|^2 = \alpha_1^2 + \alpha_2^2 + \cdots + \alpha_k^2.$$

This reveals one of several reasons why an orthonormal basis is useful.

***5.** Prove that if $U_1, U_2, \ldots, U_k$ are an orthonormal basis for a subspace $\mathscr{V}$ of $\mathscr{E}^k$, and if

$$X = \alpha_1 U_1 + \alpha_2 U_2 + \cdots + \alpha_k U_k,$$

then

$$X^{\mathsf{T}} U_i = \alpha_i, \qquad i = 1, 2, \ldots, k.$$

Hence, for each X in $\mathscr{V}$,

$$X = \sum_{i=1}^{k} (X^{\mathsf{T}} U_i) U_i.$$

Note that the associative law cannot be applied to the products in the terms of this sum. Why not?

5.10 INTERSECTION AND SUM OF TWO VECTOR SPACES

Let $\mathscr{U}$ and $\mathscr{V}$ be subspaces of a vector space $\mathscr{W}$ over a field $\mathscr{F}$. The set of all vectors which are in both $\mathscr{U}$ and $\mathscr{V}$ is called their **intersection** and is denoted by $\mathscr{U} \cap \mathscr{V}$. Since the zero vector is in every vector space, $\mathscr{U} \cap \mathscr{V}$ is never empty. We have

Theorem 5.10.1: *The intersection of two subspaces $\mathscr{U}$ and $\mathscr{V}$ of a vector space $\mathscr{W}$ is again a vector space.*

Let V_1 and V_2 belong to each of $\mathscr{U}$ and $\mathscr{V}$, hence to $\mathscr{U} \cap \mathscr{V}$. Then $V_1 + V_2$ and αV_1 are in each of $\mathscr{U}$ and $\mathscr{V}$, that is, in $\mathscr{U} \cap \mathscr{V}$, because $\mathscr{U}$ and $\mathscr{V}$ are both vector spaces. This proves that the subset $\mathscr{U} \cap \mathscr{V}$ of $\mathscr{W}$ is a vector space. It may be that $\mathscr{U} \cap \mathscr{V} = \mathscr{Z}$.

Consider now the subset of $\mathscr{W}$ consisting of all vectors of the form $\alpha U + \beta V$, where U is in $\mathscr{U}$ and V is in $\mathscr{V}$. We call this the **sum** of $\mathscr{U}$ and $\mathscr{V}$ and denote it by $\mathscr{U} + \mathscr{V}$. Note that U and V are not fixed in this definition and that α and β are arbitrary scalars of $\mathscr{F}$.

Theorem 5.10.2: *The sum of two subspaces $\mathscr{U}$ and $\mathscr{V}$ of a vector space $\mathscr{W}$ is again a vector space.*

If $\alpha_1 U_1 + \beta_1 V_1$ and $\alpha_2 U_2 + \beta_2 V_2$ are two vectors of $\mathscr{U} + \mathscr{V}$, so are $(\alpha_1 U_1 + \alpha_2 U_2) + (\beta_1 V_1 + \beta_2 V_2)$ and $(\alpha\alpha_1) U_1 + (\beta\beta_1) V_1$, so the sum is indeed a vector space.

Now let $\mathscr{U}$ and $\mathscr{V}$ be finite dimensional with dimensions u and v, respectively, let $\mathscr{U} \cap \mathscr{V}$ have dimension r, and let $\mathscr{U} + \mathscr{V}$ have dimension s. Let a basis for $\mathscr{U} \cap \mathscr{V}$ be the independent vectors

$$X_1, X_2, \ldots, X_r.$$

(If $\mathscr{U} \cap \mathscr{V} = \mathscr{Z}$, then there are no X_j's.) Starting with these, let a basis for $\mathscr{U}$ be

$$X_1, X_2, \ldots, X_r, U_{r+1}, \ldots, U_u$$

and let a basis for $\mathscr{V}$ be

$$X_1, X_2, \ldots, X_r, V_{r+1}, \ldots, V_v.$$

(How do we know such bases exist?) Then the $r + (u - r) + (v - r) = u + v - r$ vectors

$$X_1, X_2, \ldots, X_r, U_{r+1}, \ldots, U_u, V_{r+1}, \ldots, V_v$$

are linearly independent, for if

$$\sum_{i=1}^{r} \alpha_i X_i + \sum_{j=r+1}^{u} \beta_j U_j + \sum_{k=r+1}^{v} \gamma_k V_k = 0,$$

then

$$\sum \alpha_i X_i + \sum \beta_j U_j = \sum (-\gamma_k) V_k,$$

so that if some γ_k is not 0, the vector of $\mathscr{V}$ on the right is equal to a vector of $\mathscr{U}$ on the left. This means that $\sum (-\gamma_k)V_k$, being in both $\mathscr{U}$ and $\mathscr{V}$, must be a linear combination of $X_1, X_2, \ldots, X_r$, so that the vectors $X_1, X_2, \ldots, X_r, V_{r+1}, \ldots, V_v$ are linearly dependent, a contradiction. Therefore every γ_k must be 0 and hence, since $X_1, X_2, \ldots, X_r, U_{r+1}, \ldots, U_u$ are linearly independent, all the α's and β's must also be 0. This proves that the total set is independent, as claimed.

Now every U in $\mathscr{U}$, and every V in $\mathscr{V}$, is a linear combination of vectors in this set of $u + v - r$ independent vectors, and therefore so is every vector $\alpha U + \beta V$ of $\mathscr{U} + \mathscr{V}$. That is, these vectors are an independent set which span $\mathscr{U} + \mathscr{V}$ and hence are a basis for it. Thus the dimension of $\mathscr{U} + \mathscr{V}$ is $s = u + v - r$. We summarize in

Theorem 5.10.3: *If $\mathscr{U}$ of dimension u and $\mathscr{V}$ of dimension v are subspaces of a vector space $\mathscr{W}$, and if $\mathscr{U} \cap \mathscr{V}$ and $\mathscr{U} + \mathscr{V}$ have dimension r and s, respectively, then*

$$r + s = u + v. \tag{5.10.1}$$

This is often written

$$\dim (\mathscr{U} + \mathscr{V}) = \dim \mathscr{U} + \dim \mathscr{V} - \dim (\mathscr{U} \cap \mathscr{V}). \tag{5.10.2}$$

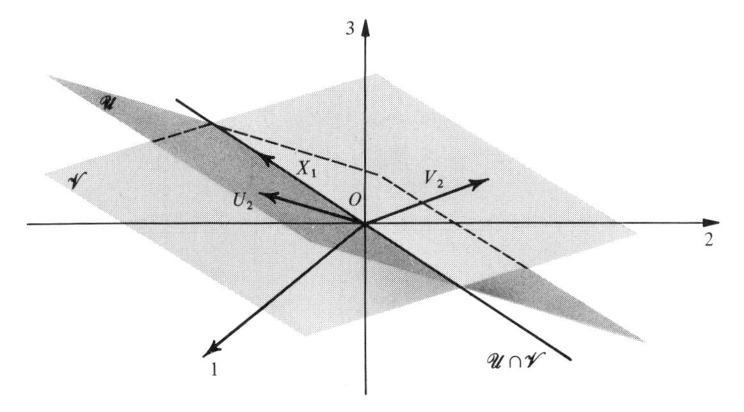

Figure 5-1. *Intersection and Sum of Two Vector Spaces.*

These results have easy geometrical interpretations. For example, let $\mathscr{U}$ and $\mathscr{V}$ be distinct two-dimensional vector spaces in $\mathscr{E}^3$, so that they can be represented by distinct planes on the origin. These two planes intersect in a line, which represents the space $\mathscr{U} \cap \mathscr{V}$ (Figure 5-1). The space $\mathscr{U} + \mathscr{V}$, spanned by X_1 in $\mathscr{U} \cap \mathscr{V}$, U_2 in $\mathscr{U}$, and V_2 in $\mathscr{V}$, is simply $\mathscr{E}^3$. Here, (5.10.1) becomes $1 + 3 = 2 + 2$.

If $\mathscr{U}$ is a 1-dimensional and $\mathscr{V}$ is a 2-dimensional vector space in $\mathscr{E}^3$, $\mathscr{U}$ not a subspace of $\mathscr{V}$, then we can represent these as a line and a plane on O, the line not lying in the plane. In this case $\mathscr{U} \cap \mathscr{V}$ is just the zero vector and a basis U_1 for $\mathscr{U}$, together with a basis V_1, V_2 for $\mathscr{V}$, is a basis for the set of all vectors in $\mathscr{E}^3$. Here (5.10.1) becomes $0 + 3 = 1 + 2$.

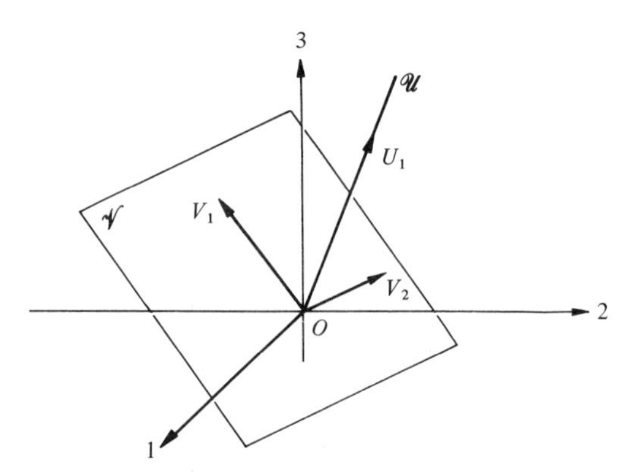

Figure 5-2. *Complementary Vector Spaces.*

When, as in this example, $\mathscr{U} \cap \mathscr{V}$ is the zero vector and $\mathscr{U} + \mathscr{V}$ is the set of all vectors in $\mathscr{E}^n$, we call $\mathscr{U}$ and $\mathscr{V}$ **complementary vector spaces** (Figure 5-2).

5.11 EXERCISES

1. Determine the dimensions of the sum and of the intersection of the vector spaces defined by the columns of these matrices:

$$\begin{bmatrix} 1 & 1 & 1 & 4 \\ 0 & 1 & 1 & 3 \\ 0 & 0 & 1 & 2 \\ 0 & 0 & -1 & -2 \end{bmatrix}, \quad \begin{bmatrix} 0 & 0 & -1 & 2 \\ 0 & 0 & -1 & 2 \\ 0 & 1 & -1 & 3 \\ 1 & 1 & 1 & 0 \end{bmatrix}.$$

2. Given that $\mathscr{U}$ and $\mathscr{V}$ are vector spaces in $\mathscr{E}^n$, of dimensions u and v, respectively, and that $u + v > n$, prove that $\mathscr{U} \cap \mathscr{V}$ is more than the zero vector.

3. Prove by induction that if $\mathscr{U}_1, \mathscr{U}_2, \ldots, \mathscr{U}_k$ in $\mathscr{E}^n$ each have dimension $n - 1$, then $\mathscr{U}_1 \cap \mathscr{U}_2 \cap \cdots \cap \mathscr{U}_k$ has dimension at least $n - k$. In particular, if $k < n$, $\mathscr{U}_1 \cap \mathscr{U}_2 \cap \cdots \cap \mathscr{U}_k$ does not consist of the zero vector alone. Illustrate the various possible cases in $\mathscr{E}^3$ with figures.

4. Given that $\mathscr{U}$ and $\mathscr{V}$ are vector spaces in $\mathscr{E}^n$, of dimensions u and v, respectively, and that $\mathscr{U} + \mathscr{V}$ is the set of all vectors of $\mathscr{E}^n$, prove that $u + v \geq n$ and that, if $u + v = n$, then $\mathscr{U} \cap \mathscr{V}$ is just the zero vector.

5. Find the dimension of the sum and of the intersection of the column spaces of the matrices

$$\begin{bmatrix} 1 & 3 & 1 \\ 2 & 2 & 14 \\ -1 & 1 & -13 \\ 4 & 0 & 40 \end{bmatrix} \quad \text{and} \quad \begin{bmatrix} 2 & 1 & 4 \\ 0 & -1 & 2 \\ 2 & 2 & 2 \\ -4 & -5 & -2 \end{bmatrix}.$$

What do the results imply?

6. The points whose coordinates satisfy a real linear equation

$$a_1x_1 + a_2x_2 + \cdots + a_nx_n = 0$$

constitute a hyperplane on the origin while the corresponding vectors constitute a vector space. How then may one interpret geometrically the set of all solutions of a system of m such linear equations in n unknowns?

5.12 ISOMORPHIC VECTOR SPACES

Recall that the set of all real quadratic polynomials of the form $ax^2 + bxy + cy^2$ is a 3-dimensional vector space over the field of real numbers. By arraying the coefficients in the form of a 3-vector, we can put these quadratic forms, as they are called, in one-to-one correspondence with the vectors of $\mathscr{E}^3$:

$$ax^2 + bxy + cy^2 \leftrightarrow \begin{bmatrix} a \\ b \\ c \end{bmatrix}.$$

The double-headed arrow means that to each quadratic form there corresponds a unique 3-vector and to each 3-vector there corresponds a unique quadratic form. Corresponding members of the two spaces are called **images** of each other under the correspondence.

If α is any scalar, then

$$\alpha(ax^2 + bxy + cy^2) = \alpha ax^2 + \alpha bxy + \alpha cy^2 \leftrightarrow \begin{bmatrix} \alpha a \\ \alpha b \\ \alpha c \end{bmatrix} = \alpha \begin{bmatrix} a \\ b \\ c \end{bmatrix};$$

that is, to a scalar multiple of a vector of either of the spaces, there corresponds the same scalar multiple of the image vector. We say that the correspondence "preserves" the operation of scalar multiplication.

Now let

$$q_i = a_i x^2 + b_i xy + c_i y^2, \qquad i = 1, 2.$$

Then

$$q_1 + q_2 = (a_1 + a_2)x^2 + (b_1 + b_2)xy + (c_1 + c_2)y^2$$

$$\leftrightarrow \begin{bmatrix} a_1 + a_2 \\ b_1 + b_2 \\ c_1 + c_2 \end{bmatrix} = \begin{bmatrix} a_1 \\ b_1 \\ c_1 \end{bmatrix} + \begin{bmatrix} a_2 \\ b_2 \\ c_2 \end{bmatrix},$$

so that the image of the sum of two vectors of either space is the sum of their images. Thus the correspondence also preserves the operation of addition.

Since these two operations are preserved by the correspondence, it follows by induction that to any linear combination of vectors of the one space there corresponds the same linear combination of the image vectors, so the linear algebraic properties of the two spaces are the same. We describe this complex of ideas by saying that the spaces are **isomorphic**.

The two vector spaces are distinct concrete representations of the same abstract entity: the 3-dimensional vector space over the field of real numbers. This space has, of course, infinitely many concrete representations, all mutually isomorphic and all isomorphic to the space of real 3-vectors. (The relation of this abstract space to its concrete representations is analogous to the relation of a positive integer to the sets of which it is the cardinal number.) Thus, when when we studied the linear algebra of $\mathscr{E}^3$, we really studied the linear algebra of all 3-dimensional vector spaces over the real field.

The preceding ideas are generalized in the following definition. Let $\mathscr{U}$ and $\mathscr{V}$ be finite-dimensional vector spaces over a field $\mathscr{F}$ whose vectors are in one-to-one correspondence, that is, such that each U in $\mathscr{U}$ has a unique image V in $\mathscr{V}$ and such that U is in turn the unique image of $V: U \leftrightarrow V$. Then $\mathscr{U}$ and $\mathscr{V}$ are said to be **isomorphic** under this correspondence if and only if, whenever $U_1 \leftrightarrow V_1$ and $U_2 \leftrightarrow V_2$,

(5.12.1) $$\alpha U_1 \leftrightarrow \alpha V_1 \qquad \text{for all scalars } \alpha \text{ in } \mathscr{F},$$

and

(5.12.2) $$U_1 + U_2 \leftrightarrow V_1 + V_2,$$

that is, if and only if the correspondence preserves the operation of multiplication by a scalar and the operation of addition.

Concerning isomorphic vector spaces, we have several basic theorems.

Theorem 5.12.1: *If $\mathscr{U}$ and $\mathscr{V}$ are isomorphic, then their zero vectors correspond.*

This follows from (5.12.1) in the case when $\alpha = 0$.

Theorem 5.12.2: *If $\mathscr{U}$ and $\mathscr{V}$ are isomorphic and if $U_1, U_2, \ldots, U_k$ correspond, respectively, to $V_1, V_2, \ldots, V_k$, then for all scalars $\alpha_1, \alpha_2, \ldots, \alpha_k$,*

$$\alpha_1 U_1 + \alpha_2 U_2 + \cdots + \alpha_k U_k \leftrightarrow \alpha_1 V_1 + \alpha_2 V_2 + \cdots + \alpha_k V_k. \tag{5.12.3}$$

This follows from (5.12.1) and (5.12.2) by induction.

Theorem 5.12.3: *If $\mathscr{U}$ and $\mathscr{V}$ are isomorphic and if $U_1, U_2, \ldots, U_k$ correspond respectively to $V_1, V_2, \ldots, V_k$, then $U_1, U_2, \ldots, U_k$ are linearly dependent (independent) if and only if $V_1, V_2, \ldots, V_k$ are linearly dependent (independent).*

The relation (5.12.3) holds. If either combination is zero with not all the α's zero, then the image combination must be zero also by Theorem 5.12.1. By this same theorem, if either combination cannot be 0 unless all the α's are 0, then neither can the image combination be 0 unless all the α's are 0. The theorem follows.

Theorem 5.12.4: *Isomorphic finite-dimensional vector spaces have the same dimension.*

Let $\mathscr{U}$ and $\mathscr{V}$ be isomorphic and finite dimensional. Let $U_1, U_2, \ldots, U_k$ be a basis for $\mathscr{U}$. Then $U_1, U_2, \ldots, U_k$ are independent and hence, by Theorem 5.12.3, so are their images $V_1, V_2, \ldots, V_k$ in $\mathscr{V}$, so the dimension of $\mathscr{V}$ is at least as great as that of $\mathscr{U}$. Similarly, the dimension of $\mathscr{U}$ is at least as great as that of $\mathscr{V}$. Hence the dimensions are the same.

We have finally

Theorem 5.12.5: *Every vector space $\mathscr{V}$ of dimension n over $\mathscr{F}$ is isomorphic to the space of n-vectors over $\mathscr{F}$.*

Choose a basis $V_1, V_2, \ldots, V_n$ for $\mathscr{V}$ and define the one-to-one correspondence

$$\alpha_1 V_1 + \alpha_2 V_2 + \cdots + \alpha_n V_n \leftrightarrow \begin{bmatrix} \alpha_1 \\ \alpha_2 \\ \vdots \\ \alpha_n \end{bmatrix}.$$

It requires only a simple computation to show that this correspondence preserves both multiplication by a scalar and addition. The theorem follows.

In view of Theorem 5.12.5, to study the linear algebra of $\mathscr{V}$, we need only study the linear algebra of the set of all n-vectors over $\mathscr{F}$. (This is somewhat analogous to our use of synthetic elimination to solve a system of linear equations.) Thus our almost exclusive attention to such n-tuples is fully justified.

5.13 THE GENERAL CONCEPT OF A FIELD

Up to now, we have restricted our attention to vectors having real or complex numbers as components and as scalar multipliers because these are the scalars most used in applications. Because our computations have involved all four of the rational operations, it has been assumed that the components and coefficients belong to a number field, a subfield of the complex number field, in fact.

There are many sets of objects other than numbers which have most of the abstract properties exhibited by number fields. These sets are called simply "fields," and the computations of Chapters 1 and 2 may be executed in the same basic fashion when the scalars belong to one of these other kinds of fields. The same holds true for the techniques of Chapters 5 through 8 except when the real number field is explicitly assumed to be the underlying field of scalars.

In general, a **field** is a set $\mathscr{F}$ of at least two objects with two operations called addition and multiplication such that

A_1: For all a and b belonging to $\mathscr{F}$, $a + b$ belongs to $\mathscr{F}$.

A_2: For all a and b belonging to $\mathscr{F}$, $a + b = b + a$.

A_3: For all a, b, and c belonging to $\mathscr{F}$, $(a + b) + c = a + (b + c)$.

A_4: There exists an element z belonging to $\mathscr{F}$, the **zero element**, such that for all a belonging to $\mathscr{F}$, $a + z = a$.

A_5: For each a belonging to $\mathscr{F}$, there exists an element $-a$, the **negative** of a, such that $a + (-a) = z$.

M_1: For all a, b belonging to $\mathscr{F}$, ab belongs to $\mathscr{F}$.

M_2: For all a, b, and c belonging to $\mathscr{F}$, $(ab)c = a(bc)$.

M_3: There exists an element u belonging to $\mathscr{F}$, the **unit element**, such that for all a belonging to $\mathscr{F}$, $au = ua = a$.

M_4: For each $a \neq z$ belonging to $\mathscr{F}$, there exists an element a^{-1} belonging to $\mathscr{F}$, the **inverse** or **reciprocal** of a, such that $a \cdot a^{-1} = a^{-1} \cdot a = u$.

D_1: For all a, b, and c belonging to $\mathscr{F}$, $a(b + c) = ab + ac$.

D_2: For all a, b, and c belonging to $\mathscr{F}$, $(a + b)c = ac + bc$.

For example, if we define $0 + 0 = 0$, $0 + 1 = 1 + 0 = 1$, $1 + 1 = 0$, and $0 \cdot 0 = 0 \cdot 1 = 1 \cdot 0 = 0$, $1 \cdot 1 = 1$, then the set $\{0, 1\}$ of two elements is a field. It is not hard to verify that all the preceding postulates hold, simply by checking all the possible cases. This finite field is called the **Boolean field**. Although this is a very simple field, it is a very important one in a variety of applications in mathematics, computer science, and electrical engineering.

Another example of a field that is not a number field is the set of expressions of the form $P(x)/Q(x)$ where $P(x)$ and $Q(x)$ are polynomials in a single variable x with coefficients in a number field $\mathscr{F}$ and where $Q(x) \not\equiv 0$. This field, called the **rational function field** over $\mathscr{F}$, is of great importance in advanced aspects of matrix theory.

Even when $\mathscr{F}$ is not a number field, the elements z and u are customarily denoted by 0 and 1, respectively.

The reader with an interest in abstract algebra may enjoy proving that in every field the elements z and u are unique, that a given element a has precisely one negative, and that a given nonzero element a has precisely one reciprocal.

5.14 EXERCISES

1. Show that in $\mathscr{E}^3$ the vector spaces defined by

$$X = t_1\begin{bmatrix}1\\0\\0\end{bmatrix} + t_2\begin{bmatrix}1\\-1\\1\end{bmatrix}, \qquad Y = s_1\begin{bmatrix}0\\0\\1\end{bmatrix} + s_2\begin{bmatrix}1\\-1\\1\end{bmatrix},$$

are isomorphic. Define a convenient one-to-one correspondence between the vectors of the two spaces such that the usual operations are preserved. Illustrate with a figure showing several corresponding pairs of vectors in the two spaces. Sketch it all on one coordinate system.

2. By defining a one-to-one correspondence such that the operations of addition and multiplication are preserved, show that the set of complex numbers $a + bi$ is isomorphic to the set of real matrices

$$\begin{bmatrix}a & -b\\b & a\end{bmatrix}.$$

What matrices correspond to 1 and to i, respectively? What matrices correspond to $c(a + bi)$, where c is real, to $-(a + bi)$, to $(a + bi)^{-1}$, to $\overline{a - bi}$? What additional operations are therefore preserved? What is the matrix equivalent of the equation $\overline{(a + bi)(c + di)} = \overline{a + bi} \cdot \overline{c + di}$? What special properties are exhibited by the matrices of this set that do not hold for real 2×2 matrices in general?

3. Prove that in the Boolean field $\{0, 1\}$, $-1 = 1$. Then solve this system of equations for the x's, the coefficients being presumed to be in the Boolean field:

$$\begin{aligned}
x_1 + x_2 \qquad\qquad\qquad &= y_1\\
x_2 + x_3 \qquad\qquad &= y_2\\
\vdots \qquad\qquad &\\
x_{n-1} + x_n &= y_{n-1}\\
x_n &= y_n.
\end{aligned}$$

CHAPTER 6

The Rank of a Matrix

6.1 THE RANK OF A MATRIX

The **column rank** of an $m \times n$ matrix A over a number field $\mathscr{F}$ is defined to be the dimension of its column space; that is, it is the maximum number of linearly independent columns in A. A simple but important example is I_n, whose columns are linearly independent, so that the column rank is n.

We have already seen in Section 5.6 how this dimension and a basis for the column space may be simultaneously computed in the case of a matrix with constant entries: We simply reduce to column echelon form or to a sufficiently close approximation thereof, using elementary column transformations only. If the entries are not constants, the column rank may vary with the values of those entries. Thus the matrix

$$\begin{bmatrix} 2 & x \\ -1 & 3 \end{bmatrix}$$

has column rank 2 unless $x = -6$. In the latter case, the column rank is 1 since the two columns are proportional, so only one is linearly independent.

The rows of A also span a vector space, the **row space** of A. Its dimension is called the **row rank** of A. The row rank of A and a basis for the row space may be determined by reducing A to echelon or near-echelon form in the usual way, operating on rows only. The remaining nonzero rows are linearly independent [the transpose of the result in (5.2.5)] and their number is the row rank. For example, the row rank of

$$\begin{bmatrix} 1 & 3 & 0 & 1 & -1 \\ 0 & 0 & 1 & 0 & 2 \\ 0 & 0 & 0 & 0 & 0 \end{bmatrix}$$

is two.

When we reduce A to column echelon form, *using elementary column transformations only*, we find both the column rank and a basis for the column space of A. If we want only to know the column rank, and do not need a basis for the column space, we can use row operations also for we have

Theorem 6.1.1: *Elementary row transformations leave the column rank of a matrix invariant and elementary column transformations leave the row rank invariant.*

Let

$$A_{n \times p} = [A_1, A_2, \ldots, A_p].$$

Suppose we interchange rows i and k of $A_{n \times p}$ and denote the resulting matrix by

$$B = [B_1, B_2, \ldots, B_p].$$

Then an equation

$$\sum_{j=1}^{p} \alpha_j A_j = 0 \tag{6.1.1}$$

holds if and only if the scalar equations

$$\sum_{j=1}^{p} \alpha_j a_{qj} = 0, \qquad q = 1, 2, \ldots, n \tag{6.1.2}$$

hold.

But by interchanging equations i and k in (6.1.2), we obtain an equivalent system of equations which is, in fact,

$$\sum_{j=1}^{p} \alpha_j b_{qj} = 0, \qquad q = 1, 2, \ldots, n$$

so (6.1.1) holds if and only if

$$\sum_{j=1}^{p} \alpha_j B_j = 0 \tag{6.1.3}$$

also holds. This implies that if any subset of columns of B is a linearly dependent set, then the corresponding subset of columns of A must be a linearly dependent set and vice versa. [Columns that do not appear in the subset have zero coefficients in equations (6.1.1) and (6.1.3).] Hence linearly *independent* sets must also correspond, so that the column rank is the same for A and B.

Next suppose we multiply the ith row of A by a nonzero constant α and call the resulting matrix

$$C_{n \times p} = [C_1, C_2, \ldots, C_p].$$

If we now multiply the ith equation of (6.1.2) by α, we obtain an equivalent system of equations which is, in fact,

$$\sum_{j=1}^{p} \alpha_j c_{qj} = 0, \qquad q = 1, 2, \ldots, n$$

so (6.1.1) holds if and only if

$$\sum_{j=1}^{p} \alpha_j C_j = 0. \tag{6.1.4}$$

By the same reasoning used in the case of the previous operation, if any columns of C are dependent (independent), the corresponding columns of A must also be dependent (independent) and conversely, so that A and C have the same column rank.

Finally, suppose we add α times the ith row of A to the kth row of A and denote the resulting matrix by

$$G_{n \times p} = [G_1, G_2, \ldots, G_p].$$

Now if we add α times the ith equation of (6.1.1) to the kth equation, we obtain an equivalent system which is, in fact,

$$\sum_{j=1}^{p} \alpha_j g_{qj} = 0, \qquad q = 1, 2, \ldots, n,$$

so (6.1.1) holds if and only if

$$\sum_{j=1}^{p} \alpha_j G_j = 0. \tag{6.1.5}$$

Thus, as before, we conclude that columns of G are linearly dependent (independent) if and only if the corresponding columns of A are dependent (independent), so that the column rank is once again invariant.

In the same way, we can show that the elementary column operations do not change the row rank, and the proof is complete.

If we now apply elementary row transformations appropriately to the column echelon form, we can sweep the nonzero columns with their leading 1's, then rearrange the rows, if necessary, so that the remaining 1's all appear on the main diagonal. The result is a matrix of the form

$$\left[\begin{array}{c|c} I_r & 0 \\ \hline 0 & 0 \end{array}\right] \tag{6.1.6}$$

in which the 0's denote zero matrices, and in which the number of linearly independent rows and the number of linearly independent columns are the same. Since none of the transformations used change either the row rank or the column rank, we have proved

Theorem 6.1.2: *The column rank and the row rank of a matrix A over a number field $\mathscr{F}$ are the same.*

This result makes it proper to refer to the common value of the row rank and the column rank of a matrix A as the **rank** of A. We denote the rank of A by the symbol $r(A)$. The matrix (6.1.6) to which A is reduced is called the **rank normal form** of A. At times some or all of the zero matrices bordering the identity matrix I_r are absent and we have one of the following instead:

$$\text{(6.1.7)} \qquad [I_r \quad 0], \quad \begin{bmatrix} I_r \\ 0 \end{bmatrix}, \quad I_r.$$

Let the symbol "$\sim$" mean "has the same rank and order as." Then an example illustrating the above theorems is the following:

$$\begin{bmatrix} 2 & -1 & 3 \\ 1 & 2 & 0 \end{bmatrix} \sim \begin{bmatrix} 1 & 2 & 0 \\ 2 & -1 & 3 \end{bmatrix} \quad \text{(interchange rows)}$$

$$\sim \begin{bmatrix} 1 & 0 & 0 \\ 2 & -5 & 3 \end{bmatrix} \quad \text{(subtract twice first column from second)}$$

$$\sim \begin{bmatrix} 1 & 0 & 0 \\ 0 & -5 & 3 \end{bmatrix} \quad \text{(subtract twice first row from second)}$$

$$\sim \begin{bmatrix} 1 & 0 & 0 \\ 0 & 1 & 3 \end{bmatrix} \quad \text{(divide second column by } -5\text{)}$$

$$\sim \begin{bmatrix} 1 & 0 & 0 \\ 0 & 1 & 0 \end{bmatrix} \quad \text{(subtract three times second column from third).}$$

The rank is 2 and the final result is the rank normal form.

6.2 BASIC THEOREMS ABOUT THE RANK OF A MATRIX

Suppose an $n \times n$ matrix A over a number field $\mathscr{F}$ has rank n so that the columns of A, as well as the rows of A, are linearly independent. Since the columns of A then form a basis for the set of all n-vectors over $\mathscr{F}$, each vector E_j is a unique linear combination of the columns of A, that is, each equation $AX_j = E_j$ has a unique solution for X_j. Then, since corresponding columns are equal,

$$[AX_1, AX_2, \ldots, AX_n] = [E_1, E_2, \ldots, E_n]$$

or, if

$$X = [X_1, X_2, \ldots, X_n],$$

$$AX = I_n. \tag{6.2.1}$$

The rows of A are the columns of A^{T} and are independent, so that each vector E_j is also a unique linear combination of the columns of A^{T}. That is, each equation $A^{\mathsf{T}} Y_j = E_j$ has a unique solution for Y_j. Then

$$[A^{\mathsf{T}} Y_1, A^{\mathsf{T}} Y_2, \ldots, A^{\mathsf{T}} Y_n] = [E_1, E_2, \ldots, E_n]$$

or, if

$$Y = [Y_1, Y_2, \ldots, Y_n],$$
$$A^{\mathsf{T}} Y = I_n,$$

so that

$$Y^{\mathsf{T}} A = I_n. \tag{6.2.2}$$

Then, from (6.2.1) and (6.2.2), we have

$$Y^{\mathsf{T}} = Y^{\mathsf{T}}(AX) = (Y^{\mathsf{T}}A)X = X,$$

so that (6.2.1) and (6.2.2) combine to yield

$$AX = XA = I,$$

from which $X = A^{-1}$.

Suppose, conversely, that A^{-1} exists and put

$$A^{-1} = [C_1, C_2, \ldots, C_n].$$

From the equation

$$A[C_1, C_2, \ldots, C_n] = I_n = [E_1, E_2, \ldots, E_n]$$

we have then

$$AC_1 = E_1, AC_2 = E_2, \ldots, AC_n = E_n.$$

Thus each of the n linearly independent vectors $E_1, E_2, \ldots, E_n$ is a linear combination of the columns of A, which are therefore independent also, so that A has rank n. We have thus proved

Theorem 6.2.1: *An $n \times n$ matrix A over a number field $\mathscr{F}$ has rank n if and only if A^{-1} exists, that is, if and only if A is nonsingular.*

Now let

$$A_{m \times n} B_{n \times p} = C_{m \times p}.$$

If we write A as a matrix of columns $[A_1, A_2, \ldots, A_n]$, then

$$AB = [(b_{11}A_1 + b_{21}A_2 + \cdots + b_{n1}A_n), \ldots, (b_{1p}A_1 + b_{2p}A_2 + \cdots + b_{np}A_n)],$$

so that the columns of C are linear combinations of the columns of A. Hence the columns of C all belong to the column space of A, and the maximum number of independent columns in C therefore does not exceed the maximum number of independent columns of A. That is, the rank of C cannot exceed the rank of A. Similarly, if we write B as a matrix of rows,

$$\begin{bmatrix} B^{(1)} \\ B^{(2)} \\ \vdots \\ B^{(n)} \end{bmatrix},$$

then

$$AB = \begin{bmatrix} a_{11}B^{(1)} + a_{12}B^{(2)} + \cdots + a_{1n}B^{(n)} \\ \vdots \\ a_{m1}B^{(1)} + a_{m2}B^{(2)} + \cdots + a_{mn}B^{(n)} \end{bmatrix},$$

so the rows of the product C are in the row space of B. Hence the rank of C cannot exceed the rank of B. Combining these results, we have

Theorem 6.2.2: *The rank of the product of two matrices cannot exceed the rank of either factor.*

For example, if A and B both have rank two, we could have

$$\begin{bmatrix} 1 & 0 & 0 & 0 \\ 0 & 0 & 1 & 0 \end{bmatrix} \begin{bmatrix} 0 & 0 \\ 1 & 0 \\ 0 & 0 \\ 0 & 1 \end{bmatrix} = \begin{bmatrix} 0 & 0 \\ 0 & 0 \end{bmatrix},$$

or

$$\begin{bmatrix} 1 & 0 & 0 \\ 0 & 1 & 0 \end{bmatrix} \begin{bmatrix} 1 & 0 \\ 0 & 0 \\ 0 & 1 \end{bmatrix} = \begin{bmatrix} 1 & 0 \\ 0 & 0 \end{bmatrix},$$

or

$$\begin{bmatrix} 1 & 0 & 0 \\ 0 & 1 & 0 \end{bmatrix} \begin{bmatrix} 1 & 0 \\ 0 & 1 \\ 0 & 0 \end{bmatrix} = \begin{bmatrix} 1 & 0 \\ 0 & 1 \end{bmatrix},$$

so the rank of the product may be 0, 1, or 2.

Now suppose that A^{-1} exists and that

$$AB = M, \qquad \text{so that } B = A^{-1}M.$$

Then, by the preceding theorem,

$$r(M) \leq r(B) \qquad \text{and} \qquad r(B) \leq r(M).$$

A similar argument applies when A is the second factor, so that we have

Theorem 6.2.3: *If A is nonsingular, the rank of the product AB is the same as the rank of B, and the rank of CA is the same as the rank of C.*

Next, suppose that

$$(6.2.2) \qquad AB = I_n,$$

where A and B are both $n \times n$. By Theorem 6.2.2, since $r(I_n) = n$, we have $r(A) = r(B) = n$. Then, by Theorem 6.2.1, A^{-1} exists. Multiplying (6.2.1) by A^{-1}, we get $B = A^{-1}$. Similarly, B^{-1} exists and $A = B^{-1}$. In summary, we have

Theorem 6.2.4: *If $A_{n \times n}B_{n \times n} = I_n$, then A and B are nonsingular and $A^{-1} = B$, $B^{-1} = A$.*

We defined B to be an inverse of A if and only if $AB = BA = I$. This theorem says that it suffices to show that $AB = I$ in order to show that B is the inverse of A.

Concerning the rank of the sum of two matrices, we have

Theorem 6.2.5: *If A and B are both $m \times n$ matrices, then*

$$(6.2.3) \qquad |r(A) - r(B)| \leq r(A + B) \leq r(A) + r(B).$$

Indeed, every column of $A + B$ is a linear combination of the set of columns of both A and B. In this set, at most $r(A) + r(B)$ columns are linearly independent. Hence at most this many columns of $A + B$ are linearly independent. That is, $r(A + B) \leq r(A) + r(B)$.

Now let $A + B = C$, so that $A = C + (-B)$. Then, since $r(-B) = r(B)$, what was just proved implies that $r(A) \leq r(C) + r(B)$, so that

$$r(A) - r(B) \leq r(C) = r(A + B).$$

Also, $B = C + (-A)$. Now we have $r(B) \leq r(C) + r(A)$, so that

$$r(B) - r(A) \leq r(C) = r(A + B).$$

Combining the two inequalities just established, we have

$$|r(A) - r(B)| \leq r(A + B),$$

and the proof of the theorem is complete.

6.3 MATRIX REPRESENTATION OF ELEMENTARY TRANSFORMATIONS

It is interesting and useful to represent the elementary row and column transformations in matrix form.

Consider the facts that

$$E_i^{\mathsf{T}}A = A^{(i)}, \qquad AE_j = A_j,$$

where E_i is the ith elementary unit vector, where $A^{(i)}$ denotes the ith row of A, and where A_j denotes its jth column (write out the details in full). From these facts, we see that to interchange two rows (columns) of a matrix A, we may simply interchange the corresponding rows (columns) of a conformable identity matrix and premultiply (postmultiply) A by the result. For example,

$$\begin{bmatrix} 0 & 1 & 0 \\ 1 & 0 & 0 \\ 0 & 0 & 1 \end{bmatrix} \begin{bmatrix} a_1 & a_2 \\ b_1 & b_2 \\ c_1 & c_2 \end{bmatrix} = \begin{bmatrix} b_1 & b_2 \\ a_1 & a_2 \\ c_1 & c_2 \end{bmatrix}$$

and

$$\begin{bmatrix} a_1 & a_2 & a_3 \\ b_1 & b_2 & b_3 \end{bmatrix} \begin{bmatrix} 1 & 0 & 0 \\ 0 & 0 & 1 \\ 0 & 1 & 0 \end{bmatrix} = \begin{bmatrix} a_1 & a_3 & a_2 \\ b_1 & b_3 & b_2 \end{bmatrix}.$$

From the facts

$$(\alpha E_i^{\mathsf{T}})A = \alpha A^{(i)}, \qquad A(\alpha E_i) = \alpha A_i,$$

or from the multiplicative behavior of a diagonal matrix (Section 2.10), we conclude that to multiply the ith row (column) of A by the scalar α, we may multiply the ith row (column) of a conformable identity matrix by α and premultiply (postmultiply) A by the result. For example,

$$\begin{bmatrix} 1 & 0 & 0 \\ 0 & \alpha & 0 \\ 0 & 0 & 1 \end{bmatrix} \begin{bmatrix} a_1 & a_2 \\ b_1 & b_2 \\ c_1 & c_2 \end{bmatrix} = \begin{bmatrix} a_1 & a_2 \\ \alpha b_1 & \alpha b_2 \\ c_1 & c_2 \end{bmatrix}, \qquad \begin{bmatrix} a_1 & a_2 \\ b_1 & b_2 \\ c_1 & c_2 \end{bmatrix} \begin{bmatrix} 1 & 0 \\ 0 & \alpha \end{bmatrix} = \begin{bmatrix} a_1 & \alpha a_2 \\ b_1 & \alpha b_2 \\ c_1 & \alpha c_2 \end{bmatrix}.$$

Finally, we observe that

$$(E_i^{\mathsf{T}} + \alpha E_j^{\mathsf{T}})A = A^{(i)} + \alpha A^{(j)} \qquad \text{and} \qquad A(E_i + \alpha E_j) = A_i + \alpha A_j,$$

from which it follows that to add α times the jth row (column) of A to the ith row (column) of A, we may first do the same to an identity matrix and then premultiply (postmultiply) A by the result. For example,

$$\begin{bmatrix} 1 & \alpha & 0 \\ 0 & 1 & 0 \\ 0 & 0 & 1 \end{bmatrix} \begin{bmatrix} a_1 & a_2 \\ b_1 & b_2 \\ c_1 & c_2 \end{bmatrix} = \begin{bmatrix} a_1 + \alpha b_1 & a_2 + \alpha b_2 \\ b_1 & b_2 \\ c_1 & c_2 \end{bmatrix}$$

and

$$\begin{bmatrix} a_1 & a_2 \\ b_1 & b_2 \\ c_1 & c_2 \end{bmatrix} \begin{bmatrix} 1 & \alpha \\ 0 & 1 \end{bmatrix} = \begin{bmatrix} a_1 & \alpha a_1 + a_2 \\ b_1 & \alpha b_1 + b_2 \\ c_1 & \alpha c_1 + c_2 \end{bmatrix}.$$

In summary, we have

Theorem 6.3.1: *To effect an elementary transformation of a matrix A, first apply the same transformation to a conformable identity matrix, then premultiply A by the result if the operation is on rows, postmultiply if it is on columns.*

The matrices that effect elementary transformations are called **elementary matrices**.

Every elementary transformation has an inverse; that is, there exists a second elementary transformation which precisely undoes the first. Thus if we interchange two parallel lines of A and then interchange them again, the end result is simply A. Hence an elementary transformation of this type is its own inverse. If we multiply a line of A by the scalar $\alpha \neq 0$, we can then return to A by multiplying the same line by $1/\alpha$. If we add α times a line j of A to another, parallel line i of A, we can restore A by adding $-\alpha$ times line j to line i. Thus we have

Theorem 6.3.2: *Every elementary transformation has an inverse which is an elementary transformation of the same kind.*

To get the matrix of an elementary transformation, we perform the transformation on an identity matrix. If we now write the matrix of the *inverse* transformation and apply it appropriately to the preceding elementary matrix, we must get the identity matrix as the product. That is, we have

Theorem 6.3.3: *Every elementary matrix has an inverse which is the matrix of the inverse elementary transformation.*

In view of this result, the inverse of an elementary matrix may be written by inspection. For example,

$$\begin{bmatrix} 0 & 1 & 0 \\ 1 & 0 & 0 \\ 0 & 0 & 1 \end{bmatrix}^{-1} = \begin{bmatrix} 0 & 1 & 0 \\ 1 & 0 & 0 \\ 0 & 0 & 1 \end{bmatrix}, \quad \begin{bmatrix} 1 & 0 & 0 \\ 0 & \frac{1}{2} & 0 \\ 0 & 0 & 1 \end{bmatrix}^{-1} = \begin{bmatrix} 1 & 0 & 0 \\ 0 & 2 & 0 \\ 0 & 0 & 1 \end{bmatrix},$$

$$\begin{bmatrix} 1 & 0 & -3 \\ 0 & 1 & 0 \\ 0 & 0 & 1 \end{bmatrix}^{-1} = \begin{bmatrix} 1 & 0 & 3 \\ 0 & 1 & 0 \\ 0 & 0 & 1 \end{bmatrix}.$$

Now recall the example of reduction to rank normal form given in Section 6.1. We can represent each of the elementary transformations used there in matrix form and represent the reduction as matrix multiplication. That is,

$$\underset{(3)}{\begin{bmatrix} 1 & 0 \\ -2 & 1 \end{bmatrix}} \underset{(1)}{\begin{bmatrix} 0 & 1 \\ 1 & 0 \end{bmatrix}} \begin{bmatrix} 2 & -1 & 3 \\ 1 & 2 & 0 \end{bmatrix} \underset{(2)}{\begin{bmatrix} 1 & -2 & 0 \\ 0 & 1 & 0 \\ 0 & 0 & 1 \end{bmatrix}} \underset{(4)}{\begin{bmatrix} 1 & 0 & 0 \\ 0 & -\frac{1}{5} & 0 \\ 0 & 0 & 1 \end{bmatrix}} \underset{(5)}{\begin{bmatrix} 1 & 0 & 0 \\ 0 & 1 & -3 \\ 0 & 0 & 1 \end{bmatrix}}$$

$$= \begin{bmatrix} 1 & 0 & 0 \\ 0 & 1 & 0 \end{bmatrix}.$$

We can combine the first two factors here, and the last three, to obtain

$$\begin{bmatrix} 0 & 1 \\ 1 & -2 \end{bmatrix}\begin{bmatrix} 2 & -1 & 3 \\ 1 & 2 & 0 \end{bmatrix}\begin{bmatrix} 1 & \frac{2}{5} & -\frac{6}{5} \\ 0 & -\frac{1}{5} & \frac{3}{5} \\ 0 & 0 & 1 \end{bmatrix} = \begin{bmatrix} 1 & 0 & 0 \\ 0 & 1 & 0 \end{bmatrix},$$

in which the first and third factors of the left member, being products of nonsingular matrices, are also nonsingular.

The procedure of the example is fully general. Given a matrix A over a number field $\mathcal{F}$, we can reduce A to rank normal form by row and column transformations using only numbers from $\mathcal{F}$. Next we represent these transformations in matrix form and express the reduced form of A as a matrix product. Finally, by multiplying them together in the correct order, we combine the matrices of the row operations into a single matrix and those of the column operations into another. These products are nonsingular since their factors are all nonsingular. Thus we may conclude

Theorem 6.3.4: *Given a matrix A over a number field $\mathcal{F}$, there exist nonsingular matrices B and C, also over $\mathcal{F}$, such that the product BAC is the rank normal form of A.*

When A is nonsingular, its rank normal form is just the identity matrix of the same order. In this case, there exist elementary matrices $R_1, \ldots, R_h$ and $C_1, \ldots, C_k$ such that

$$R_h \cdots R_1 A C_1 \cdots C_k = I,$$

so

$$A = R_1^{-1} \cdots R_h^{-1} C_k^{-1} \cdots C_1^{-1}.$$

The matrices on the right in this equation are also elementary matrices. Hence we have

Theorem 6.3.5: *A nonsingular matrix over a number field $\mathcal{F}$ can be expressed as a product of elementary matrices over $\mathcal{F}$.*

For example:

$$A = \begin{bmatrix} 1 & 3 & -2 \\ 0 & 1 & -1 \\ 0 & -1 & 0 \end{bmatrix} \sim \begin{bmatrix} 1 & 0 & -2 \\ 0 & 1 & -1 \\ 0 & -1 & 0 \end{bmatrix} \quad \text{(subtract three times first column from second)}$$

$$\sim \begin{bmatrix} 1 & 0 & 0 \\ 0 & 1 & -1 \\ 0 & -1 & 0 \end{bmatrix} \quad \text{(add twice first column to third)}$$

$$\sim \begin{bmatrix} 1 & 0 & 0 \\ 0 & 0 & -1 \\ 0 & -1 & 0 \end{bmatrix} \quad \text{(add third row to second)}$$

$$\sim \begin{bmatrix} 1 & 0 & 0 \\ 0 & -1 & 0 \\ 0 & 0 & -1 \end{bmatrix} \quad \text{(exchange second and third rows)}$$

$$\sim \begin{bmatrix} 1 & 0 & 0 \\ 0 & 1 & 0 \\ 0 & 0 & 1 \end{bmatrix} \quad \text{(multiply second column by } -1\text{, then multiply third column by } -1\text{).}$$

Hence

$$\begin{bmatrix} 1 & 0 & 0 \\ 0 & 0 & 1 \\ 0 & 1 & 0 \end{bmatrix}\begin{bmatrix} 1 & 0 & 0 \\ 0 & 1 & 1 \\ 0 & 0 & 1 \end{bmatrix}\begin{bmatrix} 1 & 3 & -2 \\ 0 & 1 & -1 \\ 0 & -1 & 0 \end{bmatrix}\begin{bmatrix} 1 & -3 & 0 \\ 0 & 1 & 0 \\ 0 & 0 & 1 \end{bmatrix}\begin{bmatrix} 1 & 0 & 2 \\ 0 & 1 & 0 \\ 0 & 0 & 1 \end{bmatrix}$$

$$\times \begin{bmatrix} 1 & 0 & 0 \\ 0 & -1 & 0 \\ 0 & 0 & 1 \end{bmatrix}\begin{bmatrix} 1 & 0 & 0 \\ 0 & 1 & 0 \\ 0 & 0 & -1 \end{bmatrix} = I.$$

Writing inverses of the elementary matrices by inspection and paying careful attention to the order in which factors appear, we obtain the factorization

$$A = \begin{bmatrix} 1 & 0 & 0 \\ 0 & 1 & -1 \\ 0 & 0 & 1 \end{bmatrix}\begin{bmatrix} 1 & 0 & 0 \\ 0 & 0 & 1 \\ 0 & 1 & 0 \end{bmatrix}\begin{bmatrix} 1 & 0 & 0 \\ 0 & 1 & 0 \\ 0 & 0 & -1 \end{bmatrix}\begin{bmatrix} 1 & 0 & 0 \\ 0 & -1 & 0 \\ 0 & 0 & 1 \end{bmatrix}\begin{bmatrix} 1 & 0 & -2 \\ 0 & 1 & 0 \\ 0 & 0 & 1 \end{bmatrix}\begin{bmatrix} 1 & 3 & 0 \\ 0 & 1 & 0 \\ 0 & 0 & 1 \end{bmatrix}.$$

In certain applications, it is essential to be able to express a given nonsingular matrix as a product of elementary matrices. The factorization is not unique, for there is always an unlimited number of ways in which to reduce A to rank normal form.

6.4 EXERCISES

1. Use elementary transformations to reduce

$$\text{(i)} \begin{bmatrix} 1 & 0 & 1 \\ 0 & 2 & 0 \\ 1 & 0 & 1 \end{bmatrix}, \qquad \text{(ii)} \begin{bmatrix} 2 & -1 & 3 & 4 & 0 \\ 1 & 3 & 1 & -2 & 2 \\ 4 & 1 & 0 & 5 & -3 \\ 1 & -1 & -4 & 3 & -5 \end{bmatrix}, \qquad \text{(iii)} \begin{bmatrix} 1 \\ 2 \\ -1 \\ 4 \\ 3 \end{bmatrix}$$

to (a) column-echelon form, (b) row-echelon form, and (c) rank normal form.

2. How does the rank of the matrix

$$\begin{bmatrix} 1 & 1 & t \\ 1 & t & 1 \\ t & 1 & 1 \end{bmatrix}$$

vary with the value of t?

3. Give examples to show that the rank of the sum of two matrices of ranks 2 and 3, respectively, can be 1, 2, 3, 4, or 5.

4. Factor the matrix

$$\begin{bmatrix} 2 & 5 \\ 3 & 1 \end{bmatrix}$$

into a product of elementary matrices.

5. Find nonsingular matrices A and B such that

$$A\begin{bmatrix} 2 & -1 & 4 \\ 3 & 0 & 1 \end{bmatrix}B$$

is in rank normal form.

6. Find the inverse of the product

$$\begin{bmatrix} 0 & 0 & 1 \\ 0 & 1 & 0 \\ 1 & 0 & 0 \end{bmatrix}\begin{bmatrix} 1 & 0 & -2 \\ 0 & 1 & 0 \\ 0 & 0 & 1 \end{bmatrix}\begin{bmatrix} 1 & 0 & 0 \\ 0 & -2 & 0 \\ 0 & 0 & 1 \end{bmatrix}$$

without first computing the product.

7. Factor into a product of elementary matrices, then find the inverse:

$$\text{(a)} \begin{bmatrix} 1 & a & b & c \\ 0 & 1 & 0 & 0 \\ 0 & 0 & 1 & 0 \\ 0 & 0 & 0 & 1 \end{bmatrix}, \qquad \text{(b)} \begin{bmatrix} a & 1 & 0 \\ 1 & 0 & 0 \\ b & c & 2 \end{bmatrix}.$$

8. Write the rank normal forms for matrices of rank 3 and orders (4, 3), (3, 6), (4, 5), and (3, 3), respectively.

9. Determine the ranks of the $n \times n$ matrices

$$\text{(a)} \begin{bmatrix} n-1 & 1 & \cdots & 1 \\ 1 & n-1 & \cdots & 1 \\ \vdots & & & \\ 1 & 1 & \cdots & n-1 \end{bmatrix}, \qquad \text{(b)} \begin{bmatrix} 1-n & 1 & \cdots & 1 \\ 1 & 1-n & \cdots & 1 \\ \vdots & & & \\ 1 & 1 & \cdots & 1-n \end{bmatrix}.$$

10. Show that three points (x_1, x_2), (y_1, y_2), and (z_1, z_2) in the plane are collinear if and only if the rank of the matrix

$$\begin{bmatrix} x_1 & x_2 & 1 \\ y_1 & y_2 & 1 \\ z_1 & z_2 & 1 \end{bmatrix}$$

is less than three.

***11.** Prove, using the concept of rank, that if $AB = I$, where all three matrices are of order n, then A^{-1} and B^{-1} exist.

12. Given that A is a 5×5 matrix, write the elementary matrices that accomplish each of these elementary transformations on A:

(a) adds 2 times row 3 of A to row 5 of A,
(b) multiplies row 2 of A by $\frac{1}{3}$,
(c) interchanges rows 1 and 4 of A.

Then write

(d) the single matrix that does all three operations in one step,
(e) the inverse of the matrix found in (d).

13. Under what conditions on x will the matrix

$$\begin{bmatrix} x & \sqrt{2} & 0 \\ \sqrt{2} & x & \sqrt{2} \\ 0 & \sqrt{2} & x \end{bmatrix}$$

fail to have an inverse? (Examine the rank.)

14. Given the matrix equation

$$AWB + C = DWB,$$

in which all matrices are $n \times n$ and C is nonsingular, what other matrices must have inverses and what is then the solution for W?

15. Which columns of a matrix in row echelon form are certainly independent?

16. Given the vectors

$$\begin{bmatrix} 1 \\ 1 \\ 1 \\ 0 \end{bmatrix}, \begin{bmatrix} 0 \\ -1 \\ 1 \\ 1 \end{bmatrix}, \begin{bmatrix} 2 \\ 1 \\ 0 \\ 0 \end{bmatrix},$$

find a fourth vector such that the resulting set of four vectors constitutes a basis for $\mathscr{E}^4$.

***17.** Name as many different ways as you can to prove that a given set of n real n-vectors, $A_1, A_2, \ldots, A_n$, constitutes a basis for $\mathscr{E}^n$.

18. Compute the rank of

$$\begin{bmatrix} 0 & t & 2t & 1 \\ t & 0 & 1 & t \\ t-1 & t & t+1 & t \\ t & -1 & -1 & 0 \end{bmatrix}$$

as a function of t. Here it pays to be clever rather than systematic in reducing to diagonal form by sweepout.

19. Invent simple examples of matrices that illustrate these relationships:

(a) $r(A + B) = r(A) + r(B)$,
(b) $r(C + D) = r(C) - r(D)$,
(c) $r(FG) = r(F)$ but $< r(G)$,
(d) $r(HK) < r(H)$ and $< r(K)$,
(e) $r(LM) = r(L) = r(M)$,
(f) $r(NP) < r(N)$ but $= r(P)$.

20. Given that

$$A_{n \times m} B_{m \times p} C_{p \times n} = I_n,$$

(a) what can you say about the ranks of A, B, and C and
(b) what can you say about the sizes of the integers m and p?

21. Reduce the symmetric matrix

$$A = \begin{bmatrix} 1 & 2 & 0 \\ 2 & 8 & 6 \\ 0 & 6 & 9 \end{bmatrix}$$

to rank normal form by performing *symmetrical* pairs of elementary operations, that is, by employing exactly the same elementary transformations on rows as on columns. Then represent the result as a product BAC. How are the matrices B and C related?

22. Under what conditions will k points in $\mathscr{R}^n$ determine a $(k - 1)$-flat in $\mathscr{R}^n$? (Start with $\mathscr{R}^2$ and $\mathscr{R}^3$; then generalize.)

23. The identity matrix I_3 can be obtained from a 3×3 matrix A by performing these four operations on A, in the order given: (a) add 3 times row 1 to row 3; (b) exchange columns 2 and 3; (c) multiply row 2 by 4; (d) subtract column 2 from column 1. What is the matrix A?

24. Prove that if $k < n$ and if the $n \times k$ matrix $[A_1, A_2, \ldots, A_k]$ has rank k, then one can always find a vector A_{k+1} such that $[A_1, A_2, \ldots, A_k, A_{k+1}]$ has rank $k + 1$; that is, show that a k-dimensional subspace of the set of all n-vectors over a number field $\mathscr{F}$ can always be **embedded** in a $(k + 1)$-dimensional subspace if $k < n$.

25. Prove that if $k < n$, there always exists a $(k + 1)$-flat in $\mathscr{E}^n$ which contains all the points of a given k-flat in $\mathscr{E}^n$; that is, prove that the k-flat can always be embedded in a $(k + 1)$-flat.

26. Given a k-flat in $\mathscr{E}^n$, $k \leq n - 1$, show that there always exists a hyperplane which contains all the points of the k-flat.

27. Given a k-flat in $\mathscr{E}^n$, $k \leq n - 1$, show that there always exists a hyperplane such that all the points of the k-flat are on the positive side of the hyperplane.

28. Given a set of points $P_0, P_1, \ldots, P_k$, in $\mathscr{E}^n$, show how to find the dimension of the smallest flat space which contains all these points.

6.5 HOMOGENEOUS SYSTEMS OF LINEAR EQUATIONS

Consider the homogeneous system of m linear equations in n unknowns, over a number field $\mathscr{F}$, represented by the equation

$$AX = 0. \tag{6.5.1}$$

Let the rank of A be r. Then, when we transform the system to reduced echelon form, we get an equivalent system of r nonvanishing equations

$$BX = 0. \tag{6.5.2}$$

If $r = n$, $B = I_n$ and the solution is $X = 0$. If $r < n$, the system can be solved for r of the variables in terms of the remaining $n - r$.

Substituting parameters $t_1, t_2, \ldots, t_{n-r}$ for these $n - r$ variables and rearranging in vector notation, as we have done before, we obtain the complete solution in the form

$$X = t_1X_1 + t_2X_2 + \cdots + t_{n-r}X_{n-r}, \tag{6.5.3}$$

where the vectors $X_1, X_2, \ldots, X_{n-r}$ are particular solutions of (6.5.1). The particular solution X_i is obtainable from (6.5.3) by assigning all the t's but t_i the value 0 and assigning t_i the value 1.

In view of how the X_i's are obtained, they are linearly independent, for when we combine the equations

$$\begin{aligned} x_{i_1} &= t_1 \\ x_{i_2} &= t_2 \\ &\vdots \\ x_{i_{n-r}} &= t_{n-r} \end{aligned}$$

which introduce the parameters with the equations obtained from (6.5.2) in order to accomplish the rearrangement (6.5.3), the result is that X_k has a 1 in row i_k and all other X_j's have 0's in row i_k, $k = 1, 2, \ldots, n - r$. By a familiar argument, these isolated 1's guarantee the independence of the X_j's. The next example will help to make this clear.

Combining all these observations, we obtain

Theorem 6.5.1: *The set of all solutions of a system of m homogeneous linear equations in n unknowns,*

$$AX = 0,$$

where A has rank $r < n$, is the set of all linear combinations of $n - r$ linearly independent particular solutions and thus is an $(n - r)$-dimensional vector space. If $r = n$, the solution set is the 0*-dimensional space consisting of the zero vector alone.*

The vector space of all solutions of $AX = 0$ is called the **null space** of the matrix A.

For example, consider

$$\begin{aligned} x_1 - 2x_2 + x_3 + 4x_4 - x_5 &= 0 \\ 2x_1 + x_2 - x_3 + 5x_4 + x_5 &= 0 \\ x_1 + 13x_2 - 8x_3 - 5x_4 + 8x_5 &= 0. \end{aligned}$$

Here the reduced echelon form is

$$\begin{aligned} x_1 \qquad - \tfrac{1}{5}x_3 + \tfrac{14}{5}x_4 + \tfrac{1}{5}x_5 &= 0 \\ x_2 - \tfrac{3}{5}x_3 - \tfrac{3}{5}x_4 + \tfrac{3}{5}x_5 &= 0. \end{aligned}$$

Now we put

$$\begin{aligned} x_3 &= 5t_1 \\ x_4 &= 5t_2 \\ x_5 &= 5t_3 \end{aligned}$$

(the coefficients 5 simplify matters) and get, after rearrangement of the equations,

$$\begin{bmatrix} x_1 \\ x_2 \\ x_3 \\ x_4 \\ x_5 \end{bmatrix} = t_1 \begin{bmatrix} 1 \\ 3 \\ 5 \\ 0 \\ 0 \end{bmatrix} + t_2 \begin{bmatrix} -14 \\ 3 \\ 0 \\ 5 \\ 0 \end{bmatrix} + t_3 \begin{bmatrix} -1 \\ -3 \\ 0 \\ 0 \\ 5 \end{bmatrix}.$$

It is easily checked that we have three particular solutions on the right. The last three components of each of these vectors make clear their linear independence. The complete solution represents a 3-dimensional subspace of $\mathscr{R}^5$.

Now let $V_1, V_2, \ldots, V_{n-r}$ denote *any* $n - r$ linearly independent solutions of (6.5.1). Then, by Theorem 5.5.6, these vectors also constitute a basis for the null space of A, so that every solution of (6.5.1) can be written in the form

$$X = \alpha_1 V_1 + \alpha_2 V_2 + \cdots + \alpha_{n-r} V_{n-r}.$$

That is, we have

Theorem 6.5.2: *The complete solution of a system of homogeneous linear equations $AX = 0$ where A has rank $r < n$ can be represented as the set of all linear combinations of any $n - r$ linearly independent particular solutions.*

To illustrate the use of this theorem, consider the system

$$\begin{aligned} x_1 + x_2 - 4x_3 - 2x_4 &= 0 \\ 2x_1 - 3x_2 + 2x_3 + 2x_4 &= 0. \end{aligned}$$

Note that $r = 2$, $n - r = 2$, so that we need two independent particular solutions. If we put $x_3 = 1$, $x_4 = 0$, we get the equations

$$\begin{aligned} x_1 + x_2 &= 4 \\ 2x_1 - 3x_2 &= -2, \end{aligned}$$

from which $x_1 = x_2 = 2$, so one solution is

$$\begin{bmatrix} 2 \\ 2 \\ 1 \\ 0 \end{bmatrix}.$$

Alternatively, if we put $x_1 = 0$, $x_2 = 2$, we get

$$\begin{aligned} 2x_3 + x_4 &= 1 \\ x_3 + x_4 &= 3, \end{aligned}$$

from which $x_3 = -2$, $x_4 = 5$, so another solution is

$$\begin{bmatrix} 0 \\ 2 \\ -2 \\ 5 \end{bmatrix}.$$

The complete solution is then

$$X = t_1 \begin{bmatrix} 2 \\ 2 \\ 1 \\ 0 \end{bmatrix} + t_2 \begin{bmatrix} 0 \\ 2 \\ -2 \\ 5 \end{bmatrix},$$

since the two particular solutions are linearly independent. By Theorem 6.5.2, this representation is not unique.

6.6 NONHOMOGENEOUS SYSTEMS OF LINEAR EQUATIONS

Consider now a nonhomogeneous system of linear equations represented by

$$AX = B, \tag{6.6.1}$$

where $A_{m \times n}$ and $B_{m \times 1} \neq 0$ are matrices over a number field $\mathscr{F}$. We saw in Chapter 1 how to solve such a system by what amounted to transforming the matrix

$$[A \quad B],$$

which is called the **augmented matrix** of the system, to reduced echelon form:

$$\left[\begin{array}{cccccccccccc|c} 1 & \alpha_{12} & \cdots & 0 & \alpha_{1,i_1+1} & \cdots & 0 & & \alpha_{1k} & \cdots\cdots & \alpha_{1n} & & \beta_1 \\ 0 & \cdots & 0 & 1 & \alpha_{2,i_1+1} & \cdots & 0 & & \alpha_{2k} & \cdots\cdots & \alpha_{2n} & & \beta_2 \\ \vdots & & & & & & & & & & & & \\ 0 & \cdots & \cdots & \cdots & \cdots & \cdots & 0 & 1 & \alpha_{rk} & \cdots\cdots & \alpha_{rn} & & \beta_r \\ \hline 0 & 0 & \cdots & \cdots & \cdots & \cdots & 0 & 0 & 0 & \cdots\cdots & 0 & & \beta_{r+1} \\ \vdots & & & & & & & & & & & & \\ 0 & 0 & \cdots & \cdots & \cdots & \cdots & 0 & 0 & 0 & \cdots\cdots & 0 & & \beta_m \end{array}\right].$$

The system is consistent if and only if $\beta_{r+1} = \cdots = \beta_m = 0$, in which case we can solve for r unknowns in terms of the remaining $n - r$ unknowns.

Since only row operations are used in this reduction to echelon form, so that the submatrix consisting of the first n columns is a reduced form of A, it follows that A has the rank r, where r is the number of rows of this submatrix with leading 1's.

The rank of the entire reduced matrix, including the column of β's, is the same as that of $[A \quad B]$. Hence the rank of the augmented matrix is also r if all of $\beta_{r+1}, \ldots, \beta_m$ are 0. Moreover, it will be r *only* if this is true, for if at least one of this set of β's is unequal to zero, we have *exactly one additional linearly independent row*, so that the rank of the augmented matrix must then be $r + 1$. This observation proves

Theorem 6.6.1: *A system of nonhomogeneous linear equations $AX = B$ is consistent if and only if the coefficient and augmented matrices, A and $[A \quad B]$, have the same rank.*

When the system is consistent, the common rank of A and $[A \quad B]$ is called the **rank of the system**.

Consider the system

$$\begin{aligned} x_1 + x_2 - 3x_3 + x_4 &= 0 \\ 2x_1 - x_2 + x_3 - x_4 &= 1 \\ x_1 + 4x_2 - 10x_3 + 4x_4 &= -1. \end{aligned} \tag{6.6.2}$$

Here the reduced echelon form of the system is

$$\begin{aligned} x_1 \qquad - \tfrac{2}{3}x_3 \qquad &= \tfrac{1}{3} \\ x_2 - \tfrac{7}{3}x_3 + x_4 &= -\tfrac{1}{3}. \end{aligned}$$

Thus $r = 2$, $n - r = 2$. We have, if we put $x_3 = t_1$, $x_4 = t_2$:

$$\begin{aligned} x_1 &= \tfrac{1}{3} + \tfrac{2}{3}t_1 \\ x_2 &= -\tfrac{1}{3} + \tfrac{7}{3}t_1 - t_2 \\ x_3 &= t_1 \\ x_4 &= t_2 \end{aligned}$$

or, in vector notation,

$$X = \begin{bmatrix} \frac{1}{3} \\ -\frac{1}{3} \\ 0 \\ 0 \end{bmatrix} + t_1 \begin{bmatrix} \frac{2}{3} \\ \frac{7}{3} \\ 1 \\ 0 \end{bmatrix} + t_2 \begin{bmatrix} 0 \\ -1 \\ 0 \\ 1 \end{bmatrix}. \tag{6.6.3}$$

Here the first column on the right provides a particular solution of the original system, obtainable by putting $t_1 = t_2 = 0$.

If we now replace the constant terms of the given system of equations by zeros, we get the *corresponding homogeneous system*. Solving this system in the same way, we simply drop the constant terms throughout and get

$$X = t_1\begin{bmatrix} \frac{2}{3} \\ \frac{7}{3} \\ 1 \\ 0 \end{bmatrix} + t_2\begin{bmatrix} 0 \\ -1 \\ 0 \\ 1 \end{bmatrix}, \tag{6.6.4}$$

which shows that the complete solution (6.6.3) is the sum of a particular solution of the nonhomogeneous system (6.6.2) and the complete solution (6.6.4) of the corresponding homogeneous system. This fact is completely general, as is shown by

Theorem 6.6.2: *The complete solution of a nonhomogeneous system $AX = B$ is the sum of any particular solution of the given system and the complete solution of the corresponding homogeneous system $AX = 0$.*

To prove this, let X_0 denote any particular solution of $AX = B$ and let Y denote an arbitrary solution so that $AY = B$ and $AX_0 = B$ are both true equations. Subtracting, we have $A(Y - X_0) = 0$, so that $Y - X_0$ is a solution of the corresponding homogeneous system $AX = 0$. Put $Y - X_0 = W$ so that

$$Y = X_0 + W;$$

that is, any solution Y of $AX = B$ is the sum of the particular solution X_0 of $AX = B$ and some solution W of $AX = 0$.

Now let X_0 denote the same particular solution of $AX = B$, let W denote *any* solution of $AX = 0$, and let

$$Y = X_0 + W.$$

Then, since $AW = 0$,

$$AY = AX_0 + AW = AX_0 = B;$$

that is, Y is a solution of $AX = B$.

The argument shows that if we write

$$Y = X_0 + W,$$

where X_0 is any particular solution of $AX = B$ and W is now the *complete* solution of $AX = 0$, we get *every solution and only solutions* of $AX = B$; that is, we have the complete solution of this equation, as the theorem asserts.

We illustrate with an example. Consider the system

$$\begin{aligned} x_1 + x_2 - x_3 - x_4 &= 2 \\ x_1 - x_2 + x_3 - x_4 &= 0. \end{aligned}$$

Here, if we put $x_3 = x_4 = 0$, we get the equations

$$\begin{aligned} x_1 + x_2 &= 2 \\ x_1 - x_2 &= 0, \end{aligned}$$

so $x_1 = x_2 = 1$. Thus a particular solution is

$$\begin{bmatrix} 1 \\ 1 \\ 0 \\ 0 \end{bmatrix}.$$

When we turn to the corresponding homogeneous system, we can see by inspection that two linearly independent solutions are

$$\begin{bmatrix} 1 \\ 0 \\ 0 \\ 1 \end{bmatrix} \quad \text{and} \quad \begin{bmatrix} 0 \\ 1 \\ 1 \\ 0 \end{bmatrix}.$$

Hence the complete solution of the given system is, by the theorem,

$$X = \begin{bmatrix} 1 \\ 1 \\ 0 \\ 0 \end{bmatrix} + t_1 \begin{bmatrix} 1 \\ 0 \\ 0 \\ 1 \end{bmatrix} + t_2 \begin{bmatrix} 0 \\ 1 \\ 1 \\ 0 \end{bmatrix}. \tag{6.6.5}$$

The geometrical interpretation here is this: The complete solution of the homogeneous system, namely

$$W = t_1 \begin{bmatrix} 1 \\ 0 \\ 0 \\ 1 \end{bmatrix} + t_2 \begin{bmatrix} 0 \\ 1 \\ 1 \\ 0 \end{bmatrix}, \tag{6.6.6}$$

is a two-dimensional vector space representable as a plane on the origin in $\mathscr{E}^4$. The complete solution (6.6.5) of the nonhomogeneous system is representable as a plane parallel to the one on the origin and lying on the point (1, 1, 0, 0). Figure 6-1 gives an intuitive idea of the situation.

It is important here to realize that the vectors given by (6.6.5), that is, those vectors X whose terminal points lie on the plane on (1, 1, 0, 0), *do not* constitute a vector space, whereas those given by (6.6.6) *do* constitute a vector space. The terminal points of these latter vectors lie on the plane on the origin. The first plane is said to be obtained from the second by **translation**. The translation is

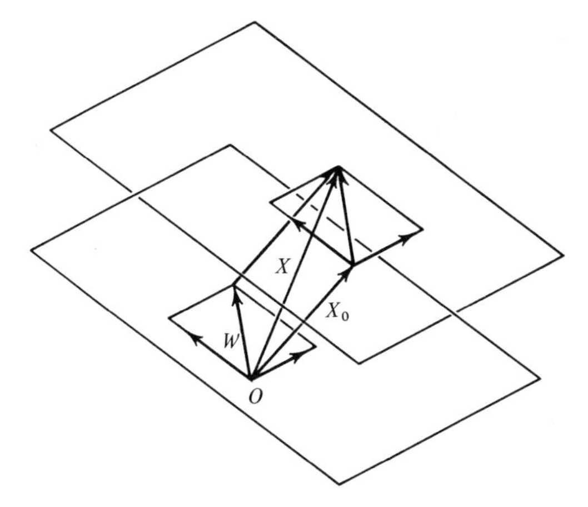

Figure 6-1. *Translation of a Vector Space.*

accomplished by adding the vector X_0 to each vector W of the space represented by the plane on the origin.

6.7 EXERCISES

1. Solve these systems of equations by the method of Theorem 6.6.2:

(a) $\begin{aligned} x_1 + x_2 - x_3 &= 1 \\ 3x_1 - x_2 + 2x_3 &= 0, \end{aligned}$

(b) $\begin{aligned} 2x_1 - x_2 + x_3 &= 4 \\ x_1 + 2x_2 - x_3 &= -1 \\ 7x_1 - 16x_2 + 11x_3 &= 29, \end{aligned}$

(c) $x_1 + x_2 + x_3 + x_4 = 1,$

(d) $\begin{aligned} -3x_1 + x_2 + x_3 + 5x_4 &= 4 \\ 2x_2 - 3x_3 + 5x_4 &= 4 \\ - x_3 + 5x_4 &= 4. \end{aligned}$

2. Find a homogeneous system of linear equations with the complete solution

$$t_1\begin{bmatrix}1\\1\\0\\0\end{bmatrix} + t_2\begin{bmatrix}0\\1\\-1\\1\end{bmatrix}.$$

3. Find a nonhomogeneous system of linear equations with the complete solution

$$\begin{bmatrix}1\\1\\1\\1\end{bmatrix} + t_1\begin{bmatrix}1\\1\\0\\0\end{bmatrix} + t_2\begin{bmatrix}0\\1\\1\\1\end{bmatrix}.$$

***4.** Show that if A is a singular matrix of order n, there always exist nonzero but necessarily singular matrices B of order n such that $AB = 0$.

5. In each case, illustrate with a figure the locus of the terminal points of the vectors of the set:

$$\text{(a)}\quad X = t_1\begin{bmatrix}1\\1\\0\end{bmatrix} + t_2\begin{bmatrix}0\\1\\1\end{bmatrix},$$

$$\text{(b)}\quad X = t_1\begin{bmatrix}1\\1\\0\end{bmatrix} + t_2\begin{bmatrix}0\\1\\1\end{bmatrix}, \quad \text{where } t_1 + t_2 = 1,$$

$$\text{(c)}\quad X = \begin{bmatrix}1\\1\\0\end{bmatrix} + t\begin{bmatrix}-1\\0\\1\end{bmatrix}.$$

***6.** Prove that if the rank of A is r, then there exists in A at least one set of r linearly independent columns and that every column of A can be written as a linear combination of these r columns. (The analogous result holds for rows, of course.)

***7.** Prove that if A has rank r, then every set of $r + 1$ columns of A is linearly dependent.

***8.** Show that if a matrix $A_{m \times n}$ has rank 1, then it can be written as the product of two matrices of rank 1:

$$A = \begin{bmatrix}\alpha_1\\ \alpha_2\\ \vdots\\ \alpha_m\end{bmatrix} \cdot [\beta_1, \beta_2, \ldots, \beta_n].$$

***9.** Show that if A has rank 2, then it can be written as the product of two matrices of rank 2:

$$A = \begin{bmatrix}\alpha_{11} & \alpha_{12}\\ \alpha_{21} & \alpha_{22}\\ \vdots & \\ \alpha_{m1} & \alpha_{m2}\end{bmatrix} \cdot \begin{bmatrix}\beta_{11} & \beta_{12} & \cdots & \beta_{1n}\\ \beta_{21} & \beta_{22} & \cdots & \beta_{2n}\end{bmatrix}.$$

Now generalize.

10. In each case, under what conditions on t will the system of equations be consistent?

$$\text{(a)}\quad \begin{aligned} 3x_1 + x_2 + x_3 &= t\\ x_1 - x_2 + 2x_3 &= 1 - t\\ x_1 + 3x_2 - 3x_3 &= 1 + t, \end{aligned}$$

$$\text{(b)}\quad \begin{aligned} x_1 + x_2 + x_3 + x_4 &= t\\ x_1 - x_2 + x_3 - x_4 &= t - 4\\ x_1 + x_2 - x_3 - x_4 &= t + 1\\ 3x_1 + x_2 + x_3 - x_4 &= 0. \end{aligned}$$

11. For what values of t will the system

$$\text{(a)}\quad \begin{aligned} x + 2y - z &= 0 \\ 2x - 3y + z &= 0 \\ tx + 7y + z &= 0, \end{aligned} \qquad \text{(b)}\quad \begin{aligned} tx_1 + x_2 \phantom{{}- x_3} &= 0 \\ x_1 + tx_2 - x_3 &= 0 \\ - x_2 + tx_3 &= 0 \end{aligned}$$

have a nontrivial solution? Why?

***12.** Show that $AX = 0$ has nontrivial solutions if and only if the columns of A are linearly dependent.

***13.** Show that $AX = B$ is consistent if and only if B is a linear combination of the columns of A.

***14.** Show that $AX = B$ is consistent if and only if every solution of $A^{\mathsf{T}}Y = 0$ is also a solution of $B^{\mathsf{T}}Y = 0$.

15. Show that the three hyperplanes in $\mathscr{E}^4$ with equations

$$\begin{aligned} x_1 - x_2 + x_3 - x_4 &= 0 \\ x_1 + x_2 - x_3 + 2x_4 &= 3 \\ 3x_1 - x_2 + x_3 - 2x_4 &= 1 \end{aligned}$$

intersect in a line. Find two points on the line and write a parametric vector equation for it.

16. Given a system of m equations in four unknowns, discuss in terms of rank and consistency the possible kinds of intersection of the corresponding hyperplanes in $\mathscr{E}^4$.

***17.** Show that the set of all vectors X in $\mathscr{E}^n$ which are orthogonal to each of the vectors $A_1, A_2, \ldots, A_k$ of $\mathscr{E}^n$ is a vector space. Assuming that the vectors $A_1, A_2, \ldots, A_k$ are linearly independent, what is the dimension of this space?

18. An orthogonal basis for $\mathscr{E}^4$ is to include the mutually orthogonal vectors

$$\begin{bmatrix} 2 \\ 1 \\ 0 \\ 0 \end{bmatrix}, \quad \begin{bmatrix} 1 \\ -2 \\ -2 \\ 1 \end{bmatrix}, \quad \begin{bmatrix} 0 \\ 0 \\ 1 \\ 2 \end{bmatrix}.$$

Find a fourth vector for the basis.

19. Prove that precisely one of the following is true:

(a) $A_{m \times n}X_{n \times 1} = B$ is consistent; (b) there exists a vector $Y_{m \times 1}$ such that $Y^{\mathsf{T}}A = 0$ and $Y^{\mathsf{T}}B = 1$. (Hint: Use the result of Problem 14.)

20. Given that $AX = B$ is a consistent system of m real equations in n unknowns, the rank of the system being r, interpret the system and its solution in geometrical language.

21. Given a k-flat in $\mathscr{E}^n$, show how to find a set of hyperplanes whose intersection is precisely the given k-flat.

6.8 THE VARIABLES ONE CAN SOLVE FOR

For the discussion in Chapter 1, it was useful to write a system of linear equations in the form

$$\begin{aligned} f_1 &= 0 \\ f_2 &= 0 \\ &\vdots \\ f_m &= 0. \end{aligned} \tag{6.8.1}$$

With the equations in this form, we say that any subset of the given set of m equations is **linearly dependent (independent)** if and only if the left members of the equations of the subset are linearly dependent (independent) functions.

Thus the equations

$$\begin{aligned} 2x - y + 3 &= 0 \\ x + 3y - 2 &= 0 \\ x + 10y - 9 &= 0 \end{aligned} \quad \text{or} \quad \begin{aligned} 2x - y &= -3 \\ x + 3y &= 2 \\ x + 10y &= 9 \end{aligned} \tag{6.8.2}$$

are linearly dependent equations because

$$1 \cdot (2x - y + 3) + (-3)(x + 3y - 2) + 1 \cdot (x + 10y - 9) \equiv 0.$$

From this identity we see that if we add the first of equations (6.8.2) to the third and subtract three times the second from the third, we get a new third equation which is just $0 = 0$ and may be dropped. The given system of three equations is thus equivalent to a system containing just two equations. This illustrates

Theorem 6.8.1: *If one equation of a linear system of equations is a linear combination of other equations of the system, it may be deleted from the system without altering the set of solutions.*

Indeed, if

$$f_i \equiv \alpha_1 f_{j_1} + \alpha_2 f_{j_2} + \cdots + \alpha_k f_{j_k},$$

then

$$f_i - \alpha_1 f_{j_1} - \alpha_2 f_{j_2} - \cdots - \alpha_k f_{j_k} \equiv 0.$$

Thus the ith equation can be reduced to the form $0 = 0$ by transformations which lead to an equivalent system so that the solution set is not altered.

As the preceding example illustrates, the procedure and the outcome are the same whether the equations are written with the constant terms on the left as in (6.8.1) or on the right as in a system represented by

$$AX = B. \tag{6.8.3}$$

Suppose (6.8.3) is obtained from (6.8.1) by transposition of constant terms and that (6.8.3) is consistent, with rank r. Then there is at least one set of r

linearly independent rows in $[A, B]$, which necessarily correspond to a set of r linearly independent equations in (6.8.1). All rows are linear combinations of the r linearly independent rows. We can therefore reduce all rows other than the r independent ones to zeros, that is, we can eliminate all but the r linearly independent equations by subtracting from each of the dependent ones suitable multiples of the independent ones. This implies, in virtue of the preceding theorem,

Theorem 6.8.2: *If the system of m linear equations in n unknowns represented by $AX = B$ is consistent and of rank r, then any r linearly independent equations of this system form an equivalent system.*

Suppose we want to solve for $x_{i_1}, x_{i_2}, \ldots, x_{i_r}$. Then we can rearrange the reduced system mentioned in the theorem in the form

$$\begin{aligned}
c_{11}x_{i_1} + c_{12}x_{i_2} + \cdots + c_{1r}x_{i_r} &= d_{11}x_{i_{r+1}} + d_{12}x_{i_{r+2}} + \cdots + d_{1,n-r}x_{i_n} + g_1 \\
c_{21}x_{i_1} + c_{22}x_{i_2} + \cdots + c_{2r}x_{i_r} &= d_{21}x_{i_{r+1}} + d_{22}x_{i_{r+2}} + \cdots + d_{2,n-r}x_{i_n} + g_2 \\
&\vdots \\
c_{r1}x_{i_1} + c_{r2}x_{i_2} + \cdots + c_{rr}x_{i_r} &= d_{r1}x_{i_{r+1}} + d_{r2}x_{i_{r+2}} + \cdots + d_{r,n-r}x_{i_n} + g_r
\end{aligned}$$

or

$$CX_1 = DX_2 + G, \tag{6.8.4}$$

where

$$X_1 = \begin{bmatrix} x_{i_1} \\ x_{i_2} \\ \vdots \\ x_{i_r} \end{bmatrix} \quad \text{and} \quad X_2 = \begin{bmatrix} x_{i_{r+1}} \\ x_{i_{r+2}} \\ \vdots \\ x_{i_n} \end{bmatrix}.$$

There exists a unique solution of (6.8.4) for $x_{i_1}, x_{i_2}, \ldots, x_{i_r}$ in terms of the remaining variables if and only if C^{-1} exists, that is, if and only if C has rank r. Now the r rows of C are rows of the $m \times r$ submatrix of A containing the r columns of coefficients of $x_{i_1}, x_{i_2}, \ldots, x_{i_r}$. This submatrix therefore has rank r if C does. Conversely, if the submatrix of coefficients of these variables has rank r, it contains r linearly independent rows. These constitute a nonsingular matrix C, which leads to a solvable equation of the form (6.8.4). We therefore have

Theorem 6.8.3: *If $AX = B$ is consistent and of rank r, we can solve for $x_{i_1}, x_{i_2}, \ldots, x_{i_r}$ in terms of the remaining variables if and only if the $m \times r$ submatrix of A which contains the coefficients of $x_{i_1}, x_{i_2}, \ldots, x_{i_r}$ has rank r. The theorem applies whether or not $B = 0$.*

For example, we can solve

$$\begin{aligned}
x_1 - 2x_2 + 3x_3 - x_4 &= 1 \\
2x_1 - 4x_2 + 5x_3 - 2x_4 &= 0
\end{aligned}$$

for x_1 and x_3, x_2 and x_3, or x_3 and x_4, but *not* for x_1 and x_2, x_1 and x_4, or x_2 and x_4.

6.9 BASIC SOLUTIONS

Assume that the system of r equations

$$AX = B$$

with real coefficients is consistent with rank r. Transforming first to the reduced echelon form, we can then write the solution in the form

$$\begin{aligned} x_{i_1} &= L_1(x_{i_{r+1}}, \ldots, x_{i_n}) + \beta_1 \\ x_{i_2} &= L_2(x_{i_{r+1}}, \ldots, x_{i_n}) + \beta_2 \\ &\vdots \\ x_{i_r} &= L_r(x_{i_{r+1}}, \ldots, x_{i_n}) + \beta_r, \end{aligned}$$

where $L_1, L_2, \ldots, L_r$ are linear functions of the indicated variables. If in these r equations we put

$$x_{i_{r+1}} = x_{i_{r+2}} = \cdots = x_{i_n} = 0,$$

we get a solution

$$\begin{aligned} x_{i_1} &= \beta_1 \\ x_{i_2} &= \beta_2 \\ &\vdots \\ x_{i_r} &= \beta_r \\ x_{i_{r+1}} &= 0 \\ &\vdots \\ x_{i_n} &= 0. \end{aligned}$$

A solution obtained by this procedure is called a **basic solution** of the given system. Such solutions are of importance in linear programming.

We can obtain as many basic solutions as there are sets of r variables for which we can solve in terms of the remaining $n - r$ variables. If we can solve for *every* set of r variables, then there will be

$$C(n, r) = \frac{n!}{r!(n-r)!}$$

basic solutions altogether. Frequently there are not this many basic solutions.

Consider, for example,

$$\begin{aligned} 3x_1 - 2x_2 - x_3 - 4x_4 &= -2 \\ 2x_1 + 5x_2 - x_3 - 4x_4 &= 5 \qquad (r = 2, n - r = 2). \end{aligned}$$

Here we can solve for x_1 and x_2, for x_1 and x_3, for x_1 and x_4, for x_2 and x_3, for x_2 and x_4, but *not* for x_3 and x_4. To get the basic solution corresponding to

x_1 and x_2, it is convenient first to put $x_3 = 0$ and $x_4 = 0$ and then solve the resulting equations,

$$\begin{aligned} 3x_1 - 2x_2 &= -2 \\ 2x_1 + 5x_2 &= 5. \end{aligned}$$

This yields the basic solution (0, 1, 0, 0). The other basic solutions are, respectively, $(-7, 0, -19, 0)$, $(-7, 0, 0, -\frac{19}{4})$, (0, 1, 0, 0), and (0, 1, 0, 0). Notice that a basic solution may have more zero values than those that were assigned. This explains why the basic solutions associated with x_1 and x_2, x_2 and x_3, and x_2 and x_4 are the same. Since there are three ways in which a pair of zeros can be picked from (0, 1, 0, 0), we have three identical basic solutions.

When more than the $n - r$ variables assigned the value 0 assume the value 0 in a basic solution, the solution is called **degenerate**. Otherwise it is **nondegenerate**.

This example suggests a question: If $AX = B$ is a set of r consistent equations and if the rank is r, which sets of r variables $x_{i_1}, x_{i_2}, \ldots, x_{i_r}$ lead to nondegenerate basic solutions?

If such a solution exists, we must first be able to solve for these variables, so there must exist an equation (6.8.4) in which C is nonsingular. Then the columns of C are linearly independent and hence are a basis for $\mathscr{E}^r$. Since $AX = B$ has r equations, the matrix G of (6.8.4) is simply B and $C = [A_{i_1}, A_{i_2}, \ldots, A_{i_r}]$. If we now put $x_{i_{r+1}} = x_{i_{r+2}} = \cdots = x_{i_n} = 0$ in (6.8.4), we get

$$CX_1 = B, \tag{6.9.1}$$

which says that

$$x_{i_1}A_{i_1} + x_{i_2}A_{i_2} + \cdots + x_{i_r}A_{i_r} = B \tag{6.9.2}$$

or, in words, that B is a linear combination of the columns of C. If the solution of (6.9.1) is a nondegenerate basic solution, no x_{i_j}, $j = 1, 2, \ldots, r$, is zero and hence B can replace any one A_{i_j}, $j = 1, 2, \ldots, r$, in a basis for $\mathscr{E}^r$. That is, if a nondegenerate basic solution exists, every set of r of the columns $A_{i_1}, A_{i_2}, \ldots, A_{i_r}, B$, is a linearly independent set.

Conversely, suppose that every such set of r columns is linearly independent. Then, since the A_{i_j}'s are independent but the $r + 1$ r-vectors $A_{i_1}, A_{i_2}, \ldots, A_{i_r}, B$ are necessarily a dependent set, B is a unique linear combination of the A_{i_j}'s; that is, there exist unique values $x_{i_1}, x_{i_2}, \ldots, x_{i_r}$ such that (6.9.2), and therefore (6.9.1), holds. Since any r of $A_{i_1}, A_{i_2}, \ldots, A_{i_r}, B$ are linearly independent, B can replace any of these A_{i_j}'s in a basis for $\mathscr{E}^r$ and hence no x_{i_j}, $j = 1, 2, \ldots, r$, can be 0. That is, we have a nondegenerate basic solution. In summary, we have

Theorem 6.9.1: *The system of r real equations in n unknowns $AX = B$ has a nondegenerate basic solution for $x_{i_1}, x_{i_2}, \ldots, x_{i_r}$ if and only if every set of r of the vectors $A_{i_1}, A_{i_2}, \ldots, A_{i_r}, B$ is a linearly independent set.*

We return to the example given earlier in this section, for which $r = 2$. The basic solution for x_1 and x_2 exists and is degenerate because, although the first two of the columns

$$\begin{bmatrix} 3 \\ 2 \end{bmatrix}, \quad \begin{bmatrix} -2 \\ 5 \end{bmatrix}, \quad \begin{bmatrix} -2 \\ 5 \end{bmatrix}$$

are linearly independent, the last two are not. The basic solution for x_1 and x_3 exists and is nondegenerate, since every pair of the columns

$$\begin{bmatrix} 3 \\ 2 \end{bmatrix}, \quad \begin{bmatrix} -1 \\ -1 \end{bmatrix}, \quad \begin{bmatrix} -2 \\ 5 \end{bmatrix}$$

is linearly independent. A basic solution for x_3 and x_4 does not exist, since the first two of the columns

$$\begin{bmatrix} -1 \\ -1 \end{bmatrix}, \quad \begin{bmatrix} -4 \\ -4 \end{bmatrix}, \quad \begin{bmatrix} -2 \\ 5 \end{bmatrix}$$

are not linearly independent.

6.10 EXERCISES

1. Determine by inspection how many basic solutions actually exist for the system

(a) $$\begin{aligned} x_1 - 2x_2 - 3x_3 + x_4 &= 2 \\ 2x_1 + x_2 - 6x_3 - 2x_4 &= 4, \end{aligned}$$

(b) $$\begin{aligned} x_1 - x_2 + x_3 - x_4 &= 1 \\ x_1 + x_2 + x_3 - x_4 &= 1, \end{aligned}$$

and state which of these are degenerate without finding the solutions.

2. Construct an example of two equations in four unknowns which has no nondegenerate basic solutions.

3. Find all basic solutions of

$$\begin{aligned} x_1 + x_2 - x_3 - 5x_4 &= 10 \\ x_1 - x_2 + x_3 - 2x_4 &= 4. \end{aligned}$$

4. For which sets of variables can one solve the following system? Which sets will lead to nondegenerate basic solutions?

$$\begin{aligned} x_1 + 2x_2 - x_3 - 2x_4 &= 1 \\ 2x_1 - x_2 + 3x_3 + x_4 &= -3 \\ 3x_1 + 2x_2 + 2x_3 - 2x_4 &= -2. \end{aligned}$$

5. Compute the basic solutions for the system of Exercise 4.

6. If you are seeking *only* nondegenerate basic solutions but obtain k additional zeros when you set some $n - r$ variables to zero, which additional combinations of variables, and how many, will you not need to set to zero?

CHAPTER 7

Determinants

7.1 THE DEFINITION OF A DETERMINANT

We saw in Chapter 2 that the inverse of the matrix

$$A = \begin{bmatrix} a_{11} & a_{12} \\ a_{21} & a_{22} \end{bmatrix}$$

exists if and only if $a_{11}a_{22} - a_{12}a_{21} \neq 0$. When the inverse exists, it is given by

$$A^{-1} = \frac{1}{a_{11}a_{22} - a_{12}a_{21}} \begin{bmatrix} a_{22} & -a_{12} \\ -a_{21} & a_{11} \end{bmatrix}. \tag{7.1.1}$$

Thus the existence of the inverse and the formula for the inverse depend on the value of a certain scalar-valued function of the matrix. The value of this function is called the **determinant** of A, abbreviated "det A." It is often denoted by vertical bars instead of brackets on the array:

$$\det A = \begin{vmatrix} a_{11} & a_{12} \\ a_{21} & a_{22} \end{vmatrix} = a_{11}a_{22} - a_{12}a_{21}. \tag{7.1.2}$$

This same function arises in many other connections. For example, the complete solution of a rank 2 system

$$a_{11}x_1 + a_{12}x_2 + a_{13}x_3 = 0$$
$$a_{21}x_1 + a_{22}x_2 + a_{23}x_3 = 0$$

involves three such determinants, each associated with a 2×2 submatrix of the coefficient matrix:

$$\text{(7.1.3)} \qquad \begin{bmatrix} x_1 \\ x_2 \\ x_3 \end{bmatrix} = t\begin{bmatrix} a_{12}a_{23} - a_{13}a_{22} \\ a_{13}a_{21} - a_{11}a_{23} \\ a_{11}a_{22} - a_{12}a_{21} \end{bmatrix} = t\begin{bmatrix} \begin{vmatrix} a_{12} & a_{13} \\ a_{22} & a_{23} \end{vmatrix} \\ -\begin{vmatrix} a_{11} & a_{13} \\ a_{21} & a_{23} \end{vmatrix} \\ \begin{vmatrix} a_{11} & a_{12} \\ a_{21} & a_{22} \end{vmatrix} \end{bmatrix}.$$

There is a simple pattern here which the reader should observe, for it generalizes to $n - 1$ independent equations in n unknowns, as we shall see later.

As another example, let us find the area of the triangle determined by the vectors

$$\begin{bmatrix} x_1 \\ x_2 \end{bmatrix} \quad \text{and} \quad \begin{bmatrix} y_1 \\ y_2 \end{bmatrix}.$$

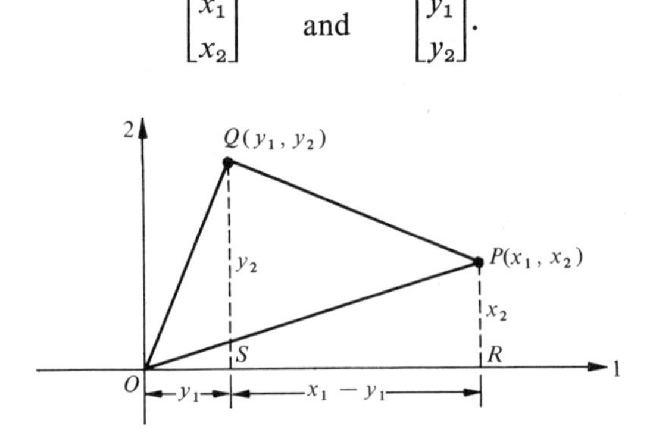

Figure 7-1. *Area of a Triangle.*

of $\mathscr{E}^2$ which terminate, respectively, in the points $P(x_1, x_2)$ and $Q(y_1, y_2)$, thus forming the triangle OPQ. In Figure 7-1, the area K of the triangle OPQ is the area of the triangle OSQ, plus the area of the trapezoid $SRPQ$, minus the area of triangle ORP, that is,

$$K = \tfrac{1}{2}y_1y_2 + \tfrac{1}{2}(x_2 + y_2)(x_1 - y_1) - \tfrac{1}{2}x_1x_2.$$

Expanding and collecting, we have

$$\text{(7.1.4)} \qquad K = \tfrac{1}{2}(x_1y_2 - x_2y_1) = \tfrac{1}{2}\begin{vmatrix} x_1 & y_1 \\ x_2 & y_2 \end{vmatrix}.$$

Here we have assumed in the figure that the directed angle θ from OP to OQ is positive. If it is negative, then the value of K computed from the formula just given will also be negative. (Can you show this?) The formula is valid no matter where P and Q are located in the plane.

These applications of determinants generalize readily to matrices of higher order, to more variables, and to higher dimensions. To accomplish such generalizations, we begin by recalling that a **permutation** of the integers $1, 2, \ldots, n$ is simply a linear arrangement of these integers in some order: $j_1, j_2, \ldots, j_{n-1}, j_n$. We can choose the first integer j_1 of such a permutation in n ways, after which we can choose the second, j_2, as any of the $n - 1$ remaining integers, then j_3 in $n - 2$ ways,..., the integer j_{n-1} in 2 ways, and finally j_n in 1 way, since it must be the one integer that remains. Thus there are $n(n - 1)(n - 2) \cdots 2 \cdot 1 = n!$ such permutations altogether. For example, the $3! = 6$ permutations of 1, 2, 3 are

$$\begin{array}{ccc} 1, 2, 3 & 2, 3, 1 & 3, 1, 2 \\ 1, 3, 2 & 2, 1, 3 & 3, 2, 1. \end{array}$$

Given a permutation π of the integers $1, 2, \ldots, n$, we associate with it a measure, denoted by $\mu(\pi)$, of its departure from the normal order, as follows: For each of the integers $j_1, j_2, \ldots, j_n$, count the number of smaller integers following it in π and add the results. The total is the **parity index** $\mu(\pi)$ of π. If $\mu(\pi)$ is even, we call π an **even permutation**. If $\mu(\pi)$ is odd, π is called an **odd permutation**. For example, if $n = 6$, we have

$$\begin{aligned} \mu(1, 2, 3, 4, 5, 6) &= 0 + 0 + 0 + 0 + 0 + 0 = 0 \\ \mu(1, 2, 3, 4, 6, 5) &= 0 + 0 + 0 + 0 + 1 + 0 = 1 \\ \mu(1, 4, 3, 2, 6, 5) &= 0 + 2 + 1 + 0 + 1 + 0 = 4 \\ \mu(6, 5, 4, 3, 2, 1) &= 5 + 4 + 3 + 2 + 1 + 0 = 15. \end{aligned}$$

The normal order is always an even permutation. Is the reverse order always odd?

Since every time one integer is followed by a smaller one, the smaller integer is preceded by the larger one, it follows that *we can also compute $\mu(\pi)$ by counting the number of larger integers preceding each of $j_1, j_2, \ldots, j_n$ and totaling the results*. This should be checked in the preceding examples.

We now define the **determinant** of a square matrix $A = [a_{ij}]_{n \times n}$, denoted by $\det A$, as follows:

$$(7.1.5) \qquad \det A = \begin{vmatrix} a_{11} & a_{12} & \cdots & a_{1n} \\ a_{21} & a_{22} & \cdots & a_{2n} \\ \vdots & & & \\ a_{n1} & a_{n2} & \cdots & a_{nn} \end{vmatrix} = \sum_{\pi} (-1)^{\mu(\pi)} a_{1j_1} a_{2j_2} \cdots a_{nj_n},$$

where $j_1, j_2, \ldots, j_n$ is a permutation π of $1, 2, \ldots, n$ and the summation extends over all $n!$ permutations π. Note that each of the $n!$ terms in this sum contains a special kind of product of n factors: The product contains one factor from each row of A since each row index appears exactly once and one factor from each column of A since each column index also appears exactly once. Moreover, every product of exactly n factors a_{ij} in which every row and every column is represented exactly once appears in this sum. Indeed, if we arrange the factors so that the row indices are in the natural order, the column indices appear in some permutation π of the natural order, so the product appears in some term of the sum (7.1.5).

The sign prefixed to a term in this sum is positive if π is an even permutation, negative if π is odd.

Because the row indices are kept in the normal order in each term, we call this the **row expansion** of det A.

We have, for example, when $n = 3$,

$$\begin{vmatrix} a_{11} & a_{12} & a_{13} \\ a_{21} & a_{22} & a_{23} \\ a_{31} & a_{32} & a_{33} \end{vmatrix} = \sum_{\pi} (-1)^{\mu(\pi)} a_{1j_1} a_{2j_2} a_{3j_3}$$

$$= a_{11}a_{22}a_{33} + a_{12}a_{23}a_{31} + a_{13}a_{21}a_{32} - a_{11}a_{23}a_{32} - a_{12}a_{21}a_{33} - a_{13}a_{22}a_{31}.$$

A particular case is

$$\begin{vmatrix} 1 & -1 & 2 \\ -1 & 0 & 3 \\ 1 & 4 & -3 \end{vmatrix} = 1 \cdot 0 \cdot (-3) + (-1) \cdot 3 \cdot 1 + 2 \cdot (-1) \cdot 4 - 1 \cdot 3 \cdot 4 - (-1)(-1)(-3) - 2 \cdot 0 \cdot 1$$

$$= 0 - 3 - 8 - 12 + 3 + 0$$

$$= -20.$$

Note that the sign factor $(-1)^{\mu(\pi)}$ appears in each term in addition to the signs carried by the individual entries of the matrix.

7.2 SOME BASIC THEOREMS

Since every row and every column of A has an entry in each term $(-1)^{\mu(\pi)} a_{1j_1} a_{2j_2} \cdots a_{nj_n}$ of det A, we have

Theorem 7.2.1: *If all the elements of one line of a matrix A are zero, then det $A = 0$.*

(Recall that "line" means "row or column.")

There are other useful formulas for det A. One appears in

Theorem 7.2.2: *If A is an $n \times n$ matrix,*

(7.2.1) $$\det A = \sum_{\pi} (-1)^{\mu(\pi)} a_{i_1 1} a_{i_2 2} \cdots a_{i_n n}.$$

That is, if we keep the column subscripts in the normal order in each term and extend the sum over all $n!$ permutations π of the row indices, we also obtain det A. This is called the **column expansion** of det A.

To prove this, note first that again we have precisely all $n!$ products of n factors a_{ij} such that each row and each column is represented exactly once in each product. (Why?) The only question that remains is whether or not a given term has the same sign in both row and column expansions. To see that it does, consider the product $a_{1j_1}a_{2j_2}\cdots a_{nj_n}$. If $j_k = n$, move the factor a_{kj_k} to the far right so that we have

$$a_{1j_1}\cdots a_{k-1,j_{k-1}}a_{k+1,j_{k+1}}\cdots a_{nj_n}a_{kj_k}.$$

Then, among the rearranged column indices, since $j_k = n$, there are precisely $n - k$ smaller integers $j_{k+1},\ldots,j_n$ that no longer follow n, so the parity index of the permutation of column indices is reduced by $n - k$. However, among row indices, the larger integers $k + 1,\ldots,n$ now all precede k, so the parity index of the permutation of row indices is increased by $n - k$.

If now j_m is $n - 1$ and if we move a_{mj_m} to the next-to-the-last position on the right, we again reduce the parity index of the permutation of column indices and increase the parity index of the permutation of row indices by the same amount in each case. Continuing thus, we eventually get a product of the form $a_{i_11}a_{i_22}\cdots a_{i_nn}$ in which the parity index of $i_1, i_2, \ldots, i_n$ is the same as that of the permutation $j_1, j_2,\ldots,j_n$ with which we began. Thus the equal products $a_{i_11}a_{i_22}\cdots a_{i_nn}$ and $a_{1j_1}a_{2j_2}\cdots a_{nj_n}$ will have the same sign factor $(-1)^{\mu(\pi)}$ in the sums (7.2.1) and (7.1.5), respectively. This completes the proof.

The preceding theorem enables us to prove

Theorem 7.2.3: *det A^{T} = det A.*

Since

$$A^{\mathsf{T}} = \begin{bmatrix} a_{11} & a_{21} & \cdots & a_{n1} \\ a_{12} & a_{22} & \cdots & a_{n2} \\ \vdots & & & \\ a_{1n} & a_{2n} & \cdots & a_{nn} \end{bmatrix},$$

the *row* expansion of det A^{T} is $\sum_\pi (-1)^{\mu(\pi)}a_{i_11}a_{i_22}\cdots a_{i_nn}$, which is precisely the *column* expansion of det A, so det A^{T} = det A, as claimed.

This theorem is important because it shows that whatever we prove about rows and/or columns of the determinant of a matrix holds equally well for columns and/or rows since transposing doesn't change the value of the determinant. Thus, when we prove one theorem, we have really proved two. We may state this as

Theorem 7.2.4: *In every theorem about determinants, it is legitimate to interchange the words "row" and "column" throughout.*

A frequently useful result is

Theorem 7.2.5: *The interchange of any two integers of a permutation results in a permutation of opposite parity.*

Consider first the effect of the interchange of two adjacent integers, j_k and j_{k+1}. If $j_k < j_{k+1}$, the value of μ is increased by 1 by the interchange but if $j_k > j_{k+1}$, the value of μ is decreased by 1. In either case the parity is changed from odd to even or from even to odd.

Consider now the interchange of any two integers j_i and j_k, where $i < k$, in the permutation

$$j_1 \cdots j_{i-1}\, j_i\, j_{i+1} \cdots j_{k-1}\, j_k\, j_{k+1} \cdots j_n.$$

We can put j_i between j_k and j_{k+1} by $k - i$ successive interchanges of adjacent integers. Thereafter, we can put j_k in the slot vacated by j_i by $(k - 1) - i$ exchanges of adjacent integers. The total number of such interchanges is $2(k - i) - 1$, which is odd. Thus the net effect on μ is to change it from odd to even or from even to odd, so the theorem is proved. We can now prove

Theorem 7.2.6: *If two parallel lines of a square matrix A are interchanged, the determinant of the resulting matrix B is $-\det A$.*

Suppose we interchange rows i and k of A, where $i < k$, to get B.

The products $a_{1j_1} \cdots a_{ij_i} \cdots a_{kj_k} \cdots a_{nj_n}$ of $\det A$ are in one-to-one correspondence with equal products $a_{1j_1} \cdots a_{kj_k} \cdots a_{ij_i} \cdots a_{nj_n}$ in $\det B$. By Theorem 7.2.5, we then have

$$\begin{aligned}\det B &= \sum (-1)^{\mu(j_1, \ldots, j_k, \ldots, j_i, \ldots j_n)}\, a_{1j_1} \cdots a_{kj_k} \cdots a_{ij_i} \cdots a_{nj_n} \\ &= \sum -(-1)^{\mu(j_1, \ldots, j_i, \ldots, j_k, \ldots, j_n)}\, a_{1j_1} \cdots a_{ij_i} \cdots a_{kj_k} \cdots a_{nj_n} \\ &= -\det A.\end{aligned}$$

The same result applies to columns by Theorem 7.3.4.

For example,

$$\det \begin{bmatrix} 1 & -1 & 2 \\ 0 & 1 & 3 \\ 0 & 0 & 1 \end{bmatrix} = -\det \begin{bmatrix} 2 & -1 & 1 \\ 3 & 1 & 0 \\ 1 & 0 & 0 \end{bmatrix} = 1.$$

This theorem yields next

Theorem 7.2.7: *If two parallel lines of a matrix A are equal, $\det A = 0$.*

For if we interchange the two equal lines, we obviously don't change the determinant. However, by Theorem 7.2.6, we change the sign of the determinant. Thus $\det A = -\det A$, so $\det A = 0$.

For example, it is easy to verify that

$$\det \begin{bmatrix} 1 & 2 & 1 \\ 0 & 5 & 0 \\ 1 & 2 & 1 \end{bmatrix} = 0.$$

Theorem 7.2.8: *If a matrix B results from a matrix A by multiplying all entries in one line of A by k, then $\det B = k \det A$.*

Suppose we multiply the ith row of A by k to get B. Then we have

$$\det B = \sum_{\pi} (-1)^{\mu(\pi)} a_{1j_1} a_{2j_2} \cdots (k a_{ij_i}) a_{i+1, j_{i+1}} \cdots a_{nj_n}.$$

Factoring out the common factor k from each term of this sum, we have

$$\det B = k\Big(\sum_{\pi} (-1)^{\mu(\pi)} a_{1j_1} a_{2j_2} \cdots a_{ij_i} \cdots a_{nj_n}\Big),$$

so

$$\det B = k \det A.$$

This is often a useful result, as the following examples show.

$$\begin{vmatrix} 12 & -24 & 36 \\ 2 & 5 & 9 \\ 4 & -2 & -6 \end{vmatrix} = 12 \begin{vmatrix} 1 & -2 & 3 \\ 2 & 5 & 9 \\ 4 & -2 & -6 \end{vmatrix}$$

$$= 12 \cdot 3 \begin{vmatrix} 1 & -2 & 1 \\ 2 & 5 & 3 \\ 4 & -2 & -2 \end{vmatrix} = 12 \cdot 3 \cdot 2 \begin{vmatrix} 1 & -2 & 1 \\ 2 & 5 & 3 \\ 2 & -1 & -1 \end{vmatrix}.$$

When we remove more than one factor, these factors are multiplied together, as is illustrated by this example.

We can also use the theorem to introduce nonzero factors:

$$\begin{vmatrix} \frac{1}{6} & \frac{1}{3} \\ \frac{1}{2} & 2 \end{vmatrix} = \frac{1}{6} \begin{vmatrix} 6 \cdot \frac{1}{6} & 6 \cdot \frac{1}{3} \\ \frac{1}{2} & 2 \end{vmatrix} = \frac{1}{6} \begin{vmatrix} 1 & 2 \\ \frac{1}{2} & 2 \end{vmatrix}$$

$$= \frac{1}{6} \cdot \frac{1}{2} \begin{vmatrix} 1 & 2 \\ 2 \cdot \frac{1}{2} & 2 \cdot 2 \end{vmatrix} = \frac{1}{12} \begin{vmatrix} 1 & 2 \\ 1 & 4 \end{vmatrix}.$$

That is, we can multiply a line by a factor $k \neq 0$ provided that we pay for it by multiplying the determinant by $1/k$.

Theorem 7.2.9: *If two parallel lines of a matrix A are proportional, then $\det A = 0$.*

Prove this, using Theorems 7.2.8 and 7.2.7.

7.3 THE COFACTOR IN det A OF AN ELEMENT OF A

In the expansion (7.1.5) of det A, we can gather all terms containing a given element a_{ij} of A and extract the common factor a_{ij}. The remaining factor is the **cofactor** A_{ij} of a_{ij}, that is, of the element in the ij-position. For example, when $n = 3$,

$$\begin{vmatrix} a_{11} & a_{12} & a_{13} \\ a_{21} & a_{22} & a_{23} \\ a_{31} & a_{32} & a_{33} \end{vmatrix} = a_{11}(a_{22}a_{33} - a_{23}a_{32}) + a_{12}(a_{23}a_{31} - a_{21}a_{33}) + a_{13}(a_{21}a_{32} - a_{22}a_{31}).$$

The expressions in parentheses are the cofactors of a_{11}, a_{12}, and a_{13}, respectively.

Since every line of A is represented exactly once by a factor in each term in the expansion of det A, the cofactor of a_{ij} contains no other element of row i of A and no other element of column j. The preceding example illustrates this. Because of this fact, we can arrange the terms of det A in disjoint sets, each having one of $a_{i1}, a_{i2}, \ldots, a_{in}$ as a common factor. Extracting these common factors, we obtain

$$\text{(7.3.1)} \qquad \det A = a_{i1}A_{i1} + a_{i2}A_{i2} + \cdots + a_{in}A_{in}.$$

This is called **the expansion of det A in terms of the elements of the ith row**. Operating the same way with the elements of the jth column, we get **the expansion of det A in terms of the elements of the jth column**:

$$\text{(7.3.2)} \qquad \det A = a_{1j}A_{1j} + a_{2j}A_{2j} + \cdots + a_{nj}A_{nj}.$$

These are particularly useful formulas for evaluating determinants, as we shall see. In words, det A is equal to the sum of the products of the elements of any line of A by their cofactors.

Now suppose, for example, that we have a 4×4 matrix B such that

$$\det B = 4A_{31} + 5A_{32} - 6A_{33} + A_{34},$$

where the A_{ij}'s are the cofactors of the elements of the third row of A. This will hold if the third row of B is $[4, 5, -6, 1]$, while the other three rows of B are the same as those of A:

$$B = \begin{bmatrix} a_{11} & a_{12} & a_{13} & a_{14} \\ a_{21} & a_{22} & a_{23} & a_{24} \\ 4 & 5 & -6 & 1 \\ a_{41} & a_{42} & a_{43} & a_{44} \end{bmatrix}.$$

More generally, by the same principle, we have

Theorem 7.3.1: *The expansions*

$$\det A = a_{i1}A_{i1} + a_{i2}A_{i2} + \cdots + a_{in}A_{in}$$

and

$$\det A = a_{1j}A_{1j} + a_{2j}A_{2j} + \cdots + a_{nj}A_{nj}$$

imply that

$$c_1A_{i1} + c_2A_{i2} + \cdots + c_nA_{in}$$

is the determinant of a matrix the same as A except that the ith *row of A has been replaced by the c's and*

$$c_1A_{1j} + c_2A_{2j} + \cdots + c_nA_{nj}$$

is the determinant of a matrix the same as A except that the jth *column has been replaced by the c's.*

For example, if $n = 3$,

$$c_1A_{13} + c_2A_{23} + c_3A_{33} = \det \begin{bmatrix} a_{11} & a_{12} & c_1 \\ a_{21} & a_{22} & c_2 \\ a_{31} & a_{32} & c_3 \end{bmatrix}.$$

This theorem enables us to prove

Theorem 7.3.2: *If $i \neq k$, $j \neq k$, then*

$$a_{i1}A_{k1} + a_{i2}A_{k2} + \cdots + a_{in}A_{kn} = 0 \tag{7.3.3}$$

and

$$a_{1j}A_{1k} + a_{2j}A_{2k} + \cdots + a_{nj}A_{nk} = 0. \tag{7.3.4}$$

In words, *the sum of the products of the elements of one line by the cofactors of the elements of a different parallel line is zero.*

To prove this, note that by Theorem 7.3.1, the expression (7.3.3) is the expansion, in terms of the elements of the kth row, of the determinant of a matrix the same as A except that the kth row has been replaced by the ith row of A, which also appears in its normal place, of course. This matrix has two identical rows and its determinant is therefore zero. A similar argument holds for (7.3.4). For example, if $i = 1$, $k = 2$, $n = 3$, we have

$$a_{11}A_{21} + a_{12}A_{22} + a_{13}A_{23} = \det \begin{bmatrix} a_{11} & a_{12} & a_{13} \\ a_{11} & a_{12} & a_{13} \\ a_{31} & a_{32} & a_{33} \end{bmatrix} = 0.$$

We can use the **Kronecker delta** symbol, defined by

$$\delta_{rs} = \begin{cases} 1 & \text{if } r = s \\ 0 & \text{if } r \neq s \end{cases}$$

to combine (7.3.1) and (7.3.3), (7.3.2) and (7.3.4), as follows:

$$a_{i1}A_{k1} + a_{i2}A_{k2} + \cdots + a_{in}A_{kn} = \delta_{ik} \det A \tag{7.3.5}$$

$$a_{1j}A_{1k} + a_{2j}A_{2k} + \cdots + a_{nj}A_{nk} = \delta_{jk} \det A. \tag{7.3.6}$$

Consider now this example:

$$\begin{aligned}
\begin{vmatrix} b_{11} + b_{12} & a_{12} & a_{13} \\ b_{21} + b_{22} & a_{22} & a_{23} \\ b_{31} + b_{32} & a_{32} & a_{33} \end{vmatrix}
&= (b_{11} + b_{12})A_{11} + (b_{21} + b_{22})A_{21} + (b_{31} + b_{32})A_{31} \\
&= (b_{11}A_{11} + b_{21}A_{21} + b_{31}A_{31}) + (b_{12}A_{11} + b_{22}A_{21} + b_{32}A_{31}) \\
&= \begin{vmatrix} b_{11} & a_{12} & a_{13} \\ b_{21} & a_{22} & a_{23} \\ b_{31} & a_{32} & a_{33} \end{vmatrix} + \begin{vmatrix} b_{12} & a_{12} & a_{13} \\ b_{22} & a_{22} & a_{23} \\ b_{32} & a_{32} & a_{33} \end{vmatrix}
\end{aligned}$$

or, in vector notation,

$$\det [B_1 + B_2, A_2, A_3] = \det [B_1, A_2, A_3] + \det [B_2, A_2, A_3].$$

This last equation suggests the following generalization, which is proved in exactly the manner suggested by the example:

Theorem 7.3.3: *If, in a square matrix A, the column A_i is a sum of columns:*

$$A_i = \sum_{k=1}^{p} B_k,$$

then

$$\det A = \sum_{k=1}^{p} \det [A_1, A_2, \ldots, A_{i-1}, B_k, A_{i+1}, \ldots, A_n].$$

A similar theorem holds for rows. For example,

$$\begin{aligned}
\begin{vmatrix} a_1 + 2b_1 & a_2 + 2b_2 & a_3 + 2b_3 \\ b_1 & b_2 & b_3 \\ c_1 & c_2 & c_3 \end{vmatrix}
&= \begin{vmatrix} a_1 & a_2 & a_3 \\ b_1 & b_2 & b_3 \\ c_1 & c_2 & c_3 \end{vmatrix} + \begin{vmatrix} 2b_1 & 2b_2 & 2b_3 \\ b_1 & b_2 & b_3 \\ c_1 & c_2 & c_3 \end{vmatrix} \\
&= \begin{vmatrix} a_1 & a_2 & a_3 \\ b_1 & b_2 & b_3 \\ c_1 & c_2 & c_3 \end{vmatrix}.
\end{aligned}$$

This example also illustrates

Theorem 7.3.4: *If a matrix B results from adding any multiple of one line of A to another parallel line of A, then $\det B = \det A$.*

By the preceding theorem and Theorem 7.2.8,

$$\begin{aligned}\det\,[A_1,\ldots,A_{i-1},A_i+kA_j,A_{i+1},\ldots,A_n] &= \det\,[A_1,\ldots,A_{i-1},A_i,A_{i+1},\ldots,A_n] \\ &\quad + k\det\,[A_1,\ldots,A_{i-1},A_j,A_{i+1},\ldots,A_n] = \det A,\end{aligned}$$

since the second determinant of the sum is 0 because of the presence of two columns A_j.

Similarly for rows. For example:

$$\begin{vmatrix} 1 & 4 & -2 \\ 2 & 5 & -4 \\ -3 & 1 & 3 \end{vmatrix} = \begin{vmatrix} 1 & 0 & 0 \\ 2 & -3 & 0 \\ -3 & 13 & -3 \end{vmatrix},$$

where the 1 in the 1,1-position has been used to sweep the first row.

7.4 COFACTORS AND THE COMPUTATION OF DETERMINANTS

The concept of a cofactor has yielded a number of useful theorems. It is now time to obtain an explicit formula for the cofactor of an element. First we determine A_{nn}. If we group all terms in $\det A$ which contain a_{nn} and factor out a_{nn}, we get

$$A_{nn}a_{nn} = \left(\sum (-1)^{\mu(j_1,\ldots,j_{n-1},n)} a_{1j_1}a_{2j_2}\cdots a_{n-1,j_{n-1}}\right)a_{nn}$$

where the summation extends over all permutations $j_1, j_2, \ldots, j_{n-1}$ of $1, 2, \ldots, n-1$. Since $\mu(j_1,\ldots,j_{n-1},n) = \mu(j_1,\ldots,j_{n-1})$, the sum in parentheses is the expansion of

$$\begin{vmatrix} a_{11} & a_{12} & \cdots & a_{1,n-1} \\ a_{21} & a_{22} & \cdots & a_{2,n-1} \\ \vdots & & & \\ a_{n-1,1} & a_{n-1,2} & \cdots & a_{n-1,n-1} \end{vmatrix},$$

and this is the formula for A_{nn}.

Now let us partition A so as to isolate a_{ij} and then move a_{ij} to the n,n-position by $n-j$ interchanges of adjacent columns and then by $n-i$ interchanges of adjacent rows, thus:

$$\det A = \begin{vmatrix} B & C & D \\ E & a_{ij} & F \\ G & H & K \end{vmatrix} = (-1)^{n-j} \begin{vmatrix} B & D & C \\ E & F & a_{ij} \\ G & K & H \end{vmatrix}$$

$$= (-1)^{n-i}(-1)^{n-j} \begin{vmatrix} B & D & C \\ G & K & H \\ E & F & a_{ij} \end{vmatrix}.$$

Now $(-1)^{n-i}(-1)^{n-j} = (-1)^{i+j}$. Also, in the last determinant written, the cofactor of a_{ij}, since it is in the n,n-position, is

$$\begin{vmatrix} B & D \\ G & K \end{vmatrix}$$

by the preceding formula for a_{nn}. Hence in det A the cofactor of a_{ij} is

$$(-1)^{i+j} \begin{vmatrix} B & D \\ G & K \end{vmatrix}.$$

Now this last determinant is the determinant of a matrix the same as A except that the ith row and the jth column have been deleted. We therefore have

Theorem 7.4.1: *The cofactor A_{ij} of a_{ij} in the expansion of det A is $(-1)^{i+j}$ times the determinant of the submatrix of A obtained by deleting the* i*th row and the* j*th column of A.*

For example, the cofactor of the element k in

$$\begin{vmatrix} 3 & 4 & -6 \\ 1 & -1 & k \\ a & 1 & 1 \end{vmatrix} \quad \text{is} \quad (-1)^{2+3} \begin{vmatrix} 3 & 4 \\ a & 1 \end{vmatrix}$$

since the element k is in row 2 and column 3. The cofactor of a_{41} in

$$\begin{vmatrix} 0 & 0 & 0 & a_{14} \\ 0 & 0 & a_{23} & 0 \\ 0 & a_{32} & 0 & 0 \\ a_{41} & 0 & 0 & 0 \end{vmatrix} \quad \text{is} \quad (-1)^{4+1} \begin{vmatrix} 0 & 0 & a_{14} \\ 0 & a_{23} & 0 \\ a_{32} & 0 & 0 \end{vmatrix}.$$

Now suppose all elements but a_{ij} in the ith row of A are zero:

$$\begin{vmatrix} a_{11} & \cdots & & & & & a_{1n} \\ \vdots & & & & & & \\ 0 & \cdots & 0 & a_{ij} & 0 & \cdots & 0 \\ \vdots & & & & & & \\ a_{n1} & \cdots & & & & & a_{nn} \end{vmatrix}.$$

Then, expanding along the ith row, we have just

$$\det A = a_{ij}A_{ij}$$

since all the other terms of the expansion are zero. We can often use this observation together with a sweepout process to compute determinants. For example, a useful result concerns triangular matrices:

$$\begin{vmatrix} a_{11} & a_{12} & \cdots & a_{1n} \\ 0 & a_{22} & \cdots & a_{2n} \\ \vdots & & & \\ 0 & 0 & \cdots & a_{nn} \end{vmatrix} = a_{11}\begin{vmatrix} a_{22} & a_{23} & \cdots & a_{2n} \\ 0 & a_{33} & \cdots & a_{3n} \\ \vdots & & & \\ 0 & 0 & \cdots & a_{nn} \end{vmatrix} = \cdots = a_{11}a_{22}\cdots a_{nn}.$$

If the zeros are not there initially, we can often create them. If we apply Theorem 7.3.4 and sweep the first row, we have

$$\begin{vmatrix} 1 & 2 & -1 \\ -2 & 1 & 4 \\ 3 & 0 & 5 \end{vmatrix} = \begin{vmatrix} 1 & 0 & 0 \\ -2 & 5 & 2 \\ 3 & -6 & 8 \end{vmatrix} = 1\cdot(-1)^{1+1}\begin{vmatrix} 5 & 2 \\ -6 & 8 \end{vmatrix} = 40 + 12 = 52.$$

In the next example, by sweeping the first column, we get

$$\begin{vmatrix} 1 & a & a^2 \\ 1 & b & b^2 \\ 1 & c & c^2 \end{vmatrix} = \begin{vmatrix} 0 & a-c & a^2-c^2 \\ 0 & b-c & b^2-c^2 \\ 1 & c & c^2 \end{vmatrix} = 1\cdot(-1)^{3+1}\begin{vmatrix} a-c & a^2-c^2 \\ b-c & b^2-c^2 \end{vmatrix}$$

$$= (a-c)(b-c)\begin{vmatrix} 1 & a+c \\ 1 & b+c \end{vmatrix} = (a-c)(b-c)(b-a).$$

The idea is to use a sweepout procedure to reduce the computation of the determinant of a matrix of order n to the computation of the determinant of a matrix of order $n - 1$, and so on, until the order becomes low enough that the determinant can be computed conveniently.

This reduction can be made fully systematic so that it can be programmed. This is illustrated by the following example.

$$\begin{vmatrix} 4 & 2 & 3 \\ 5 & -2 & 4 \\ -3 & 6 & 8 \end{vmatrix} = 4\begin{vmatrix} 1 & \frac{1}{2} & \frac{3}{4} \\ 5 & -2 & 4 \\ -3 & 6 & 8 \end{vmatrix} \qquad \text{(Create a 1 in the 1,1-position.)}$$

$$= 4\begin{vmatrix} 1 & \frac{1}{2} & \frac{3}{4} \\ 0 & -\frac{9}{2} & \frac{1}{4} \\ 0 & \frac{15}{2} & \frac{41}{4} \end{vmatrix} \qquad \text{(Sweep column 1.)}$$

$$= 4\cdot(-\tfrac{9}{2})\begin{vmatrix} 1 & \frac{1}{2} & \frac{3}{4} \\ 0 & 1 & -\frac{1}{18} \\ 0 & \frac{15}{2} & \frac{41}{4} \end{vmatrix} \qquad \text{(Create a 1 in the 2,2-position.)}$$

$$= 4\cdot(-\tfrac{9}{2})\begin{vmatrix} 1 & \frac{1}{2} & \frac{3}{4} \\ 0 & 1 & -\frac{1}{18} \\ 0 & 0 & \frac{64}{6} \end{vmatrix} \qquad \text{(Sweep column 2 below the 2,2-position.)}$$

$$= 4\cdot(-\tfrac{9}{2})\cdot\tfrac{64}{6}$$

$$= -192.$$

The reader should give a general description of this systematic process.

7.5 EXERCISES

1. Compute:

(a) $\begin{vmatrix} a-b & a+b \\ a+b & a-b \end{vmatrix}$, (b) $\begin{vmatrix} 1 & a & bc \\ 1 & b & ca \\ 1 & c & ab \end{vmatrix}$,

(c) $\begin{vmatrix} 1 & 3 & -2 & 4 \\ 1 & 4 & 3 & 0 \\ 1 & 5 & 5 & 4 \\ 1 & 6 & -2 & -3 \end{vmatrix}$, (d) $\begin{vmatrix} 2 & 1 & -3 & -1 \\ 1 & 4 & -8 & 0 \\ -1 & -1 & 4 & 2 \\ 3 & 2 & 1 & 5 \end{vmatrix}$,

(e) $\begin{vmatrix} 2 & -1 & 1 & 0 \\ -1 & 4 & 0 & 1 \\ 1 & 0 & 2 & -2 \\ 0 & 1 & -2 & 1 \end{vmatrix}$, (f) $\begin{vmatrix} 1 & 1 & 1 & 1 \\ 2 & 4 & 8 & 16 \\ 3 & 9 & 27 & 81 \\ 4 & 16 & 64 & 256 \end{vmatrix}$.

2. Compute:
(a) $\mu(2, 1, 4, 3, 6, 5, 8, 7, \ldots, 2n, 2n-1)$,
(b) $\mu(n, n-1, n-2, \ldots, 3, 2, 1)$.

***3.** Prove that if the ith row of A is a linear combination of the other rows, then det A is zero.

4. Evaluate:

$$\begin{vmatrix} 0 & 0 & 0 & a_{1n} \\ 0 & 0 & a_{2,n-1} & 0 \\ \vdots & & & \\ 0 & a_{n-1,2} & 0 & 0 \\ a_{n1} & 0 & 0 & 0 \end{vmatrix}.$$

Don't jump at a wrong conclusion here!

***5.** Prove that $\det(\lambda A) = \lambda^n \det A$, where A is $n \times n$.

6. Given that

$$A = \begin{bmatrix} 1 & 2 & -1 \\ 4 & 0 & 3 \\ 1 & 2 & 5 \end{bmatrix},$$

compute $[A_{ij}]^{\mathsf{T}}_{3\times 3}$, where A_{ij} denotes the cofactor of a_{ij}. Then compute the product $A \cdot [A_{ij}]^{\mathsf{T}}$.

***7.** If A is any $n \times n$ matrix such that $\det A \neq 0$, compute

$$A \cdot \frac{1}{\det A}[A_{ij}]^{\mathsf{T}}.$$

8. Show that $\det [\delta_{ij}a_{ij}]_{n\times n} = \prod_{j=1}^{n} a_{jj}$, where δ_{ij} is the Kronecker delta.

9. Prove that

$$\begin{vmatrix} 1-n & 1 & 1 & \cdots & 1 \\ 1 & 1-n & 1 & \cdots & 1 \\ 1 & 1 & 1-n & \cdots & 1 \\ \vdots & & & & \\ 1 & 1 & 1 & \cdots & 1-n \end{vmatrix}_{n\times n} = 0.$$

(This is useful in statistical theory, as is the next one.)

10. Prove that

$$\begin{vmatrix} x+\lambda & x & x & \cdots & x \\ x & x+\lambda & x & \cdots & x \\ \vdots & & & & \\ x & x & x & \cdots & x+\lambda \end{vmatrix}_{n\times n} = \lambda^{n-1}(nx+\lambda).$$

11. Expand and interpret geometrically:

$$\begin{vmatrix} x_1 & x_2 & x_3 & 1 \\ 2 & 0 & 0 & 1 \\ 0 & 3 & 0 & 1 \\ 0 & 0 & -4 & 1 \end{vmatrix} = 0.$$

12. Expand by successive splitting of columns and simplify:

$$\begin{vmatrix} a_1+a_2 & a_2+a_3 & a_3+a_1 \\ b_1+b_2 & b_2+b_3 & b_3+b_1 \\ c_1+c_2 & c_2+c_3 & c_3+c_1 \end{vmatrix}.$$

13. Show that elementary transformations can help you to write the expanded form of this determinant in factored form:

$$\begin{vmatrix} 1 & 1 & 1 \\ a & b & c \\ a^3 & b^3 & c^3 \end{vmatrix}.$$

14. Compute, in factored form,

$$\left|\begin{array}{cc:cc} a & b & 0 & 0 \\ c & d & 0 & 0 \\ \hdashline 0 & 0 & \alpha & \beta \\ 0 & 0 & \gamma & \delta \end{array}\right|.$$

Can you state the general rule suggested by this example?

***15** Prove, using mathematical induction, that

$$\det V \equiv \det \begin{bmatrix} x_1^{n-1} & x_1^{n-2} & \cdots & x_1 & 1 \\ x_2^{n-1} & x_2^{n-2} & \cdots & x_2 & 1 \\ \vdots & & & & \\ x_n^{n-1} & x_n^{n-2} & \cdots & x_n & 1 \end{bmatrix} = \prod_{1 \leq i < j \leq n} (x_i - x_j).$$

The matrix V is called a **Vandermonde matrix**. From the expansion it follows that V is nonsingular if and only if all the x_i's are distinct. This is an often useful fact.

7.6 THE DETERMINANT OF THE PRODUCT OF TWO MATRICES

Let

$$D = \text{diag}\,[d_1, d_2, \ldots, d_n]$$

and

$$G = \text{diag}\,[g_1, g_2, \ldots, g_n].$$

Then

$$DG = \text{diag}\,[d_1 g_1, d_2 g_2, \ldots, d_n g_n],$$

so

$$\det DG = d_1 g_1 d_2 g_2 \cdots d_n g_n = d_1 d_2 \cdots d_n \cdot g_1 g_2 \cdots g_n;$$

that is,

$$\det DG = \det D \det G. \tag{7.6.1}$$

This formula actually holds for the product of any two matrices of order n. To prove this for A and B nonsingular, recall that adding any multiple of one line of a matrix to any other parallel line of the matrix does not alter the determinant. Also, one can reduce a nonsingular matrix to a diagonal matrix with the same determinant by using row operations of this kind only, or by column operations of this kind only. For example:

$$\begin{bmatrix} 0 & 1 & 2 \\ 3 & -1 & 4 \\ 1 & 2 & 0 \end{bmatrix} \sim \begin{bmatrix} 1 & 3 & 2 \\ 0 & -7 & 4 \\ 1 & 2 & 0 \end{bmatrix}$$

Add row 3 to row 1; subtract three times row 3 from row 2.

$$\sim \begin{bmatrix} 1 & 3 & 2 \\ 0 & -7 & 4 \\ 0 & -1 & -2 \end{bmatrix}$$

Subtract row 1 from row 3.

$$\sim \begin{bmatrix} 1 & 0 & \frac{26}{7} \\ 0 & -7 & 4 \\ 0 & 0 & -\frac{18}{7} \end{bmatrix}$$

Add $\frac{3}{7}$ times row 2 to row 1; subtract $\frac{1}{7}$ times row 2 from row 3.

$$\sim \begin{bmatrix} 1 & 0 & 0 \\ 0 & -7 & 0 \\ 0 & 0 & -\frac{18}{7} \end{bmatrix}$$

Add $\frac{26}{18}$ times row 3 to row 1; add $\frac{28}{18}$ times row 3 to row 2.

The reduction is thus accomplished, and because of the type of operation used,

$$\det \begin{bmatrix} 0 & 1 & 2 \\ 3 & -1 & 4 \\ 1 & 2 & 0 \end{bmatrix} = \det \begin{bmatrix} 1 & 0 & 0 \\ 0 & -7 & 0 \\ 0 & 0 & -\frac{18}{7} \end{bmatrix} = 18.$$

A similar procedure is possible using the analogous column operations. In no case is the procedure necessarily unique.

This process can be described in a general way. It is left to the reader to do this.

Given two nonsingular matrices A and B, there exist, in view of the preceding observations, matrices $R_1, R_2, \ldots, R_k$ representing row transformations of the type named and matrices $C_1, C_2, \ldots, C_p$ representing column transformations of the type named such that

$$R_k R_{k-1} \cdots R_2 R_1 A = D_A \qquad \text{or} \qquad A = R_1^{-1} R_2^{-1} \cdots R_k^{-1} D_A$$

and

$$B C_1 C_2 \cdots C_p = D_B \qquad \text{or} \qquad B = D_B C_p^{-1} \cdots C_2^{-1} C_1^{-1},$$

where D_A and D_B are diagonal matrices such that

$$\text{(7.6.2)} \qquad \det A = \det D_A \qquad \text{and} \qquad \det B = \det D_B.$$

Hence

$$AB = R_1^{-1} \cdots R_k^{-1} D_A D_B C_p^{-1} \cdots C_1^{-1}$$

and, since the inverse of an elementary transformation is an elementary transformation of the same kind, the matrices R_i^{-1} and C_j^{-1} here do not alter the determinant either, so

$$\begin{aligned} \det (AB) &= \det (D_A D_B) \\ &= \det D_A \det D_B \qquad \text{by (7.6.1)} \end{aligned}$$

or

$$\text{(7.6.3)} \qquad \det (AB) = \det A \det B \qquad \text{by (7.6.2).}$$

If A or B is singular, that is, if $r(A) < n$ or $r(B) < n$, then $r(AB) < n$ also, so that $\det A = 0$ or $\det B = 0$ and $\det (AB) = 0$ by Exercise 3, Section 7.5. Hence, in this case also, $\det (AB) = \det A \det B$.

We have thus proved

Theorem 7.6.1: *The determinant of the product of two square matrices is the product of their determinants.*

This same technique of reduction to diagonal form by transformations that leave the determinant invariant allows us to conclude

Theorem 7.6.2: *If $A_1, A_2, \ldots, A_n$ are square matrices, then*

$$\text{(7.6.4)} \qquad \det \begin{bmatrix} A_1 & 0 & \cdots & 0 \\ 0 & A_2 & \cdots & 0 \\ \vdots & & & \\ 0 & 0 & \cdots & A_n \end{bmatrix} = \det A_1 \det A_2 \cdots \det A_n.$$

In fact, for each i, $i = 1, 2, \ldots, n$, there are elementary transformations that reduce A_i to diagonal form without affecting any other A_j. The effect of all these is to reduce the given block matrix to diagonal form. The determinant is just the product of all these diagonal elements, but the factors of this product can be grouped so that they yield the product $\det A_1 \det A_2 \cdots \det A_n$.

7.7 A FORMULA FOR A^{-1}

We begin with

Theorem 7.7.1: *A^{-1} exists if and only if $\det A \neq 0$.*

Suppose that A^{-1} exists. Then

$$AA^{-1} = I,$$

so that

$$\det A \cdot \det A^{-1} = \det I = 1.$$

Hence

$$\det A \neq 0,$$

as the theorem asserts. Of course we also have

$$\det A^{-1} \neq 0.$$

In fact,

$$\det (A^{-1}) = \frac{1}{\det A} = (\det A)^{-1}. \tag{7.7.1}$$

Note the two uses of the superscript -1: The first denotes the inverse of a matrix, the second the multiplicative inverse (reciprocal) of a scalar.

Now suppose $\det A \neq 0$ and define

$$B = \frac{1}{\det A}\begin{bmatrix} A_{11} & A_{21} & \cdots & A_{n1} \\ A_{12} & A_{22} & \cdots & A_{n2} \\ \vdots & \vdots & & \vdots \\ A_{1n} & A_{2n} & \cdots & A_{nn} \end{bmatrix} = \frac{1}{\det A}[A_{ij}]^{\mathsf{T}}, \tag{7.7.2}$$

where the A_{ij}'s are the cofactors of the a_{ij}'s in $\det A$. Then, with the aid of the identities

$$\begin{aligned} a_{i1}A_{k1} + a_{i2}A_{k2} + \cdots + a_{in}A_{kn} &= \delta_{ik} \det A \\ a_{1j}A_{1k} + a_{2j}A_{2k} + \cdots + a_{nj}A_{nk} &= \delta_{kj} \det A, \end{aligned}$$

it is not hard to prove that

$$AB = BA = I_n.$$

The reader should execute the computation. Thus, when $\det A \neq 0$, the matrix B is the unique inverse of A, and the proof of the theorem is complete.

The formula for A^{-1} that we have just derived is recorded in

Theorem 7.7.2: *If det $A \neq 0$, then*

$$A^{-1} = \frac{1}{\det A} [A_{ij}]^{\mathsf{T}}. \tag{7.7.3}$$

The matrix $[A_{ij}]^{\mathsf{T}}$ is called the **adjoint matrix**, adj A of A and we have

$$A \cdot [A_{ij}]^{\mathsf{T}} = [A_{ij}]^{\mathsf{T}} \cdot A = (\det A)I. \tag{7.7.4}$$

7.8 DETERMINANTS AND THE RANK OF A MATRIX

Given a matrix $A_{m \times n}$ with numerical entries, if $r(A) = k$, then there exists at least one set of k linearly independent rows, but no set of $k + 1$ or more linearly independent rows. In a submatrix consisting of k linearly independent rows of A, there must exist at least one set of k linearly independent columns. These columns then form a $k \times k$ submatrix of A which has an inverse, since it has rank k. Hence its determinant is not zero. Moreover, since no set of $k + 1$ or more rows is linearly independent, no square submatrix of order $k + 1$ or more has a nonzero determinant.

Conversely, if there exists in A a $k \times k$ submatrix M with determinant not zero, but no higher-order square submatrix with determinant not zero, then the rows of M are linearly independent and hence so are the corresponding rows of A. Thus the rank of A is at least k. By the preceding paragraph, it cannot be more than k, since no square submatrix of order higher than k has a nonzero determinant.

We have thus proved

Theorem 7.8.1: *The rank of a matrix A is the order of the highest-order square submatrix of A whose determinant is not zero.*

The sweepout procedure is usually easier than the determinant method of finding the rank of a matrix. However, at times the determinant rule gives a quick answer. For example, regardless of the values of a, b, c, and d, the rank of

$$\begin{bmatrix} 1 & a & b & 0 \\ 0 & c & d & 1 \\ 1 & a & b & 0 \\ 0 & c & d & 1 \end{bmatrix}$$

is 2 because

$$\det \begin{bmatrix} 1 & 0 \\ 0 & 1 \end{bmatrix} \neq 0$$

but the determinant of any third- or higher-order submatrix is zero, since it has at least two identical rows.

Consider now the case of n homogeneous equations in n unknowns, represented by $AX = 0$. We have seen in Chapter 6 that there are nontrivial solutions if and only if the rank of A is less than n. By the preceding theorem, the rank of A is less than n if and only if $\det A = 0$. We may therefore conclude

Theorem 7.8.2: *A system of n homogeneous equations in n unknowns, $AX = 0$ in matrix form, has nontrivial solutions if and only if $\det A = 0$.*

For example, the system

$$\begin{aligned} tx_1 + x_2 + x_3 &= 0 \\ x_1 + tx_2 + x_3 &= 0 \\ x_1 + x_2 + tx_3 &= 0 \end{aligned}$$

has nontrivial solutions if and only if

$$\begin{vmatrix} t & 1 & 1 \\ 1 & t & 1 \\ 1 & 1 & t \end{vmatrix} = 0,$$

that is, if and only if

$$\begin{vmatrix} t-1 & 0 & 1-t \\ 0 & t-1 & 1-t \\ 1 & 1 & t \end{vmatrix} = 0.$$

Factoring $(t - 1)$ from each of the first two rows, we have the equation

$$(t-1)^2 \begin{vmatrix} 1 & 0 & -1 \\ 0 & 1 & -1 \\ 1 & 1 & t \end{vmatrix} = 0,$$

that is, the equation

$$(t-1)^2(t+2) = 0,$$

so nontrivial solutions exist if and only if $t = 1$ or $t = -2$. When $t = 1$, the system has rank 1 so that the solution space is a two-dimensional vector space, but when $t = -2$, the rank is 2, so that the solution space is a one-dimensional vector space. The reader should solve the system in each of these two cases.

The theorem implies that if $AX = 0$ has only the trivial solution, then $\det A \neq 0$, and conversely. Hence, putting together a number of facts and definitions that we have observed at one time or another, we see that all of the following are equivalent for a square matrix of order n:

(7.8.1)
$$\begin{aligned}
&\det A \neq 0,\\
&A^{-1} \text{ exists},\\
&A \text{ is nonsingular},\\
&r(A) = n,\\
&\text{the rows of } A \text{ are linearly independent},\\
&\text{the columns of } A \text{ are linearly independent},\\
&AX = 0 \text{ has only the trivial solution},\\
&AX = Y \text{ has a unique solution for each } Y.
\end{aligned}$$

7.9 SOLUTION OF SYSTEMS OF EQUATIONS BY USING DETERMINANTS

Often one can derive a determinantal formula for the solutions of a system of equations. Such formulas are attractive and are useful in theoretical derivations. In numerical cases, they involve more computation than does the sweep-out process except in special instances, such as a small number of variables or coefficient matrices with many elements zero (**sparse matrices**).

As a first example, consider $n - 1$ homogeneous equations in n unknowns, represented by $AX = 0$, where the rank of A is $n - 1$. Then we have $n - (n - 1) = 1$ linearly independent solutions and all others are multiples of this one. To the $n - 1$ equations add an nth equation with coefficients all zero. Then the enlarged system is equivalent to the original one and has a square coefficient matrix whose determinant is zero. Hence, if $\tilde{A}$ denotes the new coefficient matrix, we have

$$a_{i1}\tilde{A}_{n1} + a_{i2}\tilde{A}_{n2} + \cdots + a_{in}\tilde{A}_{nn} = \delta_{in} \det \tilde{A} = 0, \qquad i = 1, 2, \ldots, n,$$

so $(\tilde{A}_{n1}, \tilde{A}_{n2}, \ldots, \tilde{A}_{nn})$ is a solution. Moreover, since $\tilde{A}$ has rank $n - 1$, *not all* of the cofactors $\tilde{A}_{n1}, \tilde{A}_{n2}, \ldots, \tilde{A}_{nn}$, which are formed from the elements of the first $n - 1$ rows, are zero. That is, the complete solution is

(7.9.1)
$$\begin{bmatrix} x_1 \\ x_2 \\ \vdots \\ x_n \end{bmatrix} = t \begin{bmatrix} \tilde{A}_{n1} \\ \tilde{A}_{n2} \\ \vdots \\ \tilde{A}_{nn} \end{bmatrix}.$$

For example, suppose we need a unit vector in $\mathscr{E}^4$ which is orthogonal to each of the three independent vectors

$$\begin{bmatrix} 1 \\ 1 \\ 0 \\ 1 \end{bmatrix}, \quad \begin{bmatrix} 1 \\ -1 \\ 1 \\ 0 \end{bmatrix}, \quad \begin{bmatrix} 0 \\ 1 \\ 1 \\ -1 \end{bmatrix}.$$

Then what we seek first is a solution to the system of three equations in four unknowns:

$$\begin{aligned} x_1 + x_2 \qquad + x_4 &= 0 \\ x_1 - x_2 + x_3 \qquad &= 0 \\ x_2 + x_3 - x_4 &= 0. \end{aligned}$$

By the formula (7.9.1) we determine the four cofactors of a hypothetical fourth row. These are obtained by deleting successive columns of the coefficient matrix, computing the determinant, and attaching the proper sign:

$$-\begin{vmatrix} 1 & 0 & 1 \\ -1 & 1 & 0 \\ 1 & 1 & -1 \end{vmatrix}, \quad \begin{vmatrix} 1 & 0 & 1 \\ 1 & 1 & 0 \\ 0 & 1 & -1 \end{vmatrix}, \quad -\begin{vmatrix} 1 & 1 & 1 \\ 1 & -1 & 0 \\ 0 & 1 & -1 \end{vmatrix}, \quad \begin{vmatrix} 1 & 1 & 0 \\ 1 & -1 & 1 \\ 0 & 1 & 1 \end{vmatrix},$$

so the complete solution is

$$\begin{bmatrix} x_1 \\ x_2 \\ x_3 \\ x_4 \end{bmatrix} = t \begin{bmatrix} 3 \\ 0 \\ -3 \\ -3 \end{bmatrix}.$$

The corresponding normalized vector is one of these two:

$$\pm \frac{1}{\sqrt{3}} \begin{bmatrix} 1 \\ 0 \\ -1 \\ -1 \end{bmatrix}.$$

Next consider n equations in n unknowns represented by

$$AX = B,$$

where A^{-1} exists. Then $X = A^{-1}B$, or

$$\begin{aligned} X &= \frac{1}{\det A} \begin{bmatrix} A_{11} & A_{21} & \cdots & A_{n1} \\ A_{12} & A_{22} & \cdots & A_{n2} \\ \vdots & & & \\ A_{1n} & A_{2n} & \cdots & A_{nn} \end{bmatrix} \begin{bmatrix} b_1 \\ b_2 \\ \vdots \\ b_n \end{bmatrix} \\ &= \frac{1}{\det A} \begin{bmatrix} (b_1A_{11} + b_2A_{21} + \cdots + b_nA_{n1}) \\ (b_1A_{12} + b_2A_{22} + \cdots + b_nA_{n2}) \\ \vdots \\ (b_1A_{1n} + b_2A_{2n} + \cdots + b_nA_{nn}) \end{bmatrix}. \end{aligned}$$

Thus

$$x_j = \frac{1}{\det A}(b_1A_{1j} + b_2A_{2j} + \cdots + b_nA_{nj}), \qquad j = 1, 2, \ldots, n.$$

The expression in parentheses is the determinant of a matrix the same as A except that the jth column has been replaced by the b's.

We summarize all this in **Cramer's Rule**:

Theorem 7.9.1: *Given a system of n equations in n unknowns,*

$$AX = B,$$

such that A^{-1} exists, then the unique solution is given in determinant form by

$$(7.9.2)\quad x_j = \frac{\det [A_1,\ldots, A_{j-1}, B, A_{j+1},\ldots, A_n]}{\det A}, \qquad j = 1, 2,\ldots, n.$$

For example, to solve

$$\begin{aligned} (k-1)x_1 + \quad k \quad x_2 &= \quad 1 \\ k \quad x_1 + (k-1)x_2 &= -1, \end{aligned}$$

we first compute the determinant of the coefficient matrix:

$$\begin{vmatrix} k-1 & k \\ k & k-1 \end{vmatrix} = (k-1)^2 - k^2 = 1 - 2k.$$

Then, if $1 - 2k \neq 0$, that is, if $k \neq \frac{1}{2}$,

$$x_1 = \frac{\begin{vmatrix} 1 & k \\ -1 & k-1 \end{vmatrix}}{1-2k} = \frac{2k-1}{1-2k} = -1, \qquad x_2 = \frac{\begin{vmatrix} k-1 & 1 \\ k & -1 \end{vmatrix}}{1-2k} = \frac{1-2k}{1-2k} = 1.$$

Thus every system of this one-parameter family of systems of equations, except the one for which $k = \frac{1}{2}$, has the solution $(-1, 1)$.

As another example, consider the system

$$\begin{aligned} x_1 - a_{12}x_2 \qquad\qquad &= b_1 \\ x_2 - a_{23}x_3 &= b_2 \\ -a_{31}x_1 \qquad\qquad + x_3 &= b_3. \end{aligned}$$

Here

$$\det A = \begin{vmatrix} 1 & -a_{12} & 0 \\ 0 & 1 & -a_{23} \\ -a_{31} & 0 & 1 \end{vmatrix} = 1 - a_{12}a_{23}a_{31},$$

$$x_1 = \frac{1}{\det A}\begin{vmatrix} b_1 & -a_{12} & 0 \\ b_2 & 1 & -a_{23} \\ b_3 & 0 & 1 \end{vmatrix} = \frac{1}{\det A}(b_1 + b_2a_{12} + b_3a_{12}a_{23}),$$

$$x_2 = \frac{1}{\det A}\begin{vmatrix} 1 & b_1 & 0 \\ 0 & b_2 & -a_{23} \\ -a_{31} & b_3 & 1 \end{vmatrix} = \frac{1}{\det A}(b_2 + b_3a_{23} + b_1a_{23}a_{31}).$$

$$x_3 = \frac{1}{\det A}\begin{vmatrix} 1 & -a_{12} & b_1 \\ 0 & 1 & b_2 \\ -a_{31} & 0 & b_3 \end{vmatrix} = \frac{1}{\det A}(b_3 + b_1a_{31} + b_2a_{31}a_{12}).$$

There is a fairly simple pattern here, one which suggests that there is a simple solution to a similar system of n equations in n unknowns. Can you establish the general rule?

7.10 A GEOMETRICAL APPLICATION OF DETERMINANTS

Recall Theorem 7.8.2, which says that a homogeneous system of n linear equations in n unknowns has nontrivial solutions if and only if the determinant of the coefficients is zero.

This theorem is often used in geometrical applications. For example, suppose that (a_1, a_2), (b_1, b_2) are distinct points so that they determine a line in $\mathscr{E}^2$. Then (x_1, x_2) is a point on this line if and only if the three equations

$$\begin{aligned} Ax_1 + Bx_2 + C &= 0 \\ Aa_1 + Ba_2 + C &= 0 \\ Ab_1 + Bb_2 + C &= 0 \end{aligned}$$

have a nontrivial solution for the coefficients A, B, C, that is, by Theorem 7.8.2, if and only if

$$\begin{vmatrix} x_1 & x_2 & 1 \\ a_1 & a_2 & 1 \\ b_1 & b_2 & 1 \end{vmatrix} = 0.$$

This linear equation in x_1 and x_2 is satisfied by (a_1, a_2) and by (b_1, b_2), so it is an equation of the line on these two points.

We saw in Chapter 4 that n points not in the same $(n - 2)$-flat determine a hyperplane. The equation of such a hyperplane is readily written down in determinant form. Indeed, generalizing from the preceding example, we write the equation

$$\begin{vmatrix} x_1 & x_2 & \cdots & x_n & 1 \\ a_{11} & a_{12} & \cdots & a_{1n} & 1 \\ a_{21} & a_{22} & \cdots & a_{2n} & 1 \\ \vdots & & & & \\ a_{n1} & a_{n2} & \cdots & a_{nn} & 1 \end{vmatrix} = 0, \tag{7.10.1}$$

which is satisfied by the coordinates of each of the n points

$$(a_{11}, a_{12}, \ldots, a_{1n}), (a_{21}, a_{22}, \ldots, a_{2n}), \ldots, (a_{n1}, a_{n2}, \ldots, a_{nn}),$$

since substitution results in two equal rows in each case. Thus the locus lies on the points in question. If we expand along the first row, we see that the equation is linear. Thus it represents a hyperplane unless the coefficients of *all* the x_j's are zero. In this event, the n given points all lie in a k-flat, $k < n - 1$. Consequently, given n points, we write the above determinant and expand. If a proper linear equation results, it is an equation of the hyperplane on those points. If no proper linear equation results, we know the points do not determine a hyperplane but rather lie in a space of dimension smaller than $n - 1$.

For example, the equation of the line in $\mathscr{E}^2$ on (2, 4) and (4, 2) is

$$\begin{vmatrix} x_1 & x_2 & 1 \\ 2 & 4 & 1 \\ 4 & 2 & 1 \end{vmatrix} = 0 \qquad \text{or} \qquad x_1 + x_2 = 6.$$

Again, suppose we want the x_3-intercept of the plane in $\mathscr{E}^3$ which lies on $(a, 0, 0)$, $(0, b, 0)$, and $(1, 1, 1)$. In the equation of the plane, namely

$$\begin{vmatrix} x_1 & x_2 & x_3 & 1 \\ a & 0 & 0 & 1 \\ 0 & b & 0 & 1 \\ 1 & 1 & 1 & 1 \end{vmatrix} = 0,$$

we put $x_1 = x_2 = 0$ and solve for x_3. We can do this neatly by expanding the determinant in the equation

$$\begin{vmatrix} 0 & 0 & x_3 & 1 \\ a & 0 & 0 & 1 \\ 0 & b & 0 & 1 \\ 1 & 1 & 1 & 1 \end{vmatrix} = 0$$

along the first row:

$$x_3 \begin{vmatrix} a & 0 & 1 \\ 0 & b & 1 \\ 1 & 1 & 1 \end{vmatrix} - \begin{vmatrix} a & 0 & 0 \\ 0 & b & 0 \\ 1 & 1 & 1 \end{vmatrix} = 0.$$

From this,

$$x_3 = \frac{ab}{ab - a - b},$$

provided, of course, that $ab \neq a + b$.

7.11 EXERCISES

1. For what values of t will the following matrix have no inverse? For all other values of t, what is its inverse?

$$\begin{bmatrix} 1 & t & 0 \\ 0 & 1 & -1 \\ t & 0 & 1 \end{bmatrix}.$$

2. What is the determinant of this product of matrices?

$$\begin{bmatrix} 1 & 0 & 0 \\ 0 & 2 & 0 \\ 0 & 0 & 3 \end{bmatrix} \begin{bmatrix} 2 & 0 & 1 \\ 0 & 1 & 0 \\ 1 & 0 & 1 \end{bmatrix} \begin{bmatrix} 1 & 1 & \frac{1}{3} \\ 1 & \frac{1}{2} & 0 \\ 1 & 0 & 0 \end{bmatrix}.$$

3. Given that

$$\begin{bmatrix} 1 & -2 & 0 \\ 2 & 1 & 0 \\ 0 & 0 & 1 \end{bmatrix} A \begin{bmatrix} 1 & 2 & 0 \\ -2 & 1 & 0 \\ 0 & 0 & 1 \end{bmatrix} = 5I,$$

what is $\det A$?

4. Given that $ABA = I$, show that A^{-1} and B^{-1} exist. What does $(A^2B)^{-1}$ reduce to in this case?

5. Compute the determinants of these matrices:

$$\text{(a)} \left[\begin{array}{cc|ccc} 0 & 1 & 0 & 0 & 0 \\ 1 & 0 & 0 & 0 & 0 \\ \hline a & b & 0 & 1 & 2 \\ c & d & 2 & 0 & 1 \\ e & f & 1 & 2 & 0 \end{array}\right], \qquad \text{(b)} \left[\begin{array}{c|cc|ccc} 0 & 0 & 0 & 3 & 2 & 1 \\ 0 & 0 & 0 & 2 & 1 & 0 \\ 0 & 0 & 0 & 1 & 0 & -1 \\ \hline 0 & 2 & 1 & 0 & 0 & 0 \\ 0 & 1 & 2 & 0 & 0 & 0 \\ \hline 4 & 0 & 0 & 0 & 0 & 0 \end{array}\right],$$

$$\text{(c)} \left[\begin{array}{cc|cc|cc} 0 & 0 & 0 & 0 & 1 & 2 \\ 0 & 0 & 0 & 0 & 2 & 1 \\ \hline 0 & 0 & 1 & 2 & 0 & 0 \\ 0 & 0 & 2 & 1 & 0 & 0 \\ \hline a & b & c & d & e & f \\ \alpha & \beta & \gamma & \delta & g & h \end{array}\right].$$

6. Given that $a^3 + b^3 = 1$, find by determinants the inverse of

$$\begin{bmatrix} a & b & 0 \\ 0 & a & b \\ b & 0 & a \end{bmatrix}.$$

7. Use the determinantal formula to compute A^{-1}, where

$$A = \begin{bmatrix} 0 & 0 & 1 & 1 \\ 0 & 0 & 1 & 0 \\ 1 & 1 & 0 & 0 \\ 1 & 0 & 0 & 0 \end{bmatrix}.$$

***8.** Give an example to show that $\det(A + B) \neq \det A + \det B$ in general. Then find a special case where the equality does hold.

9. Given that

$$A_j = \begin{bmatrix} a_j & d & g \\ b_j & e & h \\ c_j & f & k \end{bmatrix}, \qquad j = 1, 2, \ldots, p,$$

prove that

$$\det\left[\sum_{j=1}^{p} A_j\right] = p^2 \sum_{j=1}^{p} (\det A_j).$$

10. Find the equation of the plane in $\mathscr{E}^3$ on the points (0, 0, 0), (2, 1, 1), and (−2, 1, 0).

11. Interpret geometrically:

(a) $$\begin{vmatrix} x_1 - a_{n1} & x_2 - a_{n2} & \cdots & x_n - a_{nn} \\ a_{11} - a_{n1} & a_{12} - a_{n2} & \cdots & a_{1n} - a_{nn} \\ \vdots & & & \\ a_{n-1,1} - a_{n1} & a_{n-1,2} - a_{n2} & \cdots & a_{n-1,n} - a_{nn} \end{vmatrix} = 0,$$

(b) $$\begin{vmatrix} x_1 & x_2 & \cdots & x_n \\ a_{11} & a_{12} & \cdots & a_{1n} \\ \vdots & & & \\ a_{n-1,1} & a_{n-1,2} & \cdots & a_{n-1,n} \end{vmatrix} = 0.$$

12. Interpret geometrically:

$$\begin{vmatrix} x_1^2 + x_2^2 & x_1 & x_2 & 1 \\ a_1^2 + a_2^2 & a_1 & a_2 & 1 \\ b_1^2 + b_2^2 & b_1 & b_2 & 1 \\ c_1^2 + c_2^2 & c_1 & c_2 & 1 \end{vmatrix} = 0.$$

13. Show that no nonzero vector is orthogonal to all these vectors:

$$\begin{bmatrix} 1 \\ 1 \\ 0 \end{bmatrix}, \quad \begin{bmatrix} 1 \\ 0 \\ 1 \end{bmatrix}, \quad \begin{bmatrix} 0 \\ 1 \\ 1 \end{bmatrix}.$$

***14.** Show that

$$\det \begin{bmatrix} x_1 & x_2 & x_3 \\ y_1 & y_2 & y_3 \end{bmatrix} \begin{bmatrix} x_1 & y_1 \\ x_2 & y_2 \\ x_3 & y_3 \end{bmatrix} = \begin{vmatrix} x_1 & x_2 \\ y_1 & y_2 \end{vmatrix}^2 + \begin{vmatrix} x_1 & x_3 \\ y_1 & y_3 \end{vmatrix}^2 + \begin{vmatrix} x_2 & x_3 \\ y_2 & y_3 \end{vmatrix}^2$$

$$= X^{\mathsf{T}}X \cdot Y^{\mathsf{T}}Y - (X^{\mathsf{T}}Y)^2$$

and hence prove the Cauchy–Schwarz inequality for $\mathscr{E}^3$. Then generalize to $\mathscr{E}^n$ and $\mathscr{U}^n$.

***15.** Prove that the vectors

$$X = \begin{bmatrix} x_1 \\ x_2 \\ \vdots \\ x_n \end{bmatrix} \quad \text{and} \quad Y = \begin{bmatrix} y_1 \\ y_2 \\ \vdots \\ y_n \end{bmatrix}$$

over a number field $\mathscr{F}$ are linearly dependent if and only if $x_i y_j - x_j y_i = 0$ whenever $1 \leq i < j \leq n$. Generalize.

***16.** Compute the product $[A_{ij}]^{\mathsf{T}} A$. Use the resulting equation to obtain a formula for $\det [A_{ij}]$ in terms of $\det A$.

17. Solve by Cramer's rule:

(a)
$$\begin{aligned} x_1 + 2x_2 + 3x_3 &= 6 \\ 2x_1 - 2x_2 + 5x_3 &= 5 \\ 4x_1 - x_2 - 3x_3 &= 0, \end{aligned}$$

(b)
$$\begin{aligned} x_1 + x_2 &= 3 \\ x_2 + 2x_3 &= 2 \\ x_3 + 3x_4 &= 1 \\ 4x_1 + x_4 &= 0. \end{aligned}$$

18. For what values of t will the system

$$\begin{bmatrix} 1 & t & 0 \\ 0 & 1 & -1 \\ t & 0 & 1 \end{bmatrix} \begin{bmatrix} x_1 \\ x_2 \\ x_3 \end{bmatrix} = \begin{bmatrix} 1 \\ 1 \\ 1 \end{bmatrix}$$

have a unique solution?

19. Find the components of a vector in $\mathscr{E}^4$ orthogonal to each of

$$\begin{bmatrix} a \\ b \\ 1 \\ 0 \end{bmatrix}, \quad \begin{bmatrix} 0 \\ a \\ b \\ 1 \end{bmatrix}, \quad \begin{bmatrix} 1 \\ 0 \\ a \\ b \end{bmatrix}.$$

Are there any values of a and b for which there is no solution?

20. Show that the set of all vectors in $\mathscr{E}^4$ which are orthogonal to each of the vectors

$$\begin{bmatrix} 1 \\ 1 \\ 1 \\ 1 \end{bmatrix}, \quad \begin{bmatrix} 1 \\ -1 \\ 1 \\ -1 \end{bmatrix}, \quad \begin{bmatrix} 1 \\ 0 \\ -1 \\ 0 \end{bmatrix}$$

constitutes a vector space. (There are several ways to do this.) Then find a basis for this space.

21. Solve by Cramer's rule and determine for what values of t the solution is valid. For what real values of t is $x_3 = 0$?

$$\begin{aligned} tx_1 + x_2 - x_3 &= 1 \\ -x_1 + tx_2 + x_3 &= 1 \\ x_1 - x_2 + tx_3 &= 1. \end{aligned}$$

22. Find, by Cramer's rule, the point common to the three planes in $\mathscr{E}^3$ whose equations are

$$\begin{aligned} x_1 + x_2 + x_3 &= a \\ x_1 - x_2 \qquad &= 0 \\ x_1 + x_2 - x_3 &= 0. \end{aligned}$$

Illustrate with a figure.

23. Show that if the point (c_1, c_2, c_3) is on the straight line determined by distinct points (a_1, a_2, a_3) and (b_1, b_2, b_3) in $\mathscr{E}^3$, then the equation

$$\begin{vmatrix} x_1 & x_2 & x_3 & 1 \\ a_1 & a_2 & a_3 & 1 \\ b_1 & b_2 & b_3 & 1 \\ c_1 & c_2 & c_3 & 1 \end{vmatrix} = 0$$

reduces to $0 = 0$, which makes sense because there isn't a unique plane on the three points in this case.

24. Show that the area K of a triangle with vertices (x_1, x_2), (y_1, y_2), (z_1, z_2), listed in counterclockwise order, is given by

$$K = \tfrac{1}{2}\begin{vmatrix} x_1 & x_2 & 1 \\ y_1 & y_2 & 1 \\ z_1 & z_2 & 1 \end{vmatrix}.$$

Start with Figure 7-2. Thereafter, show that the triangle with vertices

$$(x_1 + h, x_2 + k), (y_1 + h, y_2 + k), (z_1 + h, z_2 + k)$$

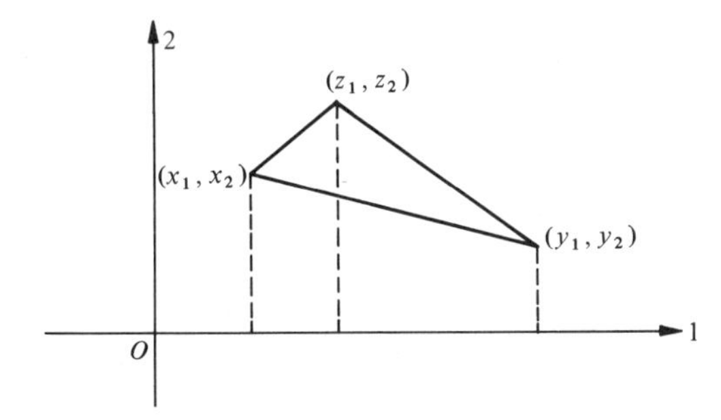

Figure 7-2. *Area of a Triangle.*

is congruent to the given one, so that having the triangle in the first quadrant is no essential limitation and so that the formula holds in any case.

25. Use Exercise 11a to solve Exercise 17, Section 5.7.

26. Use Exercise 11b to solve Exercise 18, Section 5.7.

CHAPTER 8

Linear Transformations

8.1 MAPPINGS

One of the most important ideas in mathematics is that of a **mapping** or **function**, namely a pairing of the members a of a set $\mathscr{A}$ with members b of a set $\mathscr{B}$ such that when the first member a of a pair (a, b) is given, the second member b is uniquely determined. Such a set of pairs is called a **mapping from $\mathscr{A}$ to $\mathscr{B}$** (Figure 8-1) or a **function on $\mathscr{A}$ with values in $\mathscr{B}$**.

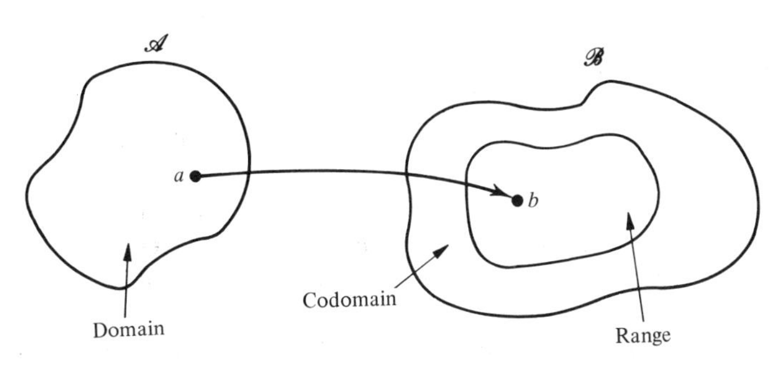

Figure 8-1. *Mapping from $\mathscr{A}$ to $\mathscr{B}$.*

For example, the equation

$$(8.1.1)\qquad \begin{bmatrix} y_1 \\ y_2 \end{bmatrix} = \begin{bmatrix} 1 & 0 & -1 \\ -1 & 1 & 0 \end{bmatrix} \begin{bmatrix} x_1 \\ x_2 \\ x_3 \end{bmatrix}$$

assigns to each vector X in $\mathscr{E}^3$ a unique corresponding vector Y in $\mathscr{E}^2$.

The member b of the pair (a, b) is called the **image** of a under the mapping. The member a of the pair is called the **counterimage** of b. If we denote the mapping by the symbol f, we write

$$a \xrightarrow{f} b$$

to mean, "a is mapped by f onto b," or, more familiarly, we write

$$b = f(a).$$

If $b = f(a)$, the pair (a, b) is said to **belong to the mapping** f.

The set $\mathscr{A}$ is called the **domain** of the mapping and the set $\mathscr{B}$ is called its **codomain**. The subset of $\mathscr{B}$ which contains all the images of the elements of $\mathscr{A}$ and no other elements is called the **range** $\mathscr{R}$ of the mapping. For example, if $\mathscr{M}$ is the set of all square matrices with elements in a number field $\mathscr{F}$, the determinant function assigns to each matrix M of $\mathscr{M}$ a unique scalar b of $\mathscr{F}$:

$$b = \det M.$$

Here we have a mapping with domain $\mathscr{M}$ and range $\mathscr{F}$.

The formula

$$(8.1.2)\qquad \begin{bmatrix} y_1 \\ y_2 \end{bmatrix} = \begin{bmatrix} 2 & 1 \\ 1 & 2 \end{bmatrix} \begin{bmatrix} x_1 \\ x_2 \end{bmatrix}$$

defines a mapping from the set of vectors in $\mathscr{E}^2$ to the set of vectors in $\mathscr{E}^2$, which illustrates the fact that the sets $\mathscr{A}$ and $\mathscr{B}$ of the definition need not be distinct. Since the coefficient matrix is nonsingular, we can rewrite this equation in the form

$$(8.1.3)\qquad \begin{bmatrix} x_1 \\ x_2 \end{bmatrix} = \frac{1}{3} \begin{bmatrix} 2 & -1 \\ -1 & 2 \end{bmatrix} \begin{bmatrix} y_1 \\ y_2 \end{bmatrix}.$$

In this case, given an image vector $\begin{bmatrix} y_1 \\ y_2 \end{bmatrix}$, we can compute its unique counterimage $\begin{bmatrix} x_1 \\ x_2 \end{bmatrix}$ simply by substituting into (8.1.3).

Whenever, as in the preceding example, each element b of the range of a mapping is the image of precisely one member of $\mathscr{A}$, the mapping is called a **one-to-one mapping**: Only one element a of $\mathscr{A}$ maps onto an element b of the range. On the other hand, the determinant function is a **many-to-one mapping**, that is, one in which more than one element a of the domain can have the same image b.

All three of the mappings used as examples so far have the property that every element of the codomain is the image of some element of the domain. In such a case, we say $\mathscr{A}$ is mapped onto $\mathscr{B}$ and we call the mapping an **onto mapping**. (Thus "onto" is used as an adjective by mathematicians, not just as a preposition.) If the range of the mapping is not all of $\mathscr{B}$ but is rather a proper subset of $\mathscr{B}$, we say that $\mathscr{A}$ is mapped **strictly into** $\mathscr{B}$ and that the mapping is a **strictly-into mapping**. If we do not know or do not wish to specify whether a mapping from $\mathscr{A}$ to $\mathscr{B}$ is onto or strictly-into, we call it an **into mapping**. Thus *into* can denote either *strictly-into* or *onto*. If we replace $\mathscr{B}$ by the range $\mathscr{R}$, the resulting mapping from $\mathscr{A}$ to $\mathscr{R}$ is, of course, onto.

An example of a strictly-into mapping is given by the determinant function applied to the set $\mathscr{A}$ of $n \times n$ matrices A with entries such that $|a_{ij}| \leq 1$, the set $\mathscr{B}$ being the set of all real numbers. Since the absolute value of every term in the expansion of $\det A$ is ≤ 1 and since $|\det A|$ is equal to or less than the sum of the absolute values of the terms of the expansion, we have at once that

$$|\det A| \leq 1 + 1 + \cdots + 1 = n!$$

Hence the mapping is a strictly-into mapping.

We have noted that the mapping from $\mathscr{E}^2$ to $\mathscr{E}^2$ defined by (8.1.2) is a one-to-one, onto mapping and that, via the rule (8.1.3), we can find for each Y in $\mathscr{E}^2$ the unique counterimage X in $\mathscr{E}^2$ with which Y is paired in the mapping. By contrast, in the case of the mapping from the set of square real matrices with $|a_{ij}| \leq 1$ to the field of real numbers via the determinant function, each member of the range arises from many members of the domain so that the mapping is not one-to-one and there is no unique counterimage. Also, since not all real numbers are the determinants of matrices in the set, some elements of the codomain have no counterimage at all. These examples illustrate the next observation.

When we have a *one-to-one* mapping f of a set $\mathscr{A}$ *onto* a set $\mathscr{B}$, and only when both these conditions are satisfied, each member of $\mathscr{B}$ has one and only one counterimage in $\mathscr{A}$. The mapping from $\mathscr{B}$ to $\mathscr{A}$ which pairs each element b of $\mathscr{B}$ with its unique counterimage in $\mathscr{A}$ is called the **inverse** of f and is denoted by f^{-1}. The pair (b, a) belongs to the mapping f^{-1} if and only if the pair (a, b) belongs to f. We write in this case

$$b \xrightarrow{f^{-1}} a \qquad \text{or} \qquad a = f^{-1}(b).$$

Note that if f^{-1} exists, f^{-1} is also one-to-one and onto and the inverse of f^{-1} is f.

An example is the mapping from the vectors of $\mathscr{E}^2$ to the vectors of $\mathscr{E}^2$ defined by

$$\begin{bmatrix} y_1 \\ y_2 \end{bmatrix} = \begin{bmatrix} 3 & 2 \\ 1 & 1 \end{bmatrix} \begin{bmatrix} x_1 \\ x_2 \end{bmatrix} + \begin{bmatrix} 1 \\ -1 \end{bmatrix},$$

which has the inverse defined by

$$\begin{bmatrix} x_1 \\ x_2 \end{bmatrix} = \begin{bmatrix} 1 & -2 \\ -1 & 3 \end{bmatrix} \begin{bmatrix} y_1 \\ y_2 \end{bmatrix} + \begin{bmatrix} -3 \\ 4 \end{bmatrix}.$$

The reader should show how this last equation was obtained and should check its correctness by substitution.

8.2 LINEAR MAPPINGS

Consider now a mapping f from a vector space $\mathscr{V}$ to a vector space $\mathscr{W}$, where the scalars in both cases belong to the same field, $\mathscr{F}$. Such a mapping f is said to be a **linear mapping** if and only if, for all vectors V_1 and V_2 in $\mathscr{V}$ and for all scalars α in $\mathscr{F}$,

$$f(V_1 + V_2) = f(V_1) + f(V_2) \tag{8.2.1}$$

and

$$f(\alpha V_1) = \alpha f(V_1). \tag{8.2.2}$$

Thus the mapping from the set $\mathscr{M}$ of all real $n \times n$ matrices to the set $\mathscr{R}$ of all real numbers defined by

$$\det A = b,$$

where A belongs to $\mathscr{M}$ and b belongs to $\mathscr{R}$ is not linear, because ordinarily

$$\det (A + B) \neq \det A + \det B.$$

The reader should provide an illustrative example. Note also that $\det(\alpha A) = \alpha^n \det A$, not $\alpha \det A$, so that, in this case, both conditions are violated.

On the other hand, the mapping from $\mathscr{M}$ to $\mathscr{M}$ defined by

$$f(A) = A^{\mathsf{T}}$$

is linear because

$$(A + B)^{\mathsf{T}} = A^{\mathsf{T}} + B^{\mathsf{T}}$$

and

$$(\alpha A)^{\mathsf{T}} = \alpha A^{\mathsf{T}}.$$

The mapping from $\mathscr{E}^n$ to $\mathscr{E}^m$ defined by

$$Y_{m \times 1} = A_{m \times n} X_{n \times 1},$$

where A is a real matrix, is also linear because, for all X_1 and X_2 in $\mathscr{E}^n$ and for all real numbers α,

$$A(X_1 + X_2) = AX_1 + AX_2$$

and

$$A(\alpha X_1) = \alpha(AX_1).$$

This example is really a little theorem. Its converse is more impressive:

Theorem 8.2.1: *Every linear mapping from $\mathscr{E}^n$ to $\mathscr{E}^m$ may be defined by an equation*

$$Y_{m\times 1} = A_{m\times n}X_{n\times 1}. \tag{8.2.3}$$

To prove the theorem, denote the mapping by f and let

$$A_j = f(E_j), \tag{8.2.4}$$

where the E_j's are the elementary unit vectors of $\mathscr{E}^n$. Then, since for every X in $\mathscr{E}^n$,

$$X = x_1E_1 + x_2E_2 + \cdots + x_nE_n,$$

by repeated use of (8.2.1) we have

$$f(X) = f(x_1E_1) + f(x_2E_2) + \cdots + f(x_nE_n),$$

so that, by applying (8.2.2) to each term, we obtain

$$f(X) = x_1f(E_1) + x_2f(E_2) + \cdots + x_nf(E_n).$$

Next, by (8.2.4), we have

$$f(X) = x_1A_1 + x_2A_2 + \cdots + x_nA_n.$$

If we now put $f(X) = Y$, we can rewrite this last equation as

$$Y = [A_1, A_2, \ldots, A_n]X, \tag{8.2.5}$$

which has the form stated in the theorem.

The proof not only establishes the required result but also proves

Corollary 8.2.2: *A linear mapping from $\mathscr{E}^n$ to $\mathscr{E}^m$ is determined uniquely by the images of the elementary unit vectors $E_1, E_2, \ldots, E_n$ of $\mathscr{E}^n$, the images of these unit vectors constituting respectively the columns $A_1, A_2, \ldots, A_n$ of the matrix A of the mapping.*

For example, the linear mapping from $\mathscr{E}^3$ to $\mathscr{E}^3$ such that

$$f(E_1) = \begin{bmatrix} 1 \\ -1 \\ 2 \end{bmatrix}, \quad f(E_2) = \begin{bmatrix} 2 \\ 1 \\ 3 \end{bmatrix}, \quad f(E_3) = \begin{bmatrix} -1 \\ 0 \\ 1 \end{bmatrix}$$

may be represented by the matrix equation

$$Y = \begin{bmatrix} 1 & 2 & -1 \\ -1 & 1 & 0 \\ 2 & 3 & 1 \end{bmatrix} X.$$

Linear mappings of vector spaces are commonly called **linear operators** or **linear transformations.**

8.3 SOME PROPERTIES OF LINEAR OPERATORS ON VECTOR SPACES

The image of a subset $\mathscr{S}$ of $\mathscr{E}^n$ by a linear operator is defined to be the set of all images of vectors of $\mathscr{S}$. We have then

Theorem 8.3.1: *Given a linear operator $Y = AX$ mapping $\mathscr{E}^n$ into $\mathscr{E}^m$, the image of a subspace $\mathscr{V}$ of $\mathscr{E}^n$ by this operator is a subspace of $\mathscr{E}^m$. This image space is spanned by the images of the vectors of any basis for $\mathscr{V}$.*

Let the subspace $\mathscr{V}$ have the basis $V_1, V_2, \ldots, V_k$. Then the vectors X of $\mathscr{V}$ are the set of all linear combinations of the form

$$X = \alpha_1 V_1 + \alpha_2 V_2 + \cdots + \alpha_k V_k.$$

The image of each such vector is given by

$$AX = \alpha_1 AV_1 + \alpha_2 AV_2 + \cdots + \alpha_k AV_k,$$

which is a linear combination of the vectors $AV_1, AV_2, \ldots, AV_k$. That is, the vectors of $\mathscr{V}$ have as images the vectors of the subspace spanned by $AV_1, AV_2, \ldots, AV_k$, and the theorem is proved.

Since, in the preceding proof, the image space is the set of linear combinations of k vectors, which may or may not be linearly independent, we have

Corollary 8.3.2: *The dimension of the image of a subspace of dimension k of $\mathscr{E}^n$ by a linear operator is not greater than k.*

Since the equation

$$Y = AX$$

can be rewritten in the form

$$Y = x_1 A_1 + x_2 A_2 + \cdots + x_n A_n, \tag{8.3.1}$$

where the A_j's are the columns of A, we have

Theorem 8.3.3: *The range of a linear operator $Y = AX$ which maps $\mathscr{E}^n$ into $\mathscr{E}^m$ is the subspace of $\mathscr{E}^m$ spanned by the columns of A. The dimension of this subspace is $r(A)$.*

As an illustration of the three preceding results, note that the linear operator

$$Y = \begin{bmatrix} 1 & 0 & 0 \\ 0 & 1 & 0 \\ 0 & 0 & 0 \end{bmatrix} X$$

maps the vectors of $\mathscr{E}^3$ onto the two-dimensional subspace spanned by

$$\begin{bmatrix}1\\0\\0\end{bmatrix}, \quad \begin{bmatrix}0\\1\\0\end{bmatrix}, \quad \begin{bmatrix}0\\0\\0\end{bmatrix},$$

that is, onto the vectors of the 1,2-plane (Figure 8-2). The image of an arbitrary vector

$$\begin{bmatrix}a_1\\a_2\\a_3\end{bmatrix}$$

by this operator is just its projection

$$\begin{bmatrix}a_1\\a_2\\0\end{bmatrix}$$

on the 12-plane (Figure 8-2). This is a many-to-one mapping. The reader should describe geometrically the set of all counterimages of a given vector

$$\begin{bmatrix}a_1\\a_2\\0\end{bmatrix}.$$

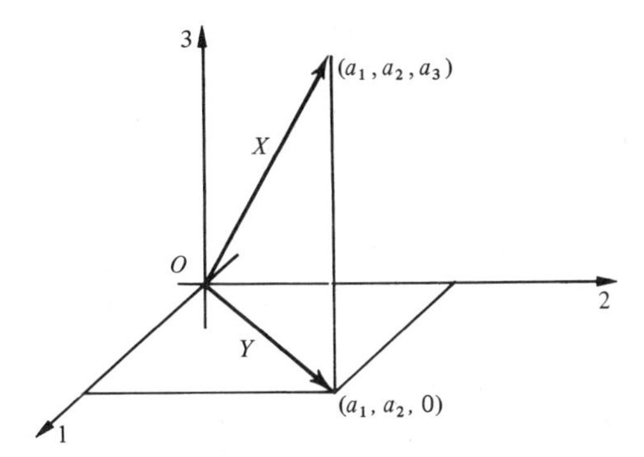

Figure 8-2. *A Projection.*

As the figure of this example suggests, we may interpret the operator as a mapping of points (the terminal points of vectors) as well as one of vectors.

Theorem 8.3.4: *If $Y = AX$ represents a linear operator mapping $\mathscr{E}^n$ into $\mathscr{E}^m$ and if $r(A) = n$, then the dimension of the image of any subspace $\mathscr{V}$ of $\mathscr{E}^n$ is the same as that of $\mathscr{V}$.*

Let $\mathscr{V}$ be of dimension $k \leq n$ and let $V_1, V_2, \ldots, V_k$ be a basis for $\mathscr{V}$ so that $AV_1, AV_2, \ldots, AV_k$ span the image space. Any equation

$$\alpha_1 AV_1 + \alpha_2 AV_2 + \cdots + \alpha_k AV_k = 0,$$

where the α's are scalars, implies

$$A(\alpha_1 V_1 + \alpha_2 V_2 + \cdots + \alpha_k V_k) = 0.$$

Since A has rank n, the only solution of $AX = 0$ is the trivial solution. That is,

$$\alpha_1 V_1 + \alpha_2 V_2 + \cdots + \alpha_k V_k = 0.$$

Since the V_j's are independent, this implies that all the α_j's are 0, which in turn proves that the vectors $AV_1, AV_2, \ldots, AV_k$ are linearly independent, so the image space also has dimension k.

A linear operator mapping $\mathscr{E}^n$ into $\mathscr{E}^n$ is called simply a **linear operator on** $\mathscr{E}^n$. Such an operator is said to be **nonsingular** if and only if the dimension of the image of *every* subspace is the same as that of the subspace itself. Otherwise it is said to be **singular**. Note that the preceding example illustrates a singular operator since the image of $\mathscr{E}^3$ itself is only two dimensional. Are there any subspaces of $\mathscr{E}^3$ whose dimension is preserved by the projection in this example?

Theorem 8.3.5: *A linear operator $Y = AX$ on $\mathscr{E}^n$ is nonsingular if and only if A is nonsingular.*

If $Y = AX$ is nonsingular, $\mathscr{E}^n$ must have $\mathscr{E}^n$, not some subspace of $\mathscr{E}^n$, as its image. Now the columns of A span the image of $\mathscr{E}^n$, namely $\mathscr{E}^n$ itself. The columns of A must therefore be linearly independent, so A must be nonsingular.

Conversely, let A be nonsingular. Then A has rank n, so by Theorem 8.3.4, the dimension of the image of every k-dimensional subspace is k, and the theorem follows.

When a linear operator on $\mathscr{E}^n$ represented by $Y = AX$ is nonsingular, so that A^{-1} exists, then the inverse mapping is given by

$$X = A^{-1}Y.$$

8.4 EXERCISES

1. Show that the range of the mapping from $\mathscr{E}^3$ to $\mathscr{E}^3$ defined by

$$\begin{bmatrix} y_1 \\ y_2 \\ y_3 \end{bmatrix} = \begin{bmatrix} 2 & -1 & 0 \\ -1 & 2 & 1 \\ -1 & -1 & -1 \end{bmatrix} \begin{bmatrix} x_1 \\ x_2 \\ x_3 \end{bmatrix}$$

is the set of vectors for which $y_1 + y_2 + y_3 = 0$. Illustrate with a figure.

2. What is the range of the mapping from $\mathscr{E}^3$ to $\mathscr{E}^3$ defined by

$$\begin{bmatrix} y_1 \\ y_2 \\ y_3 \end{bmatrix} = \begin{bmatrix} x_1 \\ x_2 \\ x_3 \end{bmatrix} + \begin{bmatrix} h_1 \\ h_2 \\ h_3 \end{bmatrix},$$

where h_1, h_2, and h_3 are fixed? Describe the mapping geometrically. Is it one-to-one? Is it onto? Is it linear? Illustrate some specific cases with figures.

3. A linear operator mapping $\mathscr{E}^3$ into $\mathscr{E}^2$ is to map

$$\begin{bmatrix} 1 \\ 0 \\ 0 \end{bmatrix} \text{ onto } \begin{bmatrix} 1 \\ 0 \end{bmatrix}, \quad \begin{bmatrix} 0 \\ 1 \\ 0 \end{bmatrix} \text{ onto } \begin{bmatrix} 0 \\ 1 \end{bmatrix}, \quad \text{and} \quad \begin{bmatrix} 0 \\ 0 \\ 1 \end{bmatrix} \text{ onto } \begin{bmatrix} 2 \\ 2 \end{bmatrix}.$$

Represent the operator in matrix form.

4. Find the range of the mapping from $\mathscr{E}^3$ to $\mathscr{E}^4$ defined by

$$\begin{bmatrix} y_1 \\ y_2 \\ y_3 \\ y_4 \end{bmatrix} = \begin{bmatrix} 1 & -1 & 2 \\ 2 & 0 & 1 \\ 3 & -1 & 3 \\ 1 & 1 & -1 \end{bmatrix} \begin{bmatrix} x_1 \\ x_2 \\ x_3 \end{bmatrix}.$$

5. Find the inverse of the linear operator on $\mathscr{E}^3$ defined by

$$\begin{bmatrix} y_1 \\ y_2 \\ y_3 \end{bmatrix} = \begin{bmatrix} \frac{1}{\sqrt{2}} & 0 & \frac{1}{\sqrt{2}} \\ -\frac{1}{\sqrt{2}} & 0 & \frac{1}{\sqrt{2}} \\ 0 & 1 & 0 \end{bmatrix} \begin{bmatrix} x_1 \\ x_2 \\ x_3 \end{bmatrix}.$$

6. Show that the image Y of a vector X by the operator of Exercise 5 has the same length as X.

***7.** Show that a linear operator on $\mathscr{E}^n$ is nonsingular if and only if the images of $E_1, E_2, \ldots, E_n$ are linearly independent vectors.

***8.** Show that the images, by a linear operator mapping $\mathscr{E}^n$ into $\mathscr{E}^m$, of dependent vectors are also dependent vectors.

***9.** Show that a linear operator mapping $\mathscr{E}^n$ into $\mathscr{E}^m$ maps every linear combination of vectors of $\mathscr{E}^n$ onto the same linear combination of the images of these vectors.

***10.** Find the unique linear operator on $\mathscr{E}^n$ which maps the linearly independent vectors $A_1, A_2, \ldots, A_n$ onto $E_1, E_2, \ldots, E_n$, respectively. (*Hint*: What is the inverse mapping?)

***11.** Show that there exists a unique nonsingular linear operator on $\mathscr{E}^n$ which maps the independent vectors $A_1, A_2, \ldots, A_n$ onto the independent vectors $B_1, B_2, \ldots, B_n$. (Use Corollary 8.2.2. and Exercise 10.)

12. Find the linear operator on $\mathscr{E}^2$ which maps

$$\begin{bmatrix}1\\1\end{bmatrix} \text{ onto } \begin{bmatrix}-2\\0\end{bmatrix} \quad \text{and} \quad \begin{bmatrix}1\\-1\end{bmatrix} \text{ onto } \begin{bmatrix}0\\2\end{bmatrix}.$$

13. A linear operator on $\mathscr{E}^3$ maps

$$\begin{bmatrix}1\\0\\0\end{bmatrix} \text{ onto } \begin{bmatrix}0\\2\\1\end{bmatrix}, \quad \begin{bmatrix}1\\1\\0\end{bmatrix} \text{ onto } \begin{bmatrix}1\\0\\2\end{bmatrix}, \quad \text{and} \quad \begin{bmatrix}1\\1\\1\end{bmatrix} \text{ onto } \begin{bmatrix}2\\1\\0\end{bmatrix}.$$

Onto what vector does it map

$$\begin{bmatrix}0\\0\\1\end{bmatrix}?$$

(Use Exercise 9.)

***14.** Extend the theorems of Section 8.3 and the starred exercises of Section 8.4 to n-vectors (and m-vectors) over an arbitrary field $\mathscr{F}$, that is, to the vectors of $\mathscr{F}^n$ (and $\mathscr{F}^m$).

8.5 SOME GEOMETRICAL PROPERTIES OF LINEAR TRANSFORMATIONS

Linear transformations, interpreted as point transformations, have many interesting and important properties, of which we can mention only a few here. As a simple example, consider the effect of the transformation of $\mathscr{E}^2$ defined by

$$\begin{aligned} y_1 &= kx_1 \\ y_2 &= \ \ x_2 \end{aligned} \quad \text{or} \quad Y = \begin{bmatrix} k & 0 \\ 0 & 1 \end{bmatrix} X, \qquad k > 0,$$

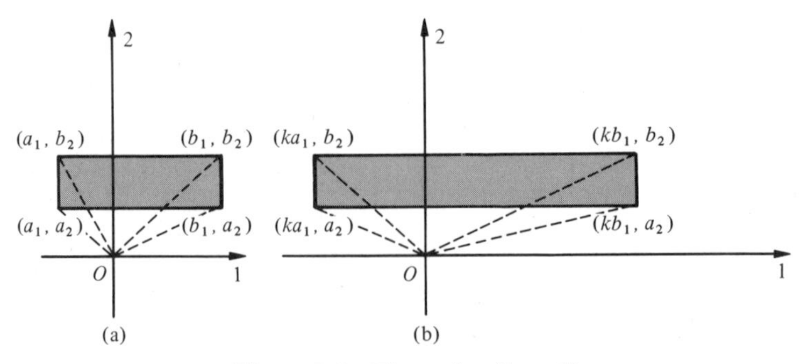

Figure 8-3. *Elongation* ($k > 1$).

on an arbitrary rectangle in the plane. We plot X and Y on separate diagrams for the sake of simplicity (Figure 8-3). Let a pair of opposite vertices of the rectangle be (a_1, a_2) and (b_1, b_2). Since the horizontal dimension is stretched ($k > 1$), unaltered ($k = 1$), or shrunk ($k < 1$) by the factor k, the vertical dimension remaining unaltered, the area of the image rectangle is k times that of the counter-image. The reader should illustrate with figures of a variety of specific choices of rectangle and of the scale factor k. Such a transformation is called an **elongation** ($k > 1$) or a **contraction** ($k < 1$).

It is intuitively clear that the area enclosed by any simple closed curve can be approximated as closely as required by inscribing rectangles of the kind appearing in the preceding figure. One can use this fact to establish that the area of a circle of radius b is πb^2. The circle may be represented by the equation $x_1^2 + x_2^2 = b^2$ (Figure 8-4).

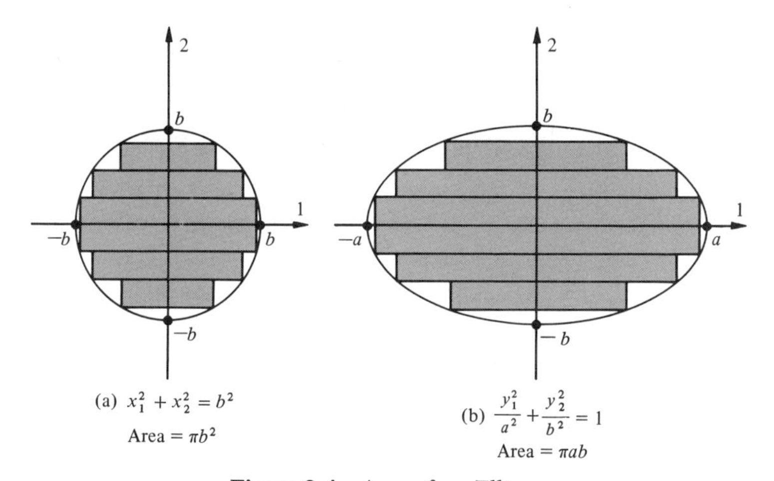

Figure 8-4. *Area of an Ellipse.*

If we now apply the transformation

$$\begin{aligned} y_1 &= \frac{a}{b} x_1 \\ y_2 &= x_2 \end{aligned} \qquad \text{or} \qquad \begin{aligned} x_1 &= \frac{b}{a} y_1 \\ x_2 &= y_2 \end{aligned} \qquad (a, b > 0),$$

which stretches the horizontal dimension by a factor $k = a/b$, the equation of the circle is transformed into the equation of an ellipse:

$$\frac{b^2 y_1^2}{a^2} + y_2^2 = b^2 \qquad \text{or} \qquad \frac{y_1^2}{a^2} + \frac{y_2^2}{b^2} = 1.$$

Now the rectangles inscribed in the circle are transformed into rectangles inscribed in the ellipse, and their areas are all multiplied by the factor a/b. Hence the area of the ellipse is a/b times that of the circle; that is, it is $\pi b^2 \cdot a/b = \pi ab$.

Note that the equation of the transformation and of the circle could be written in matrix form thus:

$$X = \begin{bmatrix} \frac{b}{a} & 0 \\ 0 & 1 \end{bmatrix} Y \qquad \text{and} \qquad X^{\mathsf{T}} I_2 X = b^2.$$

Substituting from the first equation into the second, we have

$$Y^{\mathsf{T}} \begin{bmatrix} \frac{b}{a} & 0 \\ 0 & 1 \end{bmatrix} I_2 \begin{bmatrix} \frac{b}{a} & 0 \\ 0 & 1 \end{bmatrix} Y = b^2,$$

so the equation of the ellipse is

$$Y^{\mathsf{T}} \begin{bmatrix} \frac{b^2}{a^2} & 0 \\ 0 & 1 \end{bmatrix} Y = b^2.$$

Multiplying through by the scalar a^2, we can rewrite this as

$$Y^{\mathsf{T}} \begin{bmatrix} b^2 & 0 \\ 0 & a^2 \end{bmatrix} Y = a^2 b^2,$$

which corresponds to the familiar scalar form

$$b^2 y_1^2 + a^2 y_2^2 = a^2 b^2.$$

As another example, consider the transformation of $\mathscr{E}^2$ defined by

$$\begin{aligned} y_1 &= x_1 + kx_2 \\ y_2 &= x_2 \end{aligned} \qquad \text{or} \qquad Y = \begin{bmatrix} 1 & k \\ 0 & 1 \end{bmatrix} X,$$

from which

$$X = \begin{bmatrix} 1 & -k \\ 0 & 1 \end{bmatrix} Y.$$

The vertical line with equation $x_1 = a$ is transformed into the line with equation $y_1 = ky_2 + a$. Figure 8-5 illustrates the case $a = 2$, $k = \frac{1}{2}$. The image of the line $x_1 = 2$ has the same 1-intercept but is tilted: At each vertical distance y_2 from (2, 0), there is a horizontal displacement $\frac{1}{2}y_2$.

A transformation of this type is called a **shear** (Figure 8-6) and is important in the theory of deformations.

The effect of the transformation

$$\begin{aligned} y_1 &= x_1 + \tfrac{1}{2}x_2 \\ y_2 &= \phantom{x_1 + \tfrac{1}{2}}x_2 \end{aligned} \qquad \text{or} \qquad X = \begin{bmatrix} 1 & -\frac{1}{2} \\ 0 & 1 \end{bmatrix} Y$$

on the circle $X^{\mathsf{T}} I_2 X = 4$ is to transform it into the ellipse (Figure 8-6) with equation

$$Y^{\mathsf{T}} \begin{bmatrix} 1 & 0 \\ -\frac{1}{2} & 1 \end{bmatrix} I_2 \begin{bmatrix} 1 & -\frac{1}{2} \\ 0 & 1 \end{bmatrix} Y = 4 \qquad \text{or} \qquad Y^{\mathsf{T}} \begin{bmatrix} 1 & -\frac{1}{2} \\ -\frac{1}{2} & \frac{5}{4} \end{bmatrix} Y = 4.$$

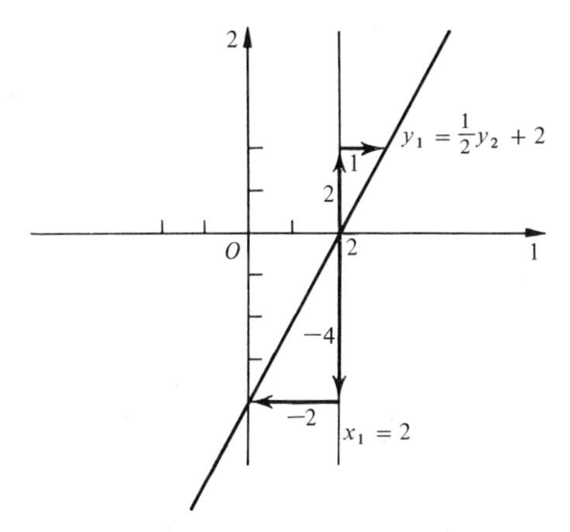

Figure 8-5. *Shear Transformation.*

Note how the shear has pushed the square into the shape of a nonrectangular parallelogram and the circle into the shape of an ellipse. The inverse substitution, applied to the equation of the ellipse, restores the equation of the circle, of course.

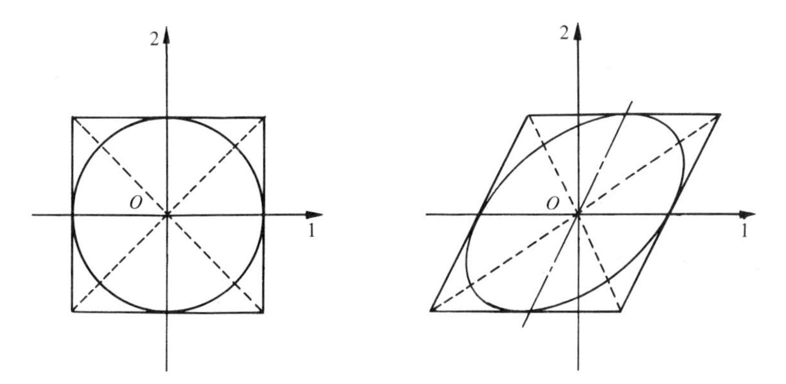

Figure 8-6. *Shear Transformation.*

A third simple sort of transformation is illustrated by

$$\begin{aligned} y_1 &= -x_1 \\ y_2 &= x_2 \end{aligned} \qquad \text{or} \qquad Y = \begin{bmatrix} -1 & 0 \\ 0 & 1 \end{bmatrix} X.$$

This transformation simply changes the sign of the first coordinate of every point and is called a **reflection** (Figure 8-7) in the 2-axis. A reflection in the 1-axis is defined similarly.

It is not hard to show that every nonsingular linear transformation, $Y = AX$, of $\mathscr{E}^2$ can be expressed as a succession of transformations of the three preceding types.

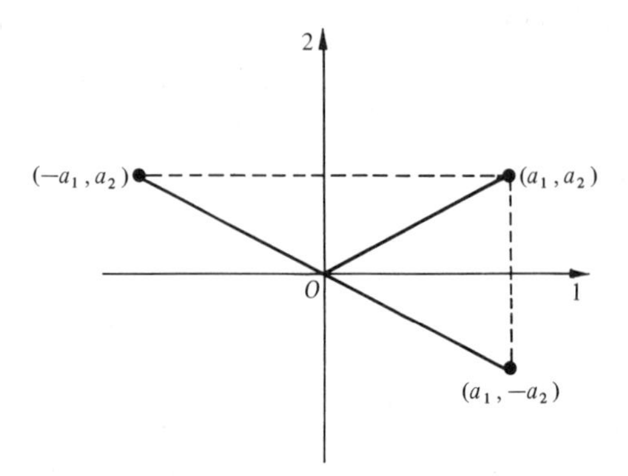

Figure 8-7. *Reflections.*

A special case of each of the first two types is the transformation

$$\begin{aligned} y_1 &= x_1 \\ y_2 &= x_2 \end{aligned} \qquad \text{or} \qquad Y = I_2 X.$$

This transformation leaves every figure unchanged and is called the identity transformation. Its importance will appear in later sections.

These and other geometrical ideas can be generalized to an arbitrary number of dimensions.

8.6 INVARIANTS OF TRANSFORMATIONS

A **figure** in $\mathscr{E}^2$ is any set of points in $\mathscr{E}^2$. A figure is **invariant** under a transformation of $\mathscr{E}^2$ if and only if the image of the figure is the figure itself (Figure 8-8). Thus the **translation**

$$\begin{aligned} y_1 &= x_1 + h_1 \\ y_2 &= x_2 + h_2 \end{aligned}$$

or

$$Y = X + H,$$

which shifts each point P a directed distance h_1 units parallel to the 1-axis and h_2 units parallel to the 2-axis, leaves invariant every line which has the same direction as the vector $\begin{bmatrix} h_1 \\ h_2 \end{bmatrix}$. It also leaves invariant every infinite set of lines orthogonal to this vector and spaced at a distance $(1/n)|H|$, where n is any positive integer. Can you think of other invariant figures?

A horizontal shear such as was used in the previous section leaves invariant every horizontal line and every set of such lines. A reflection in a given line leaves invariant every figure which is symmetric with respect to that line. These ideas are readily extended to $\mathscr{E}^n$ for any n.

Another type of invariance is that of the form of a function of the coordinates

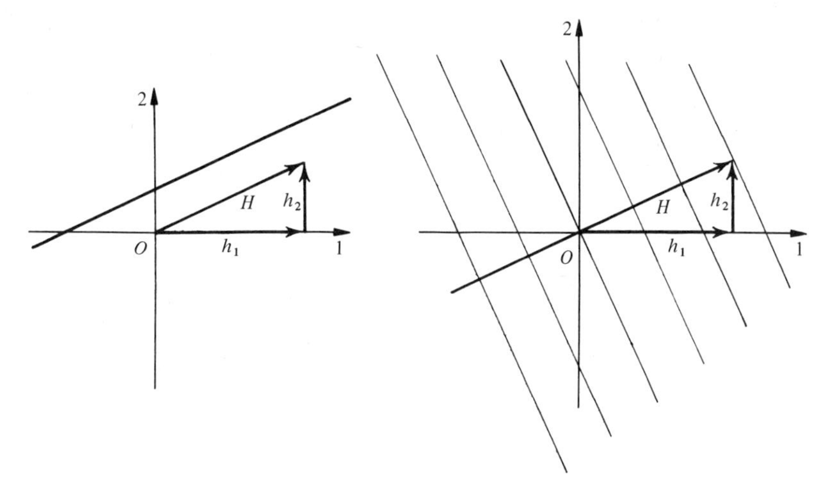

Figure 8-8. *Figures Invariant Under Translation.*

of a set of points (or of the components of a set of vectors), with respect to all the members of a certain set of transformations. For example, in $\mathscr{E}^n$, the set of all **translations**, that is, transformations of the form

$$Y = X + H,$$

where H is a fixed vector, leaves invariant the distance function so that the distance between the images of two points is the same as the distance between the points themselves. Indeed, if $Y_1 = X_1 + H$ and $Y_2 = X_2 + H$, then the distance between the endpoints of Y_1 and Y_2 is

$$\begin{aligned}\sqrt{(Y_2 - Y_1)^{\mathsf{T}}(Y_2 - Y_1)} &= \sqrt{(X_2 + H - X_1 - H)^{\mathsf{T}}(X_2 + H - X_1 - H)} \\ &= \sqrt{(X_2 - X_1)^{\mathsf{T}}(X_2 - X_1)}.\end{aligned}$$

The situation is illustrated for $\mathscr{E}^3$ in Figure 8-9.

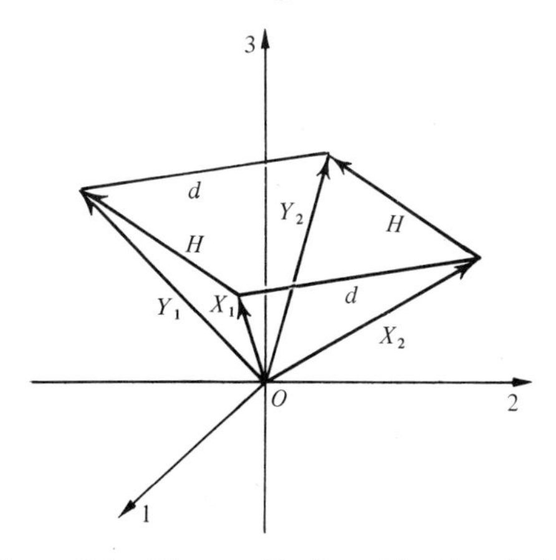

Figure 8-9. *Distance Unaltered by Translation.*

We summarize these observations by saying that *distance in $\mathscr{E}^n$ is an invariant of the set of all translations.*

In the same way, we can show that in $\mathscr{E}^n$ distance is an invariant of all transformations of the form

$$Y = D[\alpha_1, \alpha_2, \ldots, \alpha_n]X,$$

where each α_i is ± 1. These transformations are generalizations of the reflections in the coordinate axes in $\mathscr{E}^2$.

Consider, in $\mathscr{E}^2$, a rotation about the origin through a directed angle θ. Let (x_1, x_2) be any point of $\mathscr{E}^2$ and let (y_1, y_2) be its image. These points are at the same distance d from 0. From Figure 8-10 we see that when $(x_1, x_2) \neq (0, 0)$,

$$\frac{y_1}{d} = \cos(\theta + \alpha) = \cos\theta\cos\alpha - \sin\theta\sin\alpha = \frac{x_1}{d}\cos\theta - \frac{x_2}{d}\sin\theta$$

and

$$\frac{y_2}{d} = \sin(\theta + \alpha) = \sin\theta\cos\alpha + \cos\theta\sin\alpha = \frac{x_1}{d}\sin\theta + \frac{x_2}{d}\cos\theta,$$

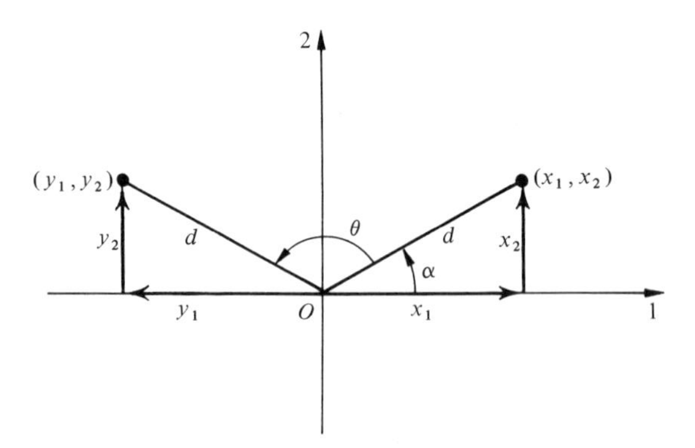

Figure 8-10. *Rotation in $\mathscr{E}^2$.*

so the rotation is defined by

$$\begin{aligned} y_1 &= x_1\cos\theta - x_2\sin\theta \\ y_2 &= x_1\sin\theta + x_2\cos\theta \end{aligned} \qquad \text{or} \qquad Y = \begin{bmatrix} \cos\theta & -\sin\theta \\ \sin\theta & \cos\theta \end{bmatrix} X.$$

These last equations hold also when $(x_1, x_2) = (0, 0)$.

Now let X_1 and X_2 be vectors associated with two arbitrary points and let Y_1 and Y_2 be their images by the rotation. Then

$$\sqrt{(Y_2 - Y_1)^{\mathsf{T}}(Y_2 - Y_1)}$$
$$= \sqrt{(X_2 - X_1)^{\mathsf{T}} \begin{bmatrix} \cos\theta & \sin\theta \\ -\sin\theta & \cos\theta \end{bmatrix} \begin{bmatrix} \cos\theta & -\sin\theta \\ \sin\theta & \cos\theta \end{bmatrix} (X_2 - X_1)}.$$

Since

$$(8.6.1) \qquad \begin{bmatrix} \cos\theta & \sin\theta \\ -\sin\theta & \cos\theta \end{bmatrix} \begin{bmatrix} \cos\theta & -\sin\theta \\ \sin\theta & \cos\theta \end{bmatrix} = \begin{bmatrix} 1 & 0 \\ 0 & 1 \end{bmatrix},$$

we have

$$\sqrt{(Y_2 - Y_1)^{\mathsf{T}}(Y_2 - Y_1)} = \sqrt{(X_2 - X_1)^{\mathsf{T}}(X_2 - X_1)},$$

so distance is an invariant of the set of all rotations about the origin.

These rotations are a special case of the class of orthogonal transformations of $\mathscr{E}^n$, which are of particular importance in applications and which we examine in the next section.

8.7 ORTHOGONAL MATRICES AND ORTHOGONAL TRANSFORMATIONS

A real $n \times n$ matrix U is said to be **orthogonal** if and only if $U^{\mathsf{T}}U = UU^{\mathsf{T}} = I_n$, that is, if and only if $U^{\mathsf{T}} = U^{-1}$. Since we have, in this case,

$$U^{\mathsf{T}}U = \begin{bmatrix} U_1^{\mathsf{T}} \\ U_2^{\mathsf{T}} \\ \vdots \\ U_n^{\mathsf{T}} \end{bmatrix} [U_1, U_2, \ldots, U_n] = [U_i^{\mathsf{T}}U_j]_{n\times n} = I_n,$$

we see that when U is orthogonal, $U_i^{\mathsf{T}}U_j = 0$ if $i \neq j$ but $U_i^{\mathsf{T}}U_i = 1$; that is, $U_i^{\mathsf{T}}U_j = \delta_{ij}$. Conversely, $U_i^{\mathsf{T}}U_j = \delta_{ij}$ implies that $U^{\mathsf{T}}U = I_n$. From this, and from a similar examination of UU^{T}, we have

Theorem 8.7.1: *A real $n \times n$ matrix U is orthogonal if and only if its columns (rows) are mutually orthogonal unit vectors.*

The matrices associated with rotations about the origin in $\mathscr{E}^2$ are a case in point, as Equation (8.6.1) demonstrates.

Since $\det U^{\mathsf{T}} = \det U$, the equation $U^{\mathsf{T}}U = I_n$ implies that $(\det U)^2 = 1$. Hence we have

Theorem 8.7.2: *If U is orthogonal, $\det U = \pm 1$.*

Any set of n mutually orthogonal, unit vectors of $\mathscr{E}^n$ is an **orthonormal reference system** or an **orthonormal basis** for the vectors of $\mathscr{E}^n$ (Chapter 5). Hence from Theorem 8.7.1 we have

Theorem 8.7.3: *The columns of a real, orthogonal, $n \times n$ matrix U constitute an orthonormal basis for the vectors of $\mathscr{E}^n$.*

We now define an **orthogonal transformation** of $\mathscr{E}^n$ to be an operator of the form

$$Y = UX,$$

where U is an orthogonal matrix. When $\det U = 1$, the transformation is called a **proper orthogonal transformation**. When $\det U = -1$, the transformation is **improper**. The proper orthogonal transformations of $\mathscr{E}^n$ are analogous to rotations about the origin in $\mathscr{E}^2$ and $\mathscr{E}^3$. The improper orthogonal transformations are the result of following proper orthogonal transformations by reflections.

The importance of orthogonal transformations is indicated by

Theorem 8.7.4: *A linear transformation $Y = AX$ of $\mathscr{E}^n$ leaves the length of all vectors invariant if and only if it is an orthogonal transformation.*

We only need to show that $Y^\mathsf{T} Y = X^\mathsf{T} X$ for all X, where $Y = AX$, if and only if A is orthogonal. Substituting for Y, we find that we must show that $X^\mathsf{T} A^\mathsf{T} A X = X^\mathsf{T} X$ for all X if and only if A is orthogonal. Assume first that the relation holds for all X, let $X = E_j$, $j = 1, 2, \ldots, n$, and note that

$$E_j^\mathsf{T} A^\mathsf{T} A E_j = (A^\mathsf{T} A)_{jj} = E_j^\mathsf{T} E_j = 1,$$

where $(A^\mathsf{T} A)_{jj}$ denotes the j,j-entry of $A^\mathsf{T} A$. Thus the diagonal entries of $A^\mathsf{T} A$ are all 1.

Next we put $X = E_j + E_k$, $j \neq k$, and get

$$(E_j^\mathsf{T} + E_k^\mathsf{T}) A^\mathsf{T} A (E_j + E_k) = (E_j^\mathsf{T} + E_k^\mathsf{T})(E_j + E_k).$$

Using the facts that $E_j^\mathsf{T} A^\mathsf{T} A E_j = 1$ and $E_j^\mathsf{T} E_k = 0$ if $j \neq k$, we find that

$$(A^\mathsf{T} A)_{jk} + (A^\mathsf{T} A)_{kj} = 0 \qquad \text{or} \qquad (A^\mathsf{T} A)_{jk} = -(A^\mathsf{T} A)_{kj}.$$

Since $A^\mathsf{T} A$ is symmetric, we have, however,

$$(A^\mathsf{T} A)_{jk} = (A^\mathsf{T} A)_{kj}.$$

The last two equations imply that $(A^\mathsf{T} A)_{jk} = 0$ if $j \neq k$. Summarizing, we have

$$A^\mathsf{T} A = I_n,$$

so A is orthogonal.

Conversely, if A is orthogonal, $A^\mathsf{T} A = I_n$, so $X^\mathsf{T} A^\mathsf{T} A X = X^\mathsf{T} X$ for all X. The theorem is thus completely proved. We can also prove

Theorem 8.7.5: *The scalar product $X^\mathsf{T} Y$ is an invariant of the set of all orthogonal transformations of $\mathscr{E}^n$.*

Suppose that U is orthogonal and that $V = UX$, $W = UY$. Then $V^\mathsf{T} W = X^\mathsf{T} U^\mathsf{T} U Y = X^\mathsf{T} Y$.

We have also:

Theorem 8.7.6: *The distance between two points is an invariant of the set of all orthogonal transformations of* $\mathscr{E}^n$.

Theorem 8.7.7: *An orthogonal transformation of* $\mathscr{E}^n$ *maps orthogonal vectors onto orthogonal vectors and nonorthogonal vectors onto nonorthogonal vectors.*

The proofs of these last two theorems are left as exercises for the reader.

8.8 EXERCISES

1. What is the effect of the mapping

$$Y = \begin{bmatrix} k_1 & 0 \\ 0 & k_2 \end{bmatrix} X$$

on area? ($k_1 k_2 \neq 0$.)

2. What figures are invariant under a given rotation about the origin?

3. What effect does a nonsingular transformation

$$Y = \begin{bmatrix} k & 0 \\ 0 & k \end{bmatrix} X$$

have on figures in $\mathscr{E}^2$? Extend to $\mathscr{E}^3$. These are **similarity transformations** with the origin as the **center of similitude**.

*__4.__ Show that if the vectors $X_1, X_2, \ldots, X_k$ of $\mathscr{E}^n$ are mutually orthogonal, then they are linearly independent.

5. If $X = [X_1, X_2, \ldots, X_k]$, where $X_1, X_2, \ldots, X_k$ are mutually orthogonal, real n-vectors, what is the nature of the matrix $X^{\mathsf{T}}X$? Extend to $\mathscr{U}^n$.

6. Show that these are orthogonal matrices:

$$\text{(a)} \begin{bmatrix} \frac{1}{\sqrt{2}} & 0 & \frac{1}{\sqrt{2}} \\ \frac{1}{\sqrt{2}} & 0 & -\frac{1}{\sqrt{2}} \\ 0 & 1 & 0 \end{bmatrix}, \qquad \text{(b)} \begin{bmatrix} \frac{1}{\sqrt{6}} & \frac{2}{\sqrt{5}} & \frac{1}{\sqrt{30}} \\ -\frac{1}{\sqrt{6}} & 0 & \frac{5}{\sqrt{30}} \\ \frac{2}{\sqrt{6}} & -\frac{1}{\sqrt{5}} & \frac{2}{\sqrt{30}} \end{bmatrix}.$$

*__7.__ Determine all unit n-vectors X whose components are exclusively 0's and 1's. Then determine all $n \times n$ orthogonal matrices whose entries are exclusively 0's and 1's.

*__8.__ State and prove the converses of Theorems 8.7.5, 8.7.6, and 8.7.7.

9. Use a transformation of the type

$$Y = \begin{bmatrix} k_1 & 0 & 0 \\ 0 & k_2 & 0 \\ 0 & 0 & k_3 \end{bmatrix} X$$

to deduce the formula for the volume of an ellipsoid

$$\frac{y_1^2}{a^2} + \frac{y_2^2}{b^2} + \frac{y_3^2}{c^2} = 1$$

from the formula for the volume of the unit sphere $x_1^2 + x_2^2 + x_3^2 = 1$.

10. What is the effect of the transformation

$$Y = \begin{bmatrix} 1 & 0 & 1 \\ 0 & 1 & k \\ 0 & 0 & 1 \end{bmatrix} X \qquad (k > 0)$$

on a line parallel to the x_3-axis? On the cylinder $x_1^2 + x_2^2 = 1$? Is the end result a circular cylinder?

11. If a triangle T: (x_{11}, x_{12}), (x_{21}, x_{22}), (x_{31}, x_{32}) in $\mathscr{E}^2$ is transformed by the nonsingular linear transformation

$$Y = \begin{bmatrix} a_{11} & a_{12} \\ a_{21} & a_{22} \end{bmatrix} X,$$

how is the area of the image triangle related to that of T? How is the area of any plane figure for which area is defined affected by this transformation?

***12.** Prove that every proper orthogonal transformation of coordinates in $\mathscr{E}^2$ is a rotation of axes.

***13.** A matrix $P = [E_{j_1}, E_{j_2}, \ldots, E_{j_n}]$, where $j_1, j_2, \ldots, j_n$ is a permutation of $1, 2, \ldots, n$, is called a **permutation matrix**. Prove that every permutation matrix is orthogonal, that is, that $PP^{\mathsf{T}} = I_n$.

8.9 ORTHOGONAL VECTOR SPACES

We now extend to $\mathscr{E}^n$ some more of the geometry of $\mathscr{E}^3$.

We define a vector to be **orthogonal** to the subspace $\mathscr{V}$ of $\mathscr{E}^n$ if and only if it is orthogonal to every vector of $\mathscr{V}$. We have

Theorem 8.9.1: *A vector V is orthogonal to a subspace $\mathscr{V}$ of $\mathscr{E}^n$ if and only if it is orthogonal to every vector of a basis for $\mathscr{V}$.*

Let $A_1, A_2, \ldots, A_k$ be a basis for $\mathscr{V}$. Then if V is orthogonal to $\mathscr{V}$, $V^{\mathsf{T}}A_i = 0$, $i = 1, 2, \ldots, k$. Conversely, if $V^{\mathsf{T}}A_i = 0$, $i = 1, 2, \ldots, k$, then

$$V^{\mathsf{T}}\left(\sum_{i=1}^{k} \alpha_i A_i\right) = \sum_{i=1}^{k} \alpha_i (V^{\mathsf{T}}A_i) = 0,$$

and the theorem follows.

We next define two vector spaces in $\mathscr{E}^n$ to be **mutually orthogonal spaces** if and only if every vector in either space is orthogonal to every vector in the other. We prove next

Theorem 8.9.2: *Two subspaces of $\mathscr{E}^n$ are orthogonal if and only if every vector in a basis of either space is orthogonal to every vector in a basis of the other.*

Let the spaces be $\mathscr{V}$, with a basis $A_1, A_2, \ldots, A_m$, and $\mathscr{W}$, with a basis $B_1, B_2, \ldots, B_p$. First assume that $\mathscr{V}$ and $\mathscr{W}$ are orthogonal. Then every A_i is orthogonal to every B_j, by the definition of orthogonal spaces.

Conversely, let every A_i be orthogonal to every B_j, so that $A_i^{\mathsf{T}}B_j = 0$ for $i = 1, 2, \ldots, k; j = 1, 2, \ldots, p$. Then

$$(\alpha_1 A_1 + \alpha_2 A_2 + \cdots + \alpha_m A_m)^{\mathsf{T}}(\beta_1 B_1 + \beta_2 B_2 + \cdots + \beta_p B_p) = \sum_{i=1}^{m} \sum_{j=1}^{p} \alpha_i \beta_j A_i^{\mathsf{T}} B_j = 0.$$

Thus every vector of $\mathscr{V}$ is orthogonal to every vector of $\mathscr{W}$ and the theorem is completely proved.

A simple example is the pair of vector spaces $\mathscr{V}$ and $\mathscr{W}$, where $\mathscr{V}$ is the space of all vectors in a fixed plane on O and $\mathscr{W}$ is the space of all vectors on a fixed line on O, the line and the plane being orthogonal (Figure 8-11).

Theorem 8.9.3: *The set $\mathscr{V}^{\perp}$ (read $\mathscr{V}$-orthocomplement) of all vectors orthogonal to every vector of a given subspace $\mathscr{V}$ of $\mathscr{E}^n$ is a vector space orthogonal to $\mathscr{V}$. The sum of the dimensions of $\mathscr{V}$ and $\mathscr{V}^{\perp}$ is n.*

Let $\mathscr{V}$ have dimension k and let $A_1, A_2, \ldots, A_k$ constitute a basis for $\mathscr{V}$. Then, by Theorem 8.9.1, the vectors X belonging to the set $\mathscr{V}^{\perp}$ constitute the set of all solutions of the system of equations

$$A_1^{\mathsf{T}}X = 0, A_2^{\mathsf{T}}X = 0, \ldots, A_k^{\mathsf{T}}X = 0.$$

Since the A's are independent, the coefficient matrix $[A_1, A_2, \ldots, A_k]^{\mathsf{T}}$ of this system has rank k and hence there are $n - k$ linearly independent solutions. Thus $\mathscr{V}^{\perp}$ is a vector space of dimension $n - k$ and the theorem now follows with the aid of Theorem 8.9.2.

The vector spaces represented in Figure 8-11 illustrate this theorem.

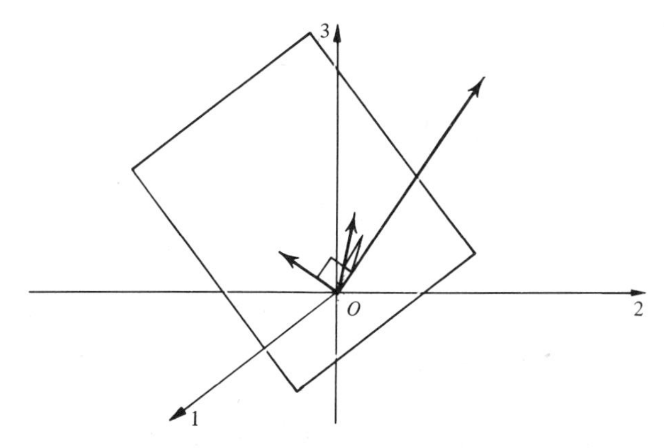

Figure 8-11. *Orthogonal Vector Spaces.*

Two mutually orthogonal subspaces of $\mathscr{E}^n$ are called **complementary orthogonal vector spaces**, briefly, **orthocomplements**, if and only if the sum of their dimensions is n. In particular, $\mathscr{V}$ and $\mathscr{V}^\perp$ are orthocomplements.

Theorem 8.9.4: *In $\mathscr{E}^n$, the sum of complementary orthogonal subspaces $\mathscr{V}$ and $\mathscr{V}^\perp$ of $\mathscr{E}^n$ is $\mathscr{E}^n$.*

Let $A_1, A_2, \ldots, A_k$ and $B_1, B_2, \ldots, B_{n-k}$ be orthonormal bases for $\mathscr{V}$ and $\mathscr{V}^\perp$. If $X = \alpha_1 A_1 + \alpha_2 A_2 + \cdots + \alpha_k A_k + \beta_1 B_1 + \beta_2 B_2 + \cdots + \beta_{n-k} B_{n-k} = 0$, then, for each i and j, $A_i^{\mathsf{T}} X = \alpha_i = 0$ and $B_j^{\mathsf{T}} X = \beta_j = 0$, so the set of n-vectors $A_1, A_2, \ldots, A_k, B_1, B_2, \ldots, B_{n-k}$ is a linearly independent set and hence constitutes a basis (actually an orthonormal basis) for $\mathscr{E}^n$. This proves the theorem.

Theorem 8.9.5: *Let $\mathscr{V}$ be the vector space consisting of all solutions of the equation $AX = 0$. Then the complementary orthogonal space $\mathscr{V}^\perp$ is spanned by the columns of A^{T}.*

Let $A^{(i)}$ denote the ith row of A, that is, the ith column of A^{T}. If X is any solution of $AX = 0$, then for each i, $A^{(i)}X = 0$, so X is orthogonal to every *column* of A^{T}. By Theorem 8.9.1, this implies that every X in the solution space of $AX = 0$ is orthogonal to every vector of the column space of A^{T}, so these two spaces are orthogonal. Moreover, their dimensions are, respectively, $n - r(A)$ and $r(A)$, so they are complementary orthogonal spaces.

For example, consider in $\mathscr{E}^4$ the subspace defined by the equation

$$AX = \begin{bmatrix} 1 & 1 & 0 & 0 \\ 0 & 1 & 1 & 0 \\ 0 & 0 & 1 & 1 \end{bmatrix} X = 0.$$

The solution space $\mathscr{V}$ has dimension $4 - 3 = 1$ and is the set of all scalar multiples of

$$\begin{bmatrix} 1 \\ -1 \\ 1 \\ -1 \end{bmatrix}.$$

The complementary orthogonal space $\mathscr{V}^\perp$ is spanned by the columns of A^{T}:

$$\begin{bmatrix} 1 \\ 1 \\ 0 \\ 0 \end{bmatrix}, \quad \begin{bmatrix} 0 \\ 1 \\ 1 \\ 0 \end{bmatrix}, \quad \begin{bmatrix} 0 \\ 0 \\ 1 \\ 1 \end{bmatrix}$$

and has dimension 3. The last four vectors written are linearly independent and hence span $\mathscr{E}^4$.

Theorem 8.9.6: *If $\mathscr{V}$ and $\mathscr{V}^\perp$ are orthocomplements in $\mathscr{E}^n$, then every vector X in $\mathscr{E}^n$ has a unique decomposition of the form $X = X_1 + X_2$, where X_1 belongs to $\mathscr{V}$ and X_2 belongs to $\mathscr{V}^\perp$.*

Let $A_1, A_2, \ldots, A_k$ and $B_1, B_2, \ldots, B_{n-k}$ be orthonormal bases for $\mathscr{V}$ and $\mathscr{V}^\perp$, respectively. Then, as in the proof of Theorem 8.9.4, these n vectors are an independent set and hence are a basis for $\mathscr{E}^n$. X therefore has a unique representation of the form

$$X = \sum_{i=1}^{k} \alpha_i A_i + \sum_{j=1}^{n-k} \beta_j B_j.$$

In this expression, the first sum represents a vector in $\mathscr{V}$ and the second represents a vector in $\mathscr{W}$, that is,

$$X_1 = \sum_{i=1}^{k} \alpha_i A_i \qquad \text{and} \qquad X_2 = \sum_{j=1}^{n-k} \beta_j B_j.$$

This completes the proof of the theorem.

8.10 EXERCISES

***1.** Show that if an n-vector X is orthogonal to each of the linearly independent vectors $A_1, A_2, \ldots, A_k$, then $X, A_1, A_2, \ldots, A_k$ are a linearly independent set.

***2.** Prove that $A_1, A_2, \ldots, A_k$ are mutually orthogonal unit n-vectors if and only if

$$[A_1, A_2, \ldots, A_k]^\mathsf{T}[A_1, A_2, \ldots, A_k] = I_k.$$

***3.** Given that $U_1, U_2, \ldots, U_k$ are mutually orthogonal unit n-vectors, prove that one can find n-vectors $U_{k+1}, \ldots, U_n$ such that $U_1, U_2, \ldots, U_n$ form an orthonormal basis for $\mathscr{E}^n$. Give examples in $\mathscr{E}^3$ for $k = 1$ and $k = 2$.

4. Given that the rows of $A_{m \times n}$ are mutually orthogonal unit vectors, show that $AX_{n \times 1} = B_{m \times 1}$ has a unique solution for X. What can you say about the relative sizes of m and n?

5. Reduce the system of equations

$$\begin{aligned} x_1 - x_2 + x_3 \qquad\quad &= 1 \\ x_1 - x_2 \qquad + 2x_4 &= 2 \\ -2x_1 \qquad + 2x_3 + x_4 &= 1 \end{aligned}$$

to an equivalent system in which the rows of coefficients of the x's are mutually orthogonal unit vectors and then solve it.

*6. Given that $A_1, A_2, \ldots, A_k$ constitute an orthonormal basis for a subspace $\mathscr{V}$ of $\mathscr{E}^n$, and that U is an orthogonal $k \times k$ matrix, what can be said about the vectors $B_1, B_2, \ldots, B_k$ determined by the relation

$$[B_1, B_2, \ldots, B_k] = [A_1, A_2, \ldots, A_k]U?$$

7. Find an orthonormal basis for the orthocomplement of the two-dimensional subspace of $\mathscr{E}^4$ spanned by the vectors

$$\begin{bmatrix} 1 \\ -1 \\ 0 \\ 1 \end{bmatrix}, \quad \begin{bmatrix} 1 \\ 0 \\ 1 \\ 0 \end{bmatrix}.$$

8. Given that

$$X = \begin{bmatrix} 0 \\ 1 \\ 0 \\ 1 \end{bmatrix},$$

find X_1 and X_2 such that $X = X_1 + X_2$ and X_1 and X_2 belong to the orthocomplementary spaces of Exercise 7.

9. Let the equation

$$[A, B]_{m \times n} X_{n \times 1} = C_{m \times 1}, \qquad n \geq m,$$

where A is an $m \times m$ orthogonal matrix, be multiplied by A^{T}. What is the effect on the corresponding system of equations?

10. Solve this system by inspection, using the idea of Exercise 9:

$$\frac{1}{\sqrt{2}} x_1 - \frac{1}{\sqrt{2}} x_2 + x_3 = 1$$
$$\frac{1}{\sqrt{2}} x_1 + \frac{1}{\sqrt{2}} x_2 - x_3 = 1.$$

*11. Prove that the system of linear equations represented by $AX = B$ is consistent if and only if B is orthogonal to every solution of the corresponding transposed homogeneous system $A^{\mathsf{T}} Y = 0$.

*12. Extend the results of Section 8.9 and the starred exercises of Section 8.10 to $\mathscr{U}^n$.

8.11 LINEAR TRANSFORMATIONS OF COORDINATES

Up to now, we have interpreted an equation $Y = AX$, where A is real, as a linear operator on $\mathscr{E}^n$, that is, as a mapping that operates on vectors of $\mathscr{E}^n$ to produce vectors of $\mathscr{E}^n$ or, equivalently, as a mapping that operates on points of $\mathscr{E}^n$ to produce points of $\mathscr{E}^n$. In this interpretation, the basis or reference system remains fixed.

If $V_1, V_2, \ldots, V_k$ constitute a basis for a vector space $\mathscr{V}$ and if

$$X = \alpha_1 V_1 + \alpha_2 V_2 + \cdots + \alpha_k V_k,$$

then $\alpha_1, \alpha_2, \ldots, \alpha_k$ are called the **coordinates** of X with respect to the given basis. Moreover, once the basis is chosen, these coordinates are uniquely determined for every X in $\mathscr{V}$. In particular, $E_1, E_2, \ldots, E_n$ are called the **standard basis** or the **natural basis** for $\mathscr{E}^n$ and since, for each X in $\mathscr{E}^n$,

$$\begin{bmatrix} x_1 \\ x_2 \\ \vdots \\ x_n \end{bmatrix} = x_1 E_1 + x_2 E_2 + \cdots + x_n E_n,$$

the components of X are its coordinates with respect to the standard basis.

Note that, given the origin, a given ordered basis for $\mathscr{E}^n$ may be used to determine a coordinate system for $\mathscr{E}^n$ in the following way: The j-axis has the direction of the jth basis vector, which identifies the positive sense and the unit of length on this axis. For example, the vectors E_1, E_2, E_3 generate the familiar, orthogonal coordinate system in $\mathscr{E}^3$ (Figure 8-12).

Equally well, we can choose any three linearly independent vectors A_1, A_2, A_3 of $\mathscr{E}^3$ as basis vectors and represent X as a linear combination,

$$X = y_1 A_1 + y_2 A_2 + y_3 A_3, \tag{8.11.1}$$

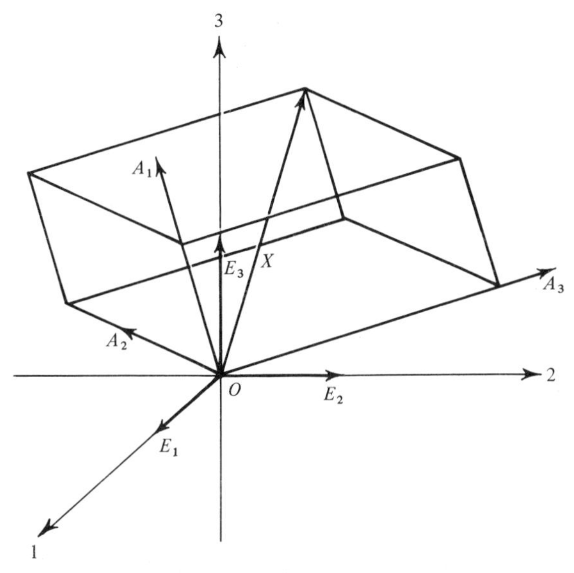

Figure 8-12. $X = x_1 E_1 + x_2 E_2 + x_3 E_3 = y_1 A_1 + y_2 A_2 + y_3 A_3$.

which may also be written

(8.11.2) $$X = AY,$$

where $A = [A_1, A_2, A_3]$. In this equation, the columns of A are the new basis vectors and, given any vector X in $\mathscr{E}^3$, the equivalent equation

(8.11.3) $$Y = A^{-1}X$$

yields explicitly the coordinates y_1, y_2, y_3, of X with respect to the basis A_1, A_2, A_3. We have computed these coordinates in earlier chapters by solving the system of equations represented by (8.11.1) by sweepout.

This interpretation generalizes at once to $\mathscr{E}^n$. Let X be any vector of $\mathscr{E}^n$, referred to the standard basis. Let $A_1, A_2, \ldots, A_n$ be an arbitrary basis for $\mathscr{E}^n$, referred to the standard reference system, so that there exist $y_1, y_2, \ldots, y_n$ such that

(8.11.4) $$X = y_1A_1 + y_2A_2 + \cdots + y_nA_n$$

or

(8.11.5) $$X = AY,$$

where

$$A = [A_1, A_2, \ldots, A_n],$$

that is, where *the columns of A are the old coordinates of the vectors constituting the new basis*. Then, for each X, the equation

(8.11.6) $$Y = A^{-1}X$$

yields the coordinates $y_1, y_2, \ldots, y_n$ of X with respect to the new basis.

We thus have an alternative geometrical interpretation of a linear transformation $Y = AX$: We regard vectors and points of $\mathscr{E}^n$ as remaining fixed while the reference system is being changed. The transformation equations now define the relationship between the coordinates of any vector X with respect to the standard basis and its coordinates with respect to the new basis. A linear transformation interpreted this way is called a **linear transformation of coordinates** or a **change of basis**.

To illustrate, consider the linear transformation of coordinates in $\mathscr{E}^2$ whose equations are

$$X = \begin{bmatrix} 5 & -1 \\ -3 & 6 \end{bmatrix} Y, \qquad Y = \tfrac{1}{27}\begin{bmatrix} 6 & 1 \\ 3 & 5 \end{bmatrix} X.$$

The new basis vectors are

$$A_1 = \begin{bmatrix} 5 \\ -3 \end{bmatrix} \quad \text{and} \quad A_2 = \begin{bmatrix} -1 \\ 6 \end{bmatrix}$$

in the standard reference system. The corresponding Y's are, of course, $\begin{bmatrix} 1 \\ 0 \end{bmatrix}$ and $\begin{bmatrix} 0 \\ 1 \end{bmatrix}$, respectively. Now the lengths of A_1 and A_2 determine the units of

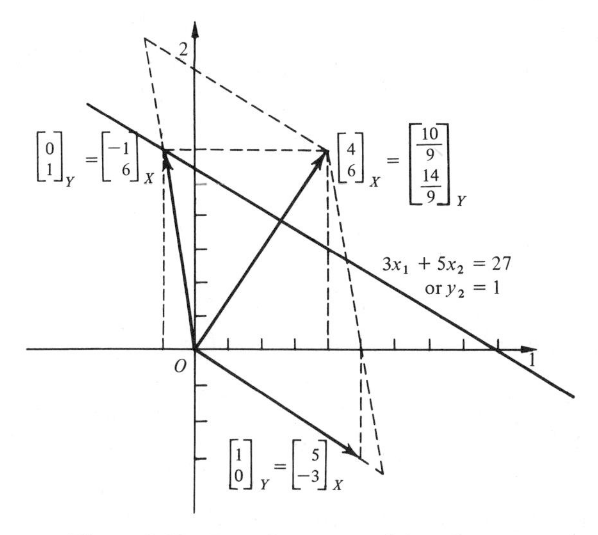

Figure 8-13. *Transformation of Coordinates.*

length on the y_1-axis and the y_2-axis, respectively, since these are the unit vectors of the new system. These units are, respectively, $\sqrt{34}$ and $\sqrt{37}$ times the original unit of length (Figure 8-13). Consider the vector $\begin{bmatrix}4\\6\end{bmatrix}$ in the standard basis. We have, for the coordinates with respect to the new basis,

$$Y = \frac{1}{27}\begin{bmatrix}6 & 1\\3 & 5\end{bmatrix}\begin{bmatrix}4\\6\end{bmatrix} = \begin{bmatrix}\frac{10}{9}\\\frac{14}{9}\end{bmatrix}.$$

Note that these smaller coordinate values are consistent with the larger units of length.

In view of the one-to-one correspondence between points and vectors, we can also answer questions such as this one: If a line has the equation

$$3x_1 + 5x_2 = 27$$

or

$$[3, 5]X = 27$$

in the standard reference system, what is its equation in the new reference system? Since

$$X = \begin{bmatrix}5 & -1\\-3 & 6\end{bmatrix}Y,$$

we have

$$[3, 5]\begin{bmatrix}5 & -1\\-3 & 6\end{bmatrix}Y = 27$$

or

$$[0, 27]Y = 27$$

or simply

$$y_2 = 1,$$

so the line is parallel to the y_1-axis and passes through the point $(0, 1)_Y$.

The preceding transformation does not preserve formulas for length or angle. But it does preserve the test for parallelism of lines. (Can you prove these statements?)

By way of contrast, consider the linear transformation of coordinates defined by

$$Y = \begin{bmatrix} \cos\theta & \sin\theta \\ -\sin\theta & \cos\theta \end{bmatrix} X \quad \text{or} \quad X = \begin{bmatrix} \cos\theta & -\sin\theta \\ \sin\theta & \cos\theta \end{bmatrix} Y,$$

where X is referred to the standard basis. The coordinates of the new unit vectors with respect to the standard basis may be obtained by substituting $\begin{bmatrix} 1 \\ 0 \end{bmatrix}$ and $\begin{bmatrix} 0 \\ 1 \end{bmatrix}$ for Y in the second of these equations and are the columns $\begin{bmatrix} \cos\theta \\ \sin\theta \end{bmatrix}$ and $\begin{bmatrix} -\sin\theta \\ \cos\theta \end{bmatrix}$, respectively. These two vectors are orthogonal unit vectors, so the new reference system is again orthonormal (Figure 8-14). Since

$$\begin{bmatrix} \cos\theta & -\sin\theta \\ \sin\theta & \cos\theta \end{bmatrix}^{\mathsf{T}} \begin{bmatrix} \cos\theta & -\sin\theta \\ \sin\theta & \cos\theta \end{bmatrix} = I_2,$$

we have $Y^{\mathsf{T}}Y = X^{\mathsf{T}}X$, so the formula for length is preserved. Also, if Y_1 corresponds to X_1 and Y_2 corresponds to X_2, we have $Y_1^{\mathsf{T}}Y_2 = X_1^{\mathsf{T}}X_2$, so the

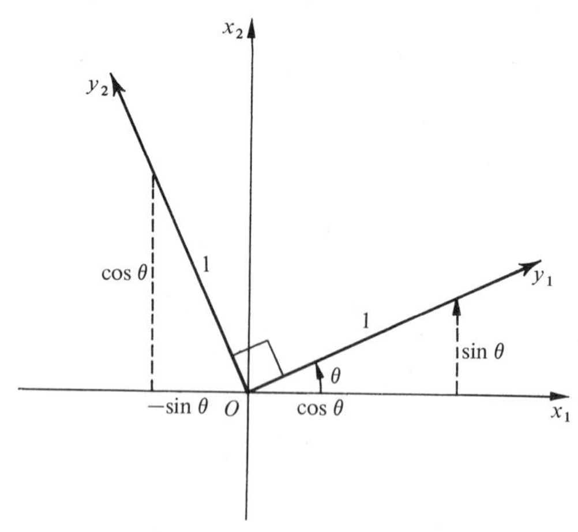

Figure 8-14. *Rotation of Axes.*

formula for angular measure and the test for orthogonality are also preserved. This rotation of the axes through the angle θ is an example of an orthogonal transformation of coordinates.

In $\mathscr{E}^n$, an orthogonal transformation of coordinates is represented by an equation $Y = UX$, where U is orthogonal. We can prove, by the same computations used in earlier sections, but employing the new interpretation,

Theorem 8.11.1: *In $\mathscr{E}^n$, a linear transformation of coordinates leaves invariant the formulas for the inner product, for the length of a vector, and for the distance between two points, if and only if it is orthogonal.*

Thus, if $Y_1 = UX_1$ and $Y_2 = UX_2$ and if U is orthogonal, then $Y_1^{\mathsf{T}} Y_2 = X_1^{\mathsf{T}} X_2$, $|Y_1| = |X_1|$, and $|Y_2 - Y_1| = |X_2 - X_1|$. These facts imply also that in the new reference system the formula for the cosine of the angle between two vectors is the same.

If the transformation of coordinates were not orthogonal, the formulas for inner product and length would not be preserved. Thus, in the case of the first example of this section,

$$X^{\mathsf{T}}X = Y^{\mathsf{T}}\begin{bmatrix} 5 & -3 \\ -1 & 6 \end{bmatrix}\begin{bmatrix} 5 & -1 \\ -3 & 6 \end{bmatrix}Y = Y^{\mathsf{T}}\begin{bmatrix} 34 & -23 \\ -23 & 37 \end{bmatrix}Y, \qquad \text{not } Y^{\mathsf{T}}Y.$$

8.12 TRANSFORMATION OF A LINEAR OPERATOR

Let the equation

$$X_2 = AX_1 \tag{8.12.1}$$

represent a linear operator on $\mathscr{E}^n$, the reference system being the standard basis (our first interpretation of a linear transformation). Let

$$X = BY, \qquad Y = B^{-1}X \tag{8.12.2}$$

be a transformation of coordinates which defines the coordinates $y_1, y_2, \ldots, y_n$ of the vector X with respect to the new basis defined by the columns of B.

What is the representation of the operator (8.12.1) in this new reference system? Since X_1 and X_2 will ordinarily have altered coordinates in the new reference system, one should expect to replace A by a different matrix in order to effect the same geometrical mapping. If the new names of X_1 and X_2 are Y_1 and Y_2, respectively, so that

$$X_1 = BY_1, \qquad X_2 = BY_2,$$

we have, from (8.12.1),

$$BY_2 = ABY_1,$$

so

$$Y_2 = (B^{-1}AB)Y_1.$$

This proves

Theorem 8.12.1: *The operator on $\mathscr{E}^n$ represented by the matrix A in the standard reference system is represented by the matrix $B^{-1}AB$ in the reference system whose basis vectors are defined, in the standard reference system, by the columns of B.*

Thus, not only does every vector of $\mathscr{E}^n$ have a particular set of coordinates in every reference system for $\mathscr{E}^n$, but also every linear operator on $\mathscr{E}^n$ has a particular representation in every reference system for $\mathscr{E}^n$.

The mapping of A onto the matrix $B^{-1}AB$ is called a **similarity transformation** of A, and A and $B^{-1}AB$ are said to be **similar**. An important problem of matrix algebra is the determination of the canonical forms to which a matrix can be reduced by similarity transformations. The particularly important case where B is orthogonal will be treated in Chapter 9.

To illustrate the theorem, let the operator be

$$X_2 = \begin{bmatrix} 1 & 2 \\ 2 & 1 \end{bmatrix} X_1$$

and let the transformation of coordinates be the orthogonal transformation defined by

$$X = \begin{bmatrix} \frac{1}{\sqrt{2}} & -\frac{1}{\sqrt{2}} \\ \frac{1}{\sqrt{2}} & \frac{1}{\sqrt{2}} \end{bmatrix} Y;$$

that is, let the axes be rotated through 45°.

In the new reference system, the operator is

$$Y_2 = \begin{bmatrix} \frac{1}{\sqrt{2}} & \frac{1}{\sqrt{2}} \\ -\frac{1}{\sqrt{2}} & \frac{1}{\sqrt{2}} \end{bmatrix} \begin{bmatrix} 1 & 2 \\ 2 & 1 \end{bmatrix} \begin{bmatrix} \frac{1}{\sqrt{2}} & -\frac{1}{\sqrt{2}} \\ \frac{1}{\sqrt{2}} & \frac{1}{\sqrt{2}} \end{bmatrix} Y_1$$

or

$$Y_2 = \begin{bmatrix} 3 & 0 \\ 0 & -1 \end{bmatrix} Y_1.$$

(See Figure 8-15).

In the new reference system, the operator is easily described geometrically. To get the image Y_2 of Y_1, one multiplies the first component of Y_1 by 3 and the second component by -1. This is a small illustration of how a properly chosen transformation can simplify and clarify a problem.

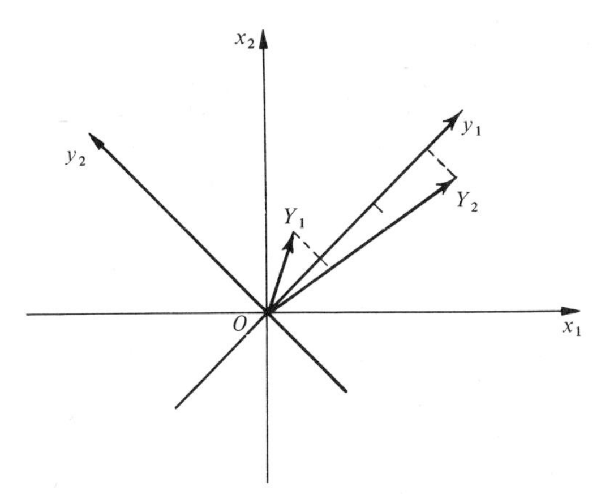

Figure 8-15. *Transformation of a Linear Operator.*

8.13 THE ALGEBRA OF LINEAR OPERATORS

Given the operators $Y = AX$ and $Z = BY$, they permit us to map X onto Z by first mapping X onto Y, then mapping Y onto Z. Substituting from the first equation into the second, we get $Z = (BA)X$, which represents an operator that maps X directly onto Z and which is called the **product** of the two first-given operators. Such a product of operators is always defined when the matrices B and A which represent them are conformable, so that the product BA is defined. Note that the matrix of the first operator appears on the *right* in this product.

If now $W = CZ$, and if we substitute for Z from the previous equation, we get $W = C(BA)X$. If, on the other hand, we first form the product of $W = CZ$ and $Z = BY$ to obtain $W = (CB)Y$, and thereafter form the product with $Y = AX$, we get $W = (CB)AX$. Since $C(BA) = (CB)A$, the two final products are the same. This fact and the fact that ordinarily $AB \neq BA$ allow us to conclude

Theorem 8.13.1: *The multiplication of linear operators is associative but is not in general commutative.*

It is important to recognize that the most essential aspect of the representation $Y = AX$ of a linear operator is the matrix A. The symbols Y and X are just convenient symbols for the vectors which correspond under the mapping. If the domain is the same for both Z and X, the equations $W = AZ$ and $Y = AX$ represent the same operator. The only occasion when the symbols for the vectors are of special significance is when we wish to form the product of the operators. In this case, these symbols serve to indicate the order in which the multiplication is to be carried out, as in the preceding discussion.

In the algebra of linear operators, the identity mapping and inverses of mappings are of particular significance. We examine them next.

The **identity operator** on $\mathscr{E}^n$ is the mapping such that for all X in $\mathscr{E}^n$, the image of X is X. (The reader should show that this mapping is linear.) If $AX = X$ for all X, by putting X successively equal to $E_1, E_2, \ldots, E_n$, we find that $A_j = E_j$, $j = 1, 2, \ldots, n$, so that $A = I_n$. Conversely, if $A = I_n$, the operator with matrix A is the identity operator. This proves

Theorem 8.13.2: *The identity operator on $\mathscr{E}^n$ is represented by the matrix I_n and by no other.*

The product of the operator represented by the matrix A and the identity operator has matrix AI or IA and hence is simply the operator represented by the matrix A.

We have already seen that if A is nonsingular, the operator represented by the matrix A^{-1} is the inverse of the operator represented by the matrix A. The product of these two operators is represented by the matrix AA^{-1} or $A^{-1}A$, namely I_n, and thus is the identity operator.

Suppose now that $Y = AX$ has an inverse. Then, like any other mapping, this transformation must be one-to-one. If A were to have rank less than n, then the equation $AX = 0$ would have more than one solution, so the vector zero would have more than one counterimage and the mapping would not be one-to-one. Hence A has rank n, A^{-1} exists, and the inverse mapping is $X = A^{-1}Y$. Thus we have

Theorem 8.13.3: *A linear operator on $\mathscr{E}^n$ represented by a matrix A has an inverse if and only if A is nonsingular, and the inverse operator is represented by A^{-1}.*

A set of mappings from a vector space $\mathscr{V}$ to itself is said to form a **group** if and only if

(a) the set is closed with respect to multiplication; that is, the product of any two operators in the set is also in the set, (b) the identity mapping belongs to the set, and (c) every mapping of the set has an inverse which also belongs to the set.

One of the simplest and most useful examples of a group of operators is the set of all nonsingular operators on $\mathscr{E}^n$.

We can, of course, discuss the algebra of linear transformations of coordinates in the same way and hence conclude that the set of all linear transformations of coordinates in $\mathscr{E}^n$ is a group.

8.14 EXERCISES

1. Find the image of the triangle with vertices $(0, 0)$, (a, b), (c, d), by the operator

$$X_2 = \begin{bmatrix} 1 & -1 \\ 1 & 1 \end{bmatrix} X_1.$$

Illustrate with a figure. What effect does this operator have on the length of an arbitrary vector? How is the area of the image triangle related to that of the original triangle? (Recall the determinant formula for the area of a triangle here.) What effect does the transformation have on areas in general? Why?

2. Note that

$$\begin{bmatrix} \sqrt{2} & 0 \\ 0 & \sqrt{2} \end{bmatrix} \begin{bmatrix} \frac{1}{\sqrt{2}} & -\frac{1}{\sqrt{2}} \\ \frac{1}{\sqrt{2}} & \frac{1}{\sqrt{2}} \end{bmatrix} = \begin{bmatrix} 1 & -1 \\ 1 & 1 \end{bmatrix}.$$

How, therefore, may the effect of the operator in Exercise 1 be described geometrically as the product of two operators whose geometrical properties are known?

3. What is the form of the operator in Exercise 1 in the reference system defined by

$$\begin{bmatrix} \cos\theta \\ \sin\theta \end{bmatrix}, \quad \begin{bmatrix} -\sin\theta \\ \cos\theta \end{bmatrix}?$$

How do you explain the result geometrically?

4. If it is desired to use

$$A_1 = \begin{bmatrix} 1 \\ 1 \\ 1 \end{bmatrix}, \quad A_2 = \begin{bmatrix} 1 \\ 1 \\ -2 \end{bmatrix}, \quad A_3 = \begin{bmatrix} -1 \\ 1 \\ 0 \end{bmatrix},$$

expressed in the standard basis, as a new basis for $\mathscr{E}^3$, what equation gives the new coordinates of a given vector X in terms of its coordinates with respect to the standard basis?

5. If a linear transformation of coordinates maps the standard basis representations

$$X_1 = \begin{bmatrix} 1 \\ 2 \\ -1 \end{bmatrix}, \quad X_2 = \begin{bmatrix} 3 \\ 1 \\ 0 \end{bmatrix}, \quad X_3 = \begin{bmatrix} -1 \\ 0 \\ 0 \end{bmatrix},$$

respectively, into

$$Y_1 = \begin{bmatrix} 1 \\ 0 \\ 0 \end{bmatrix}, \quad Y_2 = \begin{bmatrix} -1 \\ 1 \\ 0 \end{bmatrix}, \quad Y_3 = \begin{bmatrix} -2 \\ 0 \\ 1 \end{bmatrix},$$

find a matrix equation that defines the transformation and identify the new basis vectors.

***6.** Prove that an orthogonal operator on $\mathscr{E}^n$ maps any set of mutually orthogonal vectors onto a set of mutually orthogonal vectors.

7. If a transformation of coordinates in $\mathscr{E}^4$ is such that the standard basis vectors have new coordinates given by

$$\begin{bmatrix} 1 \\ 1 \\ 1 \\ 0 \end{bmatrix}, \quad \begin{bmatrix} 1 \\ 1 \\ 0 \\ 1 \end{bmatrix}, \quad \begin{bmatrix} 1 \\ 0 \\ 1 \\ 1 \end{bmatrix}, \quad \begin{bmatrix} 0 \\ 1 \\ 1 \\ 1 \end{bmatrix},$$

respectively, write a matrix equation for the transformation.

8. Show that in $\mathscr{E}^2$ the equation $x_2 = (\tan \alpha)x_1$ of a line on the origin is transformed into the equation $y_2 = (\tan(\alpha - \theta))y_1$ by a rotation of axes through the angle θ. Illustrate with a figure.

9. Let the equation

$$\begin{bmatrix} y_1 \\ y_2 \end{bmatrix} = \begin{bmatrix} 1 & 2 \\ 0 & 1 \end{bmatrix} \begin{bmatrix} x_1 \\ x_2 \end{bmatrix}$$

define a transformation of coordinates in $\mathscr{E}^2$ in the usual way. Draw a figure showing both the standard and the new reference systems. Then determine what points have the same coordinates in both systems.

10. What figures in $\mathscr{E}^2$ are invariant under a given rotation about the origin? Under *every* rotation about the origin?

11. Show that the set of all linear operators on $\mathscr{E}^n$ of the form $Y = UX$, where U is orthogonal, constitute a group (the **orthogonal group**).

12. Given that

$$A = \begin{bmatrix} a & -3b & c_1 \\ 2a & 2b & c_2 \\ -a & b & c_3 \end{bmatrix},$$

choose a, b, c_1, c_2, and c_3 so that A is orthogonal. Is the solution unique?

***13.** Let the columns of A and B define two reference systems in $\mathscr{E}^n$. Let an operator on $\mathscr{E}^n$ have the representation

$$X_2 = CX_1$$

in the reference system defined by the columns of A. What is its form in the reference system defined by the columns of B?

14. Let $X_2 = AX_1$ define a linear operator on $\mathscr{E}^n$. Let $\mathscr{V}$ be a subspace of $\mathscr{E}^n$ such that the image by this operator of every vector in $\mathscr{V}$ is still in $\mathscr{V}$. Then $\mathscr{V}$ is called an **invariant subspace** with respect to this operator. (Note that this definition does not require individual vectors to be invariant.)

(a) Invent concrete examples in $\mathscr{E}^2$ and $\mathscr{E}^3$ and illustrate with figures.

(b) Show that if $\mathscr{V}$ is invariant with respect to the operator defined by $X_2 = AX_1$, then it is also invariant with respect to every operator of the form $X_2 = f(A)X_1$, where $f(A)$ is a polynomial function of A.

15. Extend the results of Sections 8.11 and 8.12 to $\mathscr{U}^n$, so far as possible.

CHAPTER 9

The Characteristic Value Problem

9.1 DEFINITION OF THE CHARACTERISTIC VALUE PROBLEM

Given a real matrix A of order 3, consider the linear operator $X_2 = AX_1$, which maps $\mathscr{E}^3$ into $\mathscr{E}^3$. Ordinarily the vectors X_2 and X_1, where $X_2 = AX_1$, are not collinear. An important fact in many applications is that, for certain vectors $X \neq 0$, AX and X *are* collinear; that is, there exists a scalar λ such that $AX = \lambda X$.

The **characteristic value problem** or the **eigenvalue problem** is the following: Given a real or complex $n \times n$ matrix A, for what vectors $X \neq 0$ and for what scalars λ is it true that

$$AX = \lambda X; \tag{9.1.1}$$

that is, for what vectors X are AX and X proportional?

A nonzero vector X which satisfies (9.1.1) is called a **characteristic vector** (Figure 9-1) or **eigenvector** or **latent vector** of A and the associated value λ is called a **characteristic root** or **eigenvalue** or **latent root** of A.

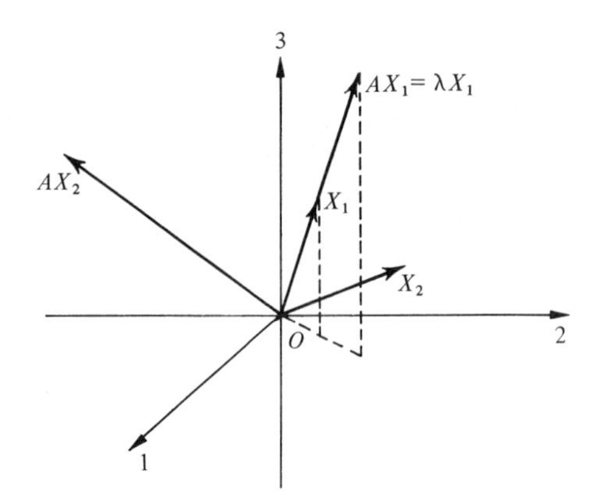

Figure 9-1. *Characteristic and Noncharacteristic Vectors.*

For example, since

$$\begin{bmatrix} 1 & 2 \\ 2 & 1 \end{bmatrix}\begin{bmatrix} 1 \\ 1 \end{bmatrix} = 3\begin{bmatrix} 1 \\ 1 \end{bmatrix},$$

$\begin{bmatrix} 1 \\ 1 \end{bmatrix}$ is a characteristic vector and 3 is the associated characteristic root of the matrix $\begin{bmatrix} 1 & 2 \\ 2 & 1 \end{bmatrix}$.

Now (9.1.1) holds if and only if

$$(A - \lambda I)X = 0, \tag{9.1.2}$$

where the scalar λ has been replaced by the scalar matrix λI in order to make it possible to extract the factor X. This equation represents a homogeneous system of n equations in n unknowns which has a nontrivial solution for X if and only if the coefficient matrix has rank $r < n$, that is, if and only if

$$\det(A - \lambda I) = 0, \tag{9.1.3}$$

or, equivalently,

$$\begin{vmatrix} a_{11} - \lambda & a_{12} & \cdots & a_{1n} \\ a_{21} & a_{22} - \lambda & \cdots & a_{2n} \\ \vdots & & & \\ a_{n1} & a_{n2} & \cdots & a_{nn} - \lambda \end{vmatrix} = 0. \tag{9.1.4}$$

This equation is called the **characteristic equation** of A.

As inspection of the main diagonal shows, this determinant will expand into a polynomial of degree n in the variable λ. If we solve the equation, we get n roots, $\lambda_1, \lambda_2, \ldots, \lambda_n$ (not necessarily all different). These are the **characteristic roots** of A. If λ_j is one of these roots, we solve the equation

$$(A - \lambda_j I)X = 0$$

for X. The nontrivial solutions are the characteristic vectors X of A that are associated with the characteristic root λ_j; that is, they are the nonzero vectors X such that

$$AX = \lambda_j X.$$

The vector space consisting of all solutions of this equation includes the zero vector in addition to the characteristic vectors associated with λ_j and is called the **characteristic subspace** associated with λ_j.

9.2 TWO EXAMPLES

Consider again the matrix

$$A = \begin{bmatrix} 2 & 1 \\ 1 & 2 \end{bmatrix}.$$

We want to determine all λ's and all X's such that

$$\begin{bmatrix} 2 & 1 \\ 1 & 2 \end{bmatrix} X = \lambda X.$$

Here we need

$$\det(A - \lambda I) = \begin{vmatrix} 2-\lambda & 1 \\ 1 & 2-\lambda \end{vmatrix} = (2-\lambda)^2 - 1 = (1-\lambda)(3-\lambda) = 0,$$

so the characteristic roots are $\lambda = 1$ and $\lambda = 3$.

Putting $\lambda = 1$ in $(A - \lambda I)X = 0$, we get

$$\begin{bmatrix} 1 & 1 \\ 1 & 1 \end{bmatrix} \begin{bmatrix} x_1 \\ x_2 \end{bmatrix} = 0$$

or, in scalar form, a system that reduces to the single equation

$$x_1 + x_2 = 0,$$

which has the complete solution

$$\begin{bmatrix} x_1 \\ x_2 \end{bmatrix} = t \begin{bmatrix} -1 \\ 1 \end{bmatrix}.$$

For any $t \neq 0$, we have a characteristic vector associated with the root $\lambda = 1$. In fact, for all t,

$$\begin{bmatrix} 2 & 1 \\ 1 & 2 \end{bmatrix} \begin{bmatrix} -t \\ t \end{bmatrix} = 1 \begin{bmatrix} -t \\ t \end{bmatrix},$$

as required. That is, associated with the root $\lambda = 1$ there is a one-dimensional characteristic subspace. In this particular instance, because $\lambda = 1$, every individual vector of this subspace is left invariant by the operator.

Now we put $\lambda = 3$ in $(A - \lambda I)X = 0$ and get

$$\begin{bmatrix} -1 & 1 \\ 1 & -1 \end{bmatrix} \begin{bmatrix} x_1 \\ x_2 \end{bmatrix} = 0,$$

which is equivalent to the single scalar equation

$$x_1 - x_2 = 0$$

with the complete solution

$$\begin{bmatrix} x_1 \\ x_2 \end{bmatrix} = t \begin{bmatrix} 1 \\ 1 \end{bmatrix}.$$

For any $t \neq 0$, we get a characteristic vector associated with the root $\lambda = 3$. In fact, for all t,

$$\begin{bmatrix} 2 & 1 \\ 1 & 2 \end{bmatrix} \begin{bmatrix} t \\ t \end{bmatrix} = 3 \begin{bmatrix} t \\ t \end{bmatrix}.$$

Associated with the characteristic root $\lambda = 3$, we thus have a one-dimensional characteristic subspace such that A maps every vector of this subspace onto a vector of the subspace (Figure 9-2). However, in this instance, except for the vector 0, individual vectors of the subspace are not left invariant by A.

Note that the subspaces determined by the roots $\lambda = 1$ and $\lambda = 3$, respectively, contain the unit characteristic vectors

$$\begin{bmatrix} -\frac{1}{\sqrt{2}} \\ \frac{1}{\sqrt{2}} \end{bmatrix} \quad \text{and} \quad \begin{bmatrix} \frac{1}{\sqrt{2}} \\ \frac{1}{\sqrt{2}} \end{bmatrix}.$$

These unit vectors determine an orthogonal transformation of coordinates,

$$X = \begin{bmatrix} -\frac{1}{\sqrt{2}} & \frac{1}{\sqrt{2}} \\ \frac{1}{\sqrt{2}} & \frac{1}{\sqrt{2}} \end{bmatrix} Y.$$

Applying this to the operator

$$X_2 = \begin{bmatrix} 2 & 1 \\ 1 & 2 \end{bmatrix} X_1,$$

we get, in the usual way,

$$Y_2 = \begin{bmatrix} -\frac{1}{\sqrt{2}} & \frac{1}{\sqrt{2}} \\ \frac{1}{\sqrt{2}} & \frac{1}{\sqrt{2}} \end{bmatrix}^{-1} \begin{bmatrix} 2 & 1 \\ 1 & 2 \end{bmatrix} \begin{bmatrix} -\frac{1}{\sqrt{2}} & \frac{1}{\sqrt{2}} \\ \frac{1}{\sqrt{2}} & \frac{1}{\sqrt{2}} \end{bmatrix} Y_1 = \begin{bmatrix} 1 & 0 \\ 0 & 3 \end{bmatrix} Y_1.$$

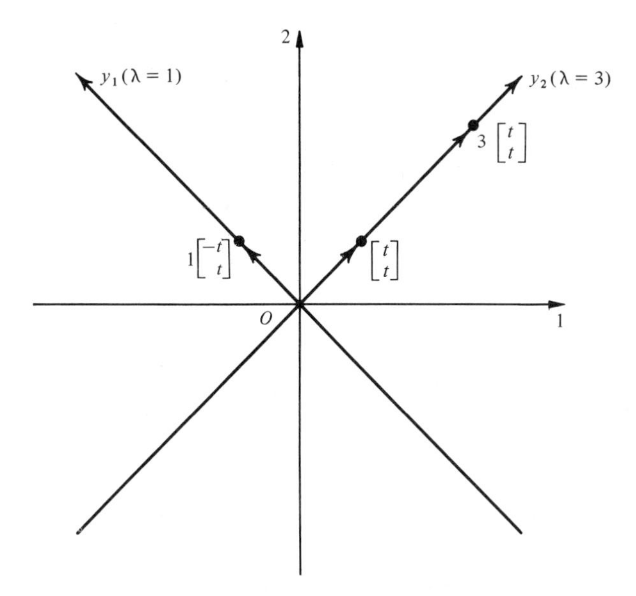

Figure 9-2. *Characteristic Subspaces of* $\begin{bmatrix} 2 & 1 \\ 1 & 2 \end{bmatrix}$.

Thus the operator has a particularly simple representation in the coordinate system determined by its unit characteristic vectors. We shall see presently when and why this happens.

As a second example, consider

$$A = \begin{bmatrix} 1 & -1 \\ 1 & 1 \end{bmatrix}, \quad \begin{vmatrix} 1-\lambda & -1 \\ 1 & 1-\lambda \end{vmatrix} = (1-\lambda)^2 + 1, \qquad \lambda = 1 \pm i.$$

Thus a real matrix may well have complex characteristic roots, and hence also complex characteristic vectors, for we cannot have A and $X \neq 0$ real in the equation $AX = \lambda X$ if λ is not also real. In this case, the operator $X_2 = AX_1$ has no nonzero, real, characteristic subspaces. This example illustrates the fact that the whole characteristic value problem is normally a problem in the complex number field. This is assumed throughout this chapter except when we treat real, symmetric matrices and the associated real quadratic forms, in which case all the characteristic roots are real, as we shall show.

9.3 TWO BASIC THEOREMS

A fundamental and often useful result is given in

Theorem 9.3.1: *Let $\lambda_1, \lambda_2, \ldots, \lambda_k$ be distinct characteristic roots of a matrix A, and let $X_1, X_2, \ldots, X_k$ be characteristic vectors associated with these roots respectively. Then $X_1, X_2, \ldots, X_k$ are linearly independent.*

If the equation

$$\alpha_1 X_1 + \alpha_2 X_2 + \cdots + \alpha_k X_k = 0 \tag{9.3.1}$$

holds, where the α's are scalars, then repeated multiplication by A, combined with the fact that $AX_i = \lambda_i X_i$, yields the additional equations

$$\begin{aligned} &\alpha_1\lambda_1 X_1 + \alpha_2\lambda_2 X_2 + \cdots + \alpha_k\lambda_k X_k = 0 \\ &\alpha_1\lambda_1^2 X_1 + \alpha_2\lambda_2^2 X_2 + \cdots + \alpha_k\lambda_k^2 X_k = 0 \\ &\vdots \\ &\alpha_1\lambda_1^{k-1} X_1 + \alpha_2\lambda_2^{k-1} X_2 + \cdots + \alpha_k\lambda_k^{k-1} X_k = 0. \end{aligned} \tag{9.3.2}$$

The equations (9.3.1) and (9.3.2) can be combined in partitioned matrix form thus:

$$[\alpha_1 X_1, \alpha_2 X_2, \ldots, \alpha_k X_k]\begin{bmatrix} 1 & \lambda_1 & \lambda_1^2 & \cdots & \lambda_1^{k-1} \\ 1 & \lambda_2 & \lambda_2^2 & \cdots & \lambda_2^{k-1} \\ \vdots \\ 1 & \lambda_k & \lambda_k^2 & \cdots & \lambda_k^{k-1} \end{bmatrix} = 0.$$

Since the λ's are all unequal, the second factor here is a nonsingular Vandermonde matrix (see Exercise 15, Section 7.5). If we multiply on the right by the inverse of this Vandermonde matrix, we conclude that

$$[\alpha_1 X_1, \alpha_2 X_2, \ldots, \alpha_k X_k] = 0,$$

which, since no X_j is 0, implies that every α_j is 0. Hence the X_j's are independent and the proof is complete.

If all n of the characteristic roots are distinct, and if

$$AV_i = \lambda_i V_i, \quad V_i \neq 0, \qquad i = 1, 2, \ldots, n,$$

then

$$[AV_1, AV_2, \ldots, AV_n] = [\lambda_1 V_1, \lambda_2 V_2, \ldots, \lambda_n V_n],$$

so that, if

$$V = [V_1, V_2, \ldots, V_n],$$

we have

$$AV = V \operatorname{diag}[\lambda_1, \lambda_2, \ldots, \lambda_n].$$

Since the V_i are linearly independent by the preceding theorem, V^{-1} exists and hence

$$V^{-1}AV = \operatorname{diag}[\lambda_1, \lambda_2, \ldots, \lambda_n].$$

This amounts to introducing the nonsingular transformation of coordinates $X = VY$ and allows us to state

Theorem 9.3.2: *When the characteristic roots of A are distinct, there exists at least one coordinate system in which the operator $X_2 = AX_1$ is represented by a diagonal matrix whose diagonal entries are the characteristic roots of A.*

In such a coordinate system, the operator has the representation

$$Y_2 = \text{diag}\,[\lambda_1, \lambda_2, \ldots, \lambda_n]\,Y_1$$

or, in scalar form,

$$y_{i2} = \lambda_i y_{i1}, \qquad i = 1, 2, \ldots, n$$

so that, if the λ_i are all real, it may be interpreted as a product of changes of scale along the several axes. This particularly simple representation is often desirable as it permits correspondingly simple analyses of a broad variety of applications in both the physical and the social sciences. In most applications, the characteristic roots are all distinct, so this kind of reduction of the operator with matrix A to diagonal form is possible. When A is a real symmetric matrix, the λ_i's are all real and the reduction is always possible, regardless of whether or not the λ_i are distinct. Moreover, the V_i may be chosen so as to constitute an orthonormal set, which makes V orthogonal. All this we prove in later sections.

9.4 EXERCISES

1. Determine the characteristic roots and the associated characteristic subspaces for these matrices:

(a) $\begin{bmatrix} 1 & 2 & 0 \\ 0 & 1 & 0 \\ 0 & 0 & 1 \end{bmatrix}$, (b) $\begin{bmatrix} 1 & 0 & 0 \\ 0 & 2 & 0 \\ 0 & 0 & 1 \end{bmatrix}$, (c) $\begin{bmatrix} 0 & 1 & 0 \\ 1 & 0 & 0 \\ 0 & 0 & 1 \end{bmatrix}$,

(d) $\begin{bmatrix} 0 & i & i \\ -i & 0 & i \\ -i & -i & 0 \end{bmatrix}$, (e) $\begin{bmatrix} 0 & 1 & 2 \\ 1 & 0 & -1 \\ 2 & -1 & 0 \end{bmatrix}$, (f) $\begin{bmatrix} \frac{1}{\sqrt{2}} & \frac{1}{\sqrt{2}} & 0 \\ \frac{1}{\sqrt{2}} & -\frac{1}{\sqrt{2}} & 0 \\ 0 & 0 & 1 \end{bmatrix}$,

(g) $\begin{bmatrix} 1 & 1 & 1 & 1 \\ 0 & 1 & 1 & 1 \\ 0 & 0 & -1 & -1 \\ 0 & 0 & 0 & -1 \end{bmatrix}$, (h) $\begin{bmatrix} 0 & 1 & 0 & 0 \\ 0 & 0 & 1 & 0 \\ 0 & 0 & 0 & 1 \\ -1 & 4 & -6 & 4 \end{bmatrix}$.

***2.** Show that $\lambda = 0$ is a characteristic root of A if and only if $\det A = 0$.

***3.** Show that the characteristic roots of a triangular matrix are just the diagonal entries of that matrix.

4. Show that if $AX = \lambda X$, $X \neq 0$, then λ^2 is a characteristic root of A^2 (with associated characteristic vector X). Then show that λ^p is a characteristic root of A^p for every positive integer p.

5. Show that if X is a characteristic vector of A associated with the characteristic root λ, then so is A^pX for every positive integer p.

6. By using the matrix

$$\begin{bmatrix} 0 & 1 & 0 & \cdots & 0 \\ 0 & 0 & 1 & \cdots & 0 \\ \vdots & & & & \\ 0 & 0 & 0 & \cdots & 1 \\ -a_0 & -a_1 & -a_2 & \cdots & -a_{n-1} \end{bmatrix}$$

show that any given polynomial $(-\lambda)^n + a_{n-1}(-\lambda)^{n-1} + \cdots + a_1(-\lambda) + a_0$ of degree n in $-\lambda$ may be regarded as the characteristic polynomial of a matrix of order n. This matrix is called the **companion matrix** of the given polynomial.

7. Show that the characteristic polynomial of the $n \times n$ matrix

$$\begin{bmatrix} 0 & 1 & 1 & \cdots & 1 \\ 1 & 0 & 1 & \cdots & 1 \\ \vdots & & & & \\ 1 & 1 & 1 & \cdots & 0 \end{bmatrix}$$

is $(-1)^{n-1}(n - 1 - \lambda)(1 + \lambda)^{n-1}$.

8. For what values of h will the matrix

$$\begin{bmatrix} h & 1 & 0 \\ 1 & h & 0 \\ 0 & 0 & 1 \end{bmatrix}$$

have multiple characteristic roots?

***9.** Prove that, if A and B are $n \times n$ matrices and if there exist n linearly independent vectors $X_1, X_2, \ldots, X_n$ and scalars $\lambda_1, \lambda_2, \ldots, \lambda_n$ such that $AX_j = \lambda_j X_j$ and $BX_j = \lambda_j X_j$, $j = 1, 2, \ldots, n$, then $A = B$.

10. Find a transformation that will reduce this operator to diagonal form:

$$X_2 = \begin{bmatrix} 4 & -1 & -2 \\ 2 & 1 & -2 \\ 1 & -1 & 1 \end{bmatrix} X_1.$$

11. The three characteristic vectors

$$\begin{bmatrix} 1 \\ -1 \\ 1 \end{bmatrix}, \quad \begin{bmatrix} 1 \\ 1 \\ 0 \end{bmatrix}, \quad \begin{bmatrix} 1 \\ -1 \\ 0 \end{bmatrix}$$

of a 3×3 matrix A are associated, respectively, with the characteristic roots $1, -1$, and 0. Find A.

9.5 MINOR DETERMINANTS OF A MATRIX

The determinant of an $r \times r$ submatrix of A is called a **minor determinant** of order r of A. Each square submatrix of A is obtained by deleting certain rows and columns of A. If A is itself square and if the rows and columns deleted have the same indices, we get a **principal minor determinant** of A. The principal minor of the form

$$p_r = \begin{vmatrix} a_{11} & \cdots & a_{1r} \\ \vdots & & \\ a_{r1} & \cdots & a_{rr} \end{vmatrix}$$

is called the **leading principal minor** of order r of A. To avoid exceptions in the statement of theorems, it is convenient to define the **minor p_0 of order zero** of an arbitrary square matrix A to be 1. Note that $p_n = \det A$.

For example, a 3×3 matrix A has these principal minors:

$$1 \qquad \text{order zero}$$

$$a_{11}, \quad a_{22}, \quad a_{33} \qquad \text{order one}$$

$$\begin{vmatrix} a_{11} & a_{12} \\ a_{21} & a_{22} \end{vmatrix}, \quad \begin{vmatrix} a_{11} & a_{13} \\ a_{31} & a_{33} \end{vmatrix}, \quad \begin{vmatrix} a_{22} & a_{23} \\ a_{32} & a_{33} \end{vmatrix} \qquad \text{order two}$$

$$\det A \qquad \text{order three.}$$

Minor determinants, in particular principal minor determinants, play an important role in the theorems of this chapter.

9.6 THE CHARACTERISTIC POLYNOMIAL AND ITS ROOTS

The polynomial

$$(9.6.1) \qquad \phi(\lambda) = \det(A - \lambda I) = \begin{vmatrix} a_{11} - \lambda & a_{12} & \cdots & a_{1n} \\ a_{21} & a_{22} - \lambda & \cdots & a_{2n} \\ \vdots & & & \\ a_{n1} & a_{n2} & \cdots & a_{nn} - \lambda \end{vmatrix}$$

is called the **characteristic polynomial** of A. The product of the diagonal entries, namely

$$(9.6.2) \qquad (a_{11} - \lambda)(a_{22} - \lambda) \cdots (a_{nn} - \lambda),$$

is a term in the expansion of the determinant. From this product, we see that the highest degree term in λ is $(-\lambda)^n$. It is convenient to write $\phi(\lambda)$ in descending powers of $-\lambda$:

$$(9.6.3) \qquad \phi(\lambda) = (-\lambda)^n + b_{n-1}(-\lambda)^{n-1} + \cdots + b_1(-\lambda) + b_0.$$

If, in the expansion of (9.6.2), we form the product of the term a_{ii} from one factor and the terms $-\lambda$ from each of the other $n - 1$ factors, then sum the results, we get

$$(a_{11} + a_{22} + \cdots + a_{nn})(-\lambda)^{n-1}.$$

There is no other way to get terms involving $(-\lambda)^{n-1}$ since every term other than (9.6.2) in the expansion of the determinant contains at most $n - 2$ of the diagonal entries. Hence

$$b_{n-1} = a_{11} + a_{22} + \cdots + a_{nn} = \text{tr } A. \tag{9.6.4}$$

By putting $\lambda = 0$ in (9.6.1) and (9.6.3), we get $\phi(0) = \det A$ and $\phi(0) = b_0$, respectively, so

$$b_0 = \det A. \tag{9.6.5}$$

These formulas for b_{n-1} and b_0 are special cases of

Theorem 9.6.1: *The coefficient of $(-\lambda)^r$ in $\phi(\lambda)$ is the sum of the principal minor determinants of order $n - r$ of A. In particular, the coefficients of $(-\lambda)^n$, $(-\lambda)^{n-1}$, and $(-\lambda)^0$ are, respectively, 1, $b_{n-1} = \text{tr } A$, and $b_0 = \det A$.*

To prove this, note that

$$A - \lambda I = [A_1 - \lambda E_1, A_2 - \lambda E_2, \ldots, A_n - \lambda E_n],$$

where A_j is the jth column of A and E_j is the jth elementary n-vector. By repeated use of Theorem 7.3.3, we see that $\det(A - \lambda I)$ is the sum of 2^n determinants, the jth column in each of which is either A_j or $-\lambda E_j$, so that

$$\begin{aligned}
\det(A - \lambda I) = {} & \det[-\lambda E_1, -\lambda E_2, \ldots, -\lambda E_n] \\
& + \det[A_1, -\lambda E_2, \ldots, -\lambda E_n] + \cdots \\
& + \det[-\lambda E_1, \ldots, -\lambda E_{n-1}, A_n] \\
& + \det[A_1, A_2, -\lambda E_3, \ldots, -\lambda E_n] + \cdots \\
& + \det[-\lambda E_1, \ldots, -\lambda E_{n-2}, A_{n-1}, A_n] \\
& + \cdots \\
& + \det[A_1, A_2, \ldots, A_{n-r}, -\lambda E_{n-r+1}, \ldots, -\lambda E_n] + \cdots \\
& + \det[-\lambda E_1, \ldots, -\lambda E_r, A_{r+1}, \ldots, A_n] \\
& + \cdots \\
& + \det[A_1, A_2, \ldots, A_n].
\end{aligned}$$

If we use Theorem 7.2.6 to factor out the coefficients $-\lambda$, what remains in each case is a principal minor determinant of A and the theorem follows. (To see this better, write out the preceding formula in full for $n = 3$.)

This theorem may be used, if we wish, to expand $\phi(\lambda)$. For example,

$$\begin{vmatrix} 2-\lambda & 0 & 1 \\ 0 & 4-\lambda & -1 \\ 1 & -1 & -2-\lambda \end{vmatrix} = (-\lambda)^3 + (2+4-2)(-\lambda)^2$$

$$+ \left(\begin{vmatrix} 2 & 0 \\ 0 & 4 \end{vmatrix} + \begin{vmatrix} 2 & 1 \\ 1 & -2 \end{vmatrix} + \begin{vmatrix} 4 & -1 \\ -1 & -2 \end{vmatrix} \right)(-\lambda)$$

$$+ \begin{vmatrix} 2 & 0 & 1 \\ 0 & 4 & -1 \\ 1 & -1 & -2 \end{vmatrix}$$

$$= (-\lambda)^3 + 4\lambda^2 + 6\lambda - 22.$$

Now let the n roots of the characteristic equation $\phi(\lambda) = 0$ be $\lambda_1, \lambda_2, \ldots, \lambda_n$. Then we can write, by the factor theorem,

(9.6.6) $$\phi(\lambda) = (\lambda_1 - \lambda)(\lambda_2 - \lambda)\cdots(\lambda_n - \lambda).$$

The coefficient of $(-\lambda)^{n-1}$ in this product will be, by the same reasoning we used to obtain (9.6.4),

$$b_{n-1} = \lambda_1 + \lambda_2 + \cdots + \lambda_n$$

so

(9.6.7) $$\text{tr } A = \lambda_1 + \lambda_2 + \cdots + \lambda_n.$$

Similarly, from (9.6.6), we find

$$\phi(0) = b_0 = \lambda_1\lambda_2\cdots\lambda_n,$$

so

(9.6.8) $$\det A = \lambda_1\lambda_2\cdots\lambda_n.$$

Making comparisons of the remaining coefficients of the expansions of the products in (9.6.4) and (9.6.6), we find that

$$b_r = \sum \lambda_{j_1}\lambda_{j_2}\cdots\lambda_{j_{n-r}}, \qquad r = 0, 1, \ldots, n-1.$$

Here the sum is extended over the $C(n, n-r)$ combinations $j_1, j_2, \ldots, j_{n-r}$ of $1, 2, \ldots, n$, taken $n - r$ at a time. For $r = 0, 1, \ldots, n-2$, these expressions are called the **higher traces** of the matrix. Thus we have proved

Theorem 9.6.2: *The trace of A is the sum of the characteristic roots of A and the determinant of A is the product of the characteristic roots of A. In general, the coefficient b_r of $\phi(\lambda)$ is the sum of the products of the roots of $\phi(\lambda)$, $n - r$ at a time.*

For example, if

$$A = \begin{bmatrix} 2 & -1 & 5 \\ -1 & -4 & 3 \\ 5 & 3 & 2 \end{bmatrix},$$

we have

$$\text{tr } A = \lambda_1 + \lambda_2 + \lambda_3 = 0$$

and

$$\det A = \lambda_1\lambda_2\lambda_3 = 34.$$

The characteristic roots $\lambda_1, \lambda_2, \ldots, \lambda_n$ may or may not be distinct. Let the distinct values be labeled $\lambda_1, \lambda_2, \ldots, \lambda_p$, with multiplicities $r_1, r_2, \ldots, r_p$, respectively, so that we may write

$$\phi(\lambda) = (\lambda_1 - \lambda)^{r_1}(\lambda_2 - \lambda)^{r_2} \cdots (\lambda_p - \lambda)^{r_p},$$

where

$$r_1 + r_2 + \cdots + r_p = n.$$

With each of the distinct values λ_i, there is associated a characteristic subspace of dimension n − rank $(A - \lambda_i I)$. Since the set of distinct characteristic roots always contains at least one member λ_1 and since rank $(A - \lambda_i I) < n$, every square matrix A has at least one characteristic subspace, whose dimension is >0. We can say more about the dimensions of the characteristic subspaces. We begin with

Lemma 9.6.3: *If λ_1 is an r_1-fold characteristic root of A, then 0 is an r_1-fold characteristic root of $A - \lambda_1 I$.*

Since $\lambda = \lambda_1$ is an r_1-fold characteristic root of A, precisely r_1 of the factors $\lambda_i - \lambda$ must be identical to $\lambda_1 - \lambda$, so we can write

$$\phi(\lambda) = (\lambda_1 - \lambda)^{r_1}\psi(\lambda), \tag{9.6.9}$$

where $\psi(\lambda)$ represents the product of the remaining factors $\lambda_j - \lambda$, so that $\psi(\lambda_1) \neq 0$. Hence the characteristic polynomial of $A - \lambda_1 I$ is

$$\begin{aligned} \det((A - \lambda_1 I) - \lambda I) &= \det(A - (\lambda_1 + \lambda)I) \\ &= \phi(\lambda_1 + \lambda) \\ &= (\lambda_1 - (\lambda_1 + \lambda))^{r_1}\psi(\lambda_1 + \lambda) \text{ by (9.6.9)} \\ &= (0 - \lambda)^{r_1}\psi(\lambda_1 + \lambda). \end{aligned} \tag{9.6.10}$$

Now when $\lambda = 0$, $\psi(\lambda_1 + \lambda)$ becomes $\psi(\lambda_1)$, which is not zero. Thus no additional factors $-\lambda$ appear in $\psi(\lambda_1 + \lambda)$ and 0 is precisely an r_1-fold characteristic root of $A - \lambda_1 I$, as was to be shown.

Next recall that in the expansion of det $((A - \lambda_1 I) - \lambda I)$, the coefficient of $(-\lambda)^{r_1}$ is the sum of the principal minor determinants of order $n - r_1$ of $A - \lambda_1 I$. Since this coefficient is also the constant term in the expansion of $\psi(\lambda_1 + \lambda)$, namely $\psi(\lambda_1) \neq 0$, there must be at least one principal minor determinant of order $n - r_1$ of $A - \lambda_1 I$ which is $\neq 0$. Hence the rank of $A = \lambda_1 I$ is *at least* $n - r_1$. It may be even higher, as this example shows:

$$A = \begin{bmatrix} 1 & 1 & 1 & 1 \\ 0 & 1 & 1 & 1 \\ 0 & 0 & 1 & 1 \\ 0 & 0 & 0 & 1 \end{bmatrix}, \qquad \begin{array}{l} \det(A - \lambda I) = (1 - \lambda)^4, \quad \lambda_1 = 1, \quad r_1 = 4, \\ n - r_1 = 0, \quad \text{rank}\,(A - 1 \cdot I) = 3 > n - r_1. \end{array}$$

In summary, we have

Theorem 9.6.4: *If λ_1 is an r_1-fold characteristic root of a matrix A, then the rank of $A - \lambda_1 I$ is not less than $n - r_1$ and the dimension of the associated characteristic subspace is not greater than $n - (n - r_1) = r_1$.*

If ρ_j is the dimension of the characteristic subspace associated with λ_j, then

$$(9.6.11) \qquad 1 \leq \rho_j \leq r_j, \qquad j = 1, 2, \ldots, p,$$

so that, summing these inequalities, we have

$$(9.6.12) \qquad p \leq \sum_1^p \rho_j \leq \sum_1^p r_j = n.$$

A useful special case follows directly from (9.6.11):

Theorem 9.6.5: *If λ_1 is a simple root of $\phi(\lambda) = 0$ so that $r_1 = 1$, the rank of $A - \lambda_1 I$ is $n - 1$ and the dimension of the associated characteristic subspace is* 1.

9.7 SIMILAR MATRICES

Given the matrices A and C, we say that C is **similar** to A if and only if there exists a matrix B such that $C = B^{-1}AB$. We saw in Chapter 8 how a change of basis gives rise to similar matrices. Since $A = I^{-1}AI$, A is similar to itself. That is, similarity is a *reflexive* relation. Since $C = B^{-1}AB$ implies $A = (B^{-1})^{-1}C(B^{-1})$, C similar to A implies A similar to C. That is, similarity is a *symmetric* relation so that we may properly use the symmetrical expression, "A and C are similar." Finally, let A be similar to C and C to F so that for some B and D, $A = B^{-1}CB$ and $C = D^{-1}FD$. Then $A = (DB)^{-1}F(DB)$ and A is similar to F, so similarity is a *transitive* relation. In summary, we have

Theorem 9.7.1: *Similarity of matrices is an equivalence relation.*

In the relation $C = B^{-1}AB$, all three matrices may be over the same number field, but this is not necessarily the case. We may have A real, B and C complex, for example, or C real, A and B complex. Such cases will arise in what follows.

Similar matrices have important properties in common, as is indicated by

Theorem 9.7.1: *Let A and C be similar matrices and let $C = B^{-1}AB$. Then A and C have identical characteristic polynomials and hence have identical characteristic roots. Moreover, if X is a characteristic vector of A associated with the root λ, then $B^{-1}X$ is a characteristic vector of C, again associated with the root λ.*

The characteristic polynomial of C is given by

$$\begin{aligned}
\det(C - \lambda I) &= \det(B^{-1}AB - \lambda I) \\
&= \det(B^{-1}AB - B^{-1}\lambda IB) \\
&= \det(B^{-1}(A - \lambda I)B) \\
&= \det B^{-1} \det(A - \lambda I) \det B \\
&= \det B^{-1} \det B \det(A - \lambda I) \quad \text{since determinants are scalars} \\
&= \det(B^{-1}B(A - \lambda I)) \\
&= \det(A - \lambda I).
\end{aligned}$$

Thus C and A have the same characteristic polynomial and therefore have precisely the same characteristic roots $\lambda_1, \lambda_2, \ldots, \lambda_n$, so that

$$\operatorname{tr} A = \lambda_1 + \lambda_2 + \cdots + \lambda_n = \operatorname{tr} C$$

and

$$\det A = \lambda_1\lambda_2 \cdots \lambda_n = \det C.$$

Now suppose that X is a characteristic vector of A associated with the root λ so that $AX = \lambda X$. Then, since $CB^{-1} = B^{-1}A$, we have

$$C(B^{-1}X) = (CB^{-1})X = (B^{-1}A)X = B^{-1}(AX) = B^{-1}(\lambda X) = \lambda(B^{-1}X),$$

so that $B^{-1}X$ is a characteristic vector of C associated with the characteristic root λ.

All this has a geometrical interpretation. Let A be the matrix of a linear operator $X_2 = AX_1$ in the standard reference system. Recall that if we introduce the transformation of coordinates $X = BY$ or $Y = B^{-1}X$, we get the representation of the operator in the new reference system: $Y_2 = (B^{-1}AB)Y_1$. The previous theorem then shows that in every reference system, the operator has the same characteristic polynomial, the same characteristic roots, the same trace, the same determinant, and (since $B^{-1}X$ and X are representations of the same vector in the new and in the old reference systems, respectively) the same characteristic subspaces. Since they are independent of the choice of reference system, the characteristic polynomial, the characteristic roots, the trace, the determinant, the higher traces, and the characteristic subspaces are called **invariants** of the operator; that is, they are properties of the operator rather than of the particular matrix used to represent it.

9.8 EXERCISES

1. If A is of order 3, if the sum of its characteristic roots is 9, if their product is 24, and if $\lambda_3 = 3$, what are the possible values for λ_1 and λ_2?

***2.** Show that $\lambda = 0$ is a characteristic root of A if and only if $\det A = 0$.

3. Prove, *without finding the characteristic roots*, that if

$$A = \begin{bmatrix} -1 & 1 & -1 \\ 1 & -1 & 1 \\ -1 & 1 & 2 \end{bmatrix},$$

then two of the characteristic roots of A differ only in sign.

4. Use Theorem 9.6.1 to find the characteristic polynomial of

$$\begin{bmatrix} 1 & 1 & 0 & 0 \\ 1 & -1 & 0 & 1 \\ 0 & 0 & 2 & 1 \\ 0 & 1 & 1 & 2 \end{bmatrix}.$$

***5.** Show that A and A^{T} have the same characteristic polynomial.

6. Show that if A is of order n, then

$$\det (A - I) = \sum_{k=0}^{n} (-1)^{n-k} \sigma_k,$$

where σ_k is the sum of the principal minors of order k of A. This includes the principal minor of order 0, which is by definition 1.

7. A linear operator on $\mathscr{E}^3$ maps

$$\begin{bmatrix} 1 \\ 2 \\ 1 \end{bmatrix} \text{ onto } \begin{bmatrix} 1 \\ 2 \\ 1 \end{bmatrix}, \quad \begin{bmatrix} 2 \\ 1 \\ 0 \end{bmatrix} \text{ onto } \begin{bmatrix} -4 \\ -2 \\ 0 \end{bmatrix}, \quad \text{and} \quad \begin{bmatrix} 1 \\ 1 \\ 1 \end{bmatrix} \text{ onto } \begin{bmatrix} 0 \\ 0 \\ 0 \end{bmatrix}.$$

Without finding A itself, find the characteristic polynomial of A and also its trace and determinant.

***8.** Show that if the rank of A is r, then at least $n - r$ characteristic roots of A are zero. Give an example to show that the words "at least" are justified.

9.9 THE CHARACTERISTIC ROOTS OF A REAL SYMMETRIC MATRIX

We have seen that a real matrix may well have complex characteristic roots. In applications, the most important matrices are the real, symmetric matrices, concerning which we now prove

Theorem 9.9.1: *The characteristic roots of every real, symmetric matrix are real.*

Let A be a real symmetric matrix, let λ be any characteristic root of A, and let Y be an associated *unit* characteristic vector. Since λ may be complex (until we prove otherwise), Y may also be complex, so the fact that it is a unit vector must be expressed by the equation $Y^*Y = 1$ (Section 4.11). Then, from the equation

$$AY = \lambda Y$$

we obtain

$$Y^*AY = \lambda Y^*Y = \lambda. \tag{9.9.1}$$

We also have, since $A^* = A$ and since Y^*AY is a scalar,

$$Y^*AY = Y^*A^*(Y^*)^* = (Y^*AY)^* = \overline{Y^*AY}.$$

Thus the scalar Y^*AY is equal to its own conjugate and hence is real. That is, by (9.9.1), λ is real.

Since all the coefficients in the system of equations represented by $(A - \lambda I)Y = 0$ are real, we have

Theorem 9.9.2: *All the characteristic subspaces of a real symmetric matrix are spanned by real vectors.*

We prove also

Theorem 9.9.3: *If X and Y are characteristic vectors associated with distinct characteristic roots λ and μ of a real, symmetric matrix A, then X and Y are orthogonal.*

Suppose, in fact, that

$$AX = \lambda X \quad \text{and} \quad AY = \mu Y, \qquad \lambda \neq \mu.$$

Then from these equations we have

$$Y^{\mathsf{T}}AX = \lambda Y^{\mathsf{T}}X \qquad \text{and} \qquad X^{\mathsf{T}}AY = \mu X^{\mathsf{T}}Y.$$

Forming transposes in the first of these equations, we have, since $A^{\mathsf{T}} = A$,

$$X^{\mathsf{T}}AY = \lambda X^{\mathsf{T}}Y.$$

Hence

$$\lambda X^{\mathsf{T}}Y = \mu X^{\mathsf{T}}Y,$$

so that, since $\lambda \neq \mu$,

$$X^{\mathsf{T}}Y = 0;$$

that is, X and Y are orthogonal. As a consequence, the characteristic subspaces associated with λ and μ are also orthogonal. This is illustrated by the first example in Section 9.2 (see Figure 9-2).

Another useful result is

Theorem 9.9.4: *If U is orthogonal and if $U^{\mathsf{T}}AU = D[\lambda_1, \lambda_2, \ldots, \lambda_n]$, then $\lambda_1, \lambda_2, \ldots, \lambda_n$ are necessarily the characteristic roots of A.*

Indeed, since A and $D[\lambda_1, \lambda_2, \ldots, \lambda_n]$ are similar, they have identical characteristic roots, and the characteristic roots of a diagonal matrix are its diagonal entries.

9.10 THE DIAGONAL FORM OF A REAL SYMMETRIC MATRIX

Recall that the notation $D[\lambda_1, \lambda_2, \ldots, \lambda_n]$ denotes a diagonal matrix with diagonal elements $\lambda_1, \lambda_2, \ldots, \lambda_n$.

We prove first an existence theorem:

Theorem 9.10.1: *If A is a real, symmetric matrix, then there exists an orthogonal matrix U such that $U^{\mathsf{T}}AU = D[\lambda_1, \lambda_2, \ldots, \lambda_n]$, where $\lambda_1, \lambda_2, \ldots, \lambda_n$ are the characteristic roots of A.*

The matrix $D[\lambda_1, \lambda_2, \ldots, \lambda_n]$ is called the **diagonal form** of A and the equation $Y_2 = D[\lambda_1, \lambda_2, \ldots, \lambda_n]Y_1$ represents the corresponding operator in diagonal form.

First recall that if A is a real symmetric matrix and if C is any matrix of order n, then $C^{\mathsf{T}}AC$ is also symmetric. From this fact it follows that if $C^{\mathsf{T}}AC = B$ and if $b_{21} = b_{31} = \cdots = b_{n1} = 0$, then $b_{12} = b_{13} = \cdots = b_{1n} = 0$.

We shall prove the theorem by induction on the order of A, and then later shall show how the matrix U may be computed more directly.

In case A is of order 2, we must show there exists an orthogonal matrix U of order 2 such that

$$U^{\mathsf{T}}AU = D[\lambda_1, \lambda_2],$$

where λ_1 and λ_2 are the characteristic roots of A. That is, since $U^{\mathsf{T}} = U^{-1}$, we must show there exists an orthogonal matrix U such that

$$\begin{bmatrix} a_{11} & a_{12} \\ a_{21} & a_{22} \end{bmatrix}\begin{bmatrix} u_{11} & u_{12} \\ u_{21} & u_{22} \end{bmatrix} = \begin{bmatrix} u_{11} & u_{12} \\ u_{21} & u_{22} \end{bmatrix}\begin{bmatrix} \lambda_1 & 0 \\ 0 & \lambda_2 \end{bmatrix}. \tag{9.10.1}$$

Forming both products and equating their first columns, we obtain the equations

$$\begin{aligned} (a_{11} - \lambda_1)u_{11} + a_{12}u_{21} &= 0 \\ a_{21}u_{11} + (a_{22} - \lambda_1)u_{21} &= 0, \end{aligned} \tag{9.10.2}$$

which necessarily have nontrivial solutions for u_{11} and u_{21}, since λ_1 is a characteristic root of A and the coefficient matrix therefore has rank less than 2. We select one of these solutions and normalize it. This vector and any unit vector orthogonal to it (How many are there?) are used as the columns of an orthogonal matrix U. Then from (9.10.2) it follows that $U^{\mathsf{T}}AU$ has as its first column the vector $[\lambda_1, 0]^{\mathsf{T}}$. Since the product $U^{\mathsf{T}}AU$ is symmetric, it follows that the 1,2-entry of this product is also zero. Then from Theorem 9.9.4 we conclude that the 2,2-entry is necessarily λ_2. This proves the theorem for $n = 2$.

Now let $n - 1$ be any integer ≥ 2 such that the theorem holds true for all symmetric matrices of order $n - 1$ and let A be any symmetric matrix of order n. Using the same approach as we did for $n = 2$, we determine first an orthogonal matrix V such that

$$V^{\mathsf{T}}AV = \begin{bmatrix} \lambda_1 & 0 \\ 0 & B \end{bmatrix},$$

that is, such that

$$AV = V\begin{bmatrix} \lambda_1 & 0 \\ 0 & B \end{bmatrix}, \tag{9.10.3}$$

where λ_1 is any given characteristic root of A and where B is a matrix of order $n - 1$ still to be determined.

Once again we compute and equate the elements of the first columns of the left and right members of this last equation and obtain the equations

$$\begin{aligned}
(a_{11} - \lambda_1)v_{11} + a_{12}v_{21} + \cdots + a_{1n}v_{n1} &= 0 \\
a_{21}v_{11} + (a_{22} - \lambda_1)v_{21} + \cdots + a_{2n}v_{n1} &= 0 \\
&\vdots \\
a_{n1}v_{11} + a_{n2}v_{21} + \cdots + (a_{nn} - \lambda_1)v_{n1} &= 0,
\end{aligned}$$

for the determination of the first column $V_1 = [v_{11}, v_{21}, \ldots, v_{n1}]^{\mathsf{T}}$ of V. Since λ_1 is a characteristic root of A, the determinant of the coefficients here is zero, so that there exist nontrivial solutions, one of which we normalize and use for V_1.

The remaining columns of V we choose in such a way that $V_1, V_2, \ldots, V_n$ form an orthonormal basis for $\mathscr{E}^n$, which is always possible.

Then the first column of $V^{\mathsf{T}}AV$ is $[\lambda_1, 0, \ldots, 0]^{\mathsf{T}}$, so that, since $V^{\mathsf{T}}AV$ is symmetric, we must have

$$V^{\mathsf{T}}AV = \left[\begin{array}{c|c} \lambda_1 & 0 \\ \hline 0 & B \end{array}\right],$$

where B is determined by our choices of $\lambda_1, V_1, V_2, \ldots, V_n$.

Since $V^{\mathsf{T}}AV$ is symmetric, B must be symmetric also. Since $V^{\mathsf{T}}AV$ is similar to A, it follows that the characteristic roots of B are the remaining characteristic roots $\lambda_2, \lambda_3, \ldots, \lambda_n$ of A. Hence, by the induction hypothesis, there exists an orthogonal matrix W of order $n - 1$ such that

$$W^{\mathsf{T}}BW = D[\lambda_2, \lambda_3, \ldots, \lambda_n].$$

It then follows that

$$\left[\begin{array}{c|c} 1 & 0 \\ \hline 0 & W^{\mathsf{T}} \end{array}\right] V^{\mathsf{T}}AV \left[\begin{array}{c|c} 1 & 0 \\ \hline 0 & W \end{array}\right] = \left[\begin{array}{c|c} 1 & 0 \\ \hline 0 & W^{\mathsf{T}} \end{array}\right]\left[\begin{array}{c|c} \lambda_1 & 0 \\ \hline 0 & B \end{array}\right]\left[\begin{array}{c|c} 1 & 0 \\ \hline 0 & W \end{array}\right]$$

$$= \left[\begin{array}{c|c} \lambda_1 & 0 \\ \hline 0 & W^{\mathsf{T}}BW \end{array}\right] = D[\lambda_1, \lambda_2, \ldots, \lambda_n].$$

Thus the orthogonal matrix

$$U = V\left[\begin{array}{c|c} 1 & 0 \\ \hline 0 & W \end{array}\right]$$

is such that

$$U^{\mathsf{T}}AU = D[\lambda_1, \lambda_2, \ldots, \lambda_n],$$

so the theorem is true for all real symmetric matrices of order n. The theorem now follows for all integers $n \geq 2$.

The geometric interpretation of this result is simple. Let A be the matrix of a real, symmetric, linear operator in the standard reference system. If U is the matrix of an orthogonal transformation of coordinates, then the matrix representing the operator in the new coordinate system is $U^{\mathsf{T}}AU$, since $U^{-1} = U^{\mathsf{T}}$ here. Because all characteristic roots of a real symmetric matrix are real, the theorem says that, given a symmetric linear operator over the real field, we can always find a reference system in $\mathscr{E}^n$ in which the matrix of that operator is diagonal. In this reference system, the effect of the operator is simply to multiply each component of a vector by a factor which is one of the characteristic roots of the operator. The first example of Section 9.3 illustrates these remarks.

Theorem 9.10.2: *The orthogonal matrix U of Theorem* 9.10.1 *may be chosen so that the characteristic roots $\lambda_1, \lambda_2, \ldots, \lambda_n$ appear in any desired order in the diagonal matrix $U^{\mathsf{T}}AU$.*

This follows from the fact that, if one were to derive U in the manner of the proof of the theorem, one would choose an arbitrary root λ at each step of the process. It also follows from the fact that the matrix $[E_{j_1}, E_{j_2}, \ldots, E_{j_n}]$, which may be used to permute the columns or rows of a matrix, is orthogonal. The reader may supply the details.

9.11 THE DIAGONALIZATION OF A REAL SYMMETRIC MATRIX

Theorem 9.10.1 now permits us to deduce a chain of theorems which show how the orthogonal matrix U of that theorem may be computed more directly.

Theorem 9.11.1: *If λ_1 is an r_1-fold characteristic root of a real symmetric matrix A of order n, then the rank of $A - \lambda_1 I$ is $n - r_1$.*

In Theorem 9.6.4 we saw that for arbitrary A the rank of $A - \lambda_1 I$ cannot be *less* than $n - r_1$. Here we show that for a real, symmetric matrix A, the rank of $A - \lambda_1 I$ is precisely $n - r_1$, that is, that the reduction in rank is maximal.

By Theorems 9.10.1 and 9.10.2, there exists an orthogonal matrix U such that

$$U^{\mathsf{T}}AU = D[\lambda_1, \lambda_1, \ldots, \lambda_1; \lambda_{r_1+1}, \ldots, \lambda_n],$$

where λ_1 occurs r_1 times and $\lambda_{r_1+1}, \ldots, \lambda_n$ are all distinct from λ_1. Since U is orthogonal, subtracting $\lambda_1 I$ from both members of this equation gives

$$U^{\mathsf{T}}(A - \lambda_1 I)U = D[0, 0, \ldots, 0; \lambda_{r_1+1} - \lambda_1, \ldots, \lambda_n - \lambda_1].$$

Since U is nonsingular, it follows that the rank of $A - \lambda_1 I$ is the same as that of the matrix on the right, which is precisely $n - r_1$ since $\lambda_{r_1+1} - \lambda_1, \ldots, \lambda_n - \lambda_1$ are all different from zero.

Since the characteristic vectors of A associated with the root λ_1 are the non-zero solutions of $(A - \lambda_1 I)X = 0$, an immediate consequence of the preceding theorem is

Theorem 9.11.2: *If λ_1 is an r_1-fold characteristic root of a real symmetric matrix A of order n, there exist r_1 linearly independent characteristic vectors associated with λ_1; that is, with λ_1 there is associated an r_1-dimensional characteristic subspace. The sum of the dimensions of all the characteristic subspaces is n.*

For each characteristic subspace we can construct an orthonormal basis of characteristic vectors, using the Gram–Schmidt process if necessary. For each A, there are altogether n vectors in the bases so constructed, by the preceding theorem. Since characteristic vectors associated with distinct characteristic values of a real symmetric matrix are orthogonal, it follows that these n basis vectors are an orthonormal set. Hence we have

Theorem 9.11.3: *With every real symmetric matrix we can associate an orthonormal set of n characteristic vectors.*

This result indicates a second way in which the diagonalization process may be effected. For let the n vectors of such an orthonormal system be $U_1, U_2, \ldots, U_n$. Then

$$AU_j = \lambda_j U_j, \qquad j = 1, 2, \ldots, n, \tag{9.11.1}$$

where $\lambda_1, \lambda_2, \ldots, \lambda_n$ are the characteristic roots of A. Equations (9.11.1) may be combined in the single equation

$$[AU_1, AU_2, \ldots, AU_n] = [\lambda_1 U_1, \lambda_2 U_2, \ldots, \lambda_n U_n]$$

or, if

$$U = [U_1, U_2, \ldots, U_n],$$

in the equation

$$AU = UD[\lambda_1, \lambda_2, \ldots, \lambda_n].$$

Since U is orthogonal by construction, this implies

$$U^{\mathsf{T}}AU = D[\lambda_1, \lambda_2, \ldots, \lambda_n].$$

We summarize in

Theorem 9.11.4: *If $U_1, U_2, \ldots, U_n$ is an orthonormal system of characteristic vectors associated, respectively, with the characteristic roots $\lambda_1, \lambda_2, \ldots, \lambda_n$ of a real symmetric matrix A, and if U is the orthogonal matrix $[U_1, U_2, \ldots, U_n]$, then*

$$U^{\mathsf{T}}AU = D[\lambda_1, \lambda_2, \ldots, \lambda_n]. \tag{9.11.2}$$

The vectors $U_1, U_2, \ldots, U_n$ are often called a set of **principal axes** of A and the transformation with matrix U used to diagonalize A is called a **principal axis transformation.**

We illustrate with the following example: Let

$$A = \begin{bmatrix} 5 & 2 & 0 & 0 \\ 2 & 2 & 0 & 0 \\ 0 & 0 & 5 & -2 \\ 0 & 0 & -2 & 2 \end{bmatrix}.$$

Then the characteristic roots are 1, 1, 6, 6. Putting $\lambda = 1$ in $AX = \lambda X$, we obtain the system of equations

$$\begin{aligned} 4x_1 + 2x_2 \qquad\qquad &= 0 \\ 2x_1 + x_2 \qquad\qquad &= 0 \\ 4x_3 - 2x_4 &= 0 \\ -2x_3 + x_4 &= 0, \end{aligned}$$

with the complete solution

$$X = k_1\begin{bmatrix} 1 \\ -2 \\ 0 \\ 0 \end{bmatrix} + k_2\begin{bmatrix} 0 \\ 0 \\ 1 \\ 2 \end{bmatrix}.$$

The basis vectors here are already orthogonal, so we need only normalize them to obtain two columns of U:

$$U_1 = \left[\frac{1}{\sqrt{5}}, \frac{-2}{\sqrt{5}}, 0, 0\right]^{\mathsf{T}}, \qquad U_2 = \left[0, 0, \frac{1}{\sqrt{5}}, \frac{2}{\sqrt{5}}\right]^{\mathsf{T}}.$$

Proceeding in the same fashion with the root $\lambda = 6$, we obtain for the other two columns of U,

$$U_3 = \left[\frac{2}{\sqrt{5}}, \frac{1}{\sqrt{5}}, 0, 0\right]^{\mathsf{T}}, \qquad U_4 = \left[0, 0, \frac{-2}{\sqrt{5}}, \frac{1}{\sqrt{5}}\right]^{\mathsf{T}}.$$

Then if $U = [U_1, U_2, U_3, U_4]$, it is easy to check that

$$U^{\mathsf{T}}AU = D[1, 1, 6, 6].$$

9.12 CHARACTERISTIC ROOTS OF A POLYNOMIAL FUNCTION OF A MATRIX

Before continuing with the study of real, symmetric matrices, we prove three theorems which are useful in a variety of applications and which apply to arbitrary real or complex matrices. We begin with

Theorem 9.12.1: *If $\lambda_1, \lambda_2, \ldots, \lambda_n$ are the characteristic roots, distinct or not, of a matrix A of order n, and if $g(A)$ is any polynomial function of A, then the characteristic roots of $g(A)$ are $g(\lambda_1), g(\lambda_2), \ldots, g(\lambda_n)$.*

We know that

$$\det(A - \lambda I_n) = (\lambda_1 - \lambda)(\lambda_2 - \lambda) \cdots (\lambda_n - \lambda),$$

and we wish to prove that, for any polynomial function $g(A)$,

$$\det(g(A) - \lambda I_n) = (g(\lambda_1) - \lambda)(g(\lambda_2) - \lambda) \cdots (g(\lambda_n) - \lambda).$$

Now suppose that $g(x)$ is of degree r in x and that, for a fixed value of λ, the roots of $g(x) - \lambda = 0$ are $x_1, x_2, \ldots, x_r$. Then we have

$$g(x) - \lambda = \alpha(x - x_1)(x - x_2) \cdots (x - x_r),$$

where α is the coefficient of x^r in $g(x)$. Hence

$$g(A) - \lambda I_n = \alpha(A - x_1 I_n)(A - x_2 I_n) \cdots (A - x_r I_n),$$

so that, if $\phi(\lambda)$ is the characteristic polynomial of A,

$$\begin{aligned}
\det(g(A) - \lambda I_n) &= \alpha^n \det(A - x_1 I_n) \det(A - x_2 I_n) \cdots \det(A - x_r I_n) \\
&= \alpha^n \phi(x_1)\phi(x_2) \cdots \phi(x_r) \\
&= \alpha^n (\lambda_1 - x_1)(\lambda_2 - x_1) \cdots (\lambda_n - x_1) \\
&\quad \cdot (\lambda_1 - x_2)(\lambda_2 - x_2) \cdots (\lambda_n - x_2) \\
&\quad \vdots \\
&\quad \cdot (\lambda_1 - x_r)(\lambda_2 - x_r) \cdots (\lambda_n - x_r).
\end{aligned}$$

By rearranging the orders of the factors, we now obtain

$$\begin{aligned}\det(g(A) - \lambda I_n) &= \alpha(\lambda_1 - x_1)(\lambda_1 - x_2)\cdots(\lambda_1 - x_r)\\ &\quad \cdot\alpha(\lambda_2 - x_1)(\lambda_2 - x_2)\cdots(\lambda_2 - x_r)\\ &\quad \vdots\\ &\quad \cdot\alpha(\lambda_n - x_1)(\lambda_n - x_2)\cdots(\lambda_n - x_r)\\ &= (g(\lambda_1) - \lambda)(g(\lambda_2) - \lambda)\cdots(g(\lambda_n) - \lambda),\end{aligned}$$

as was to be proved.

For example, since A^p, where p is a positive integer, is a polynomial function of A, the characteristic roots of A^p are $\lambda_1^p, \lambda_2^p, \ldots, \lambda_n^p$.

One might expect a similar result to hold for the roots of A^{-1}, when A^{-1} exists. However, the preceding theorem does not apply, since A^{-1} is not a polynomial function of A. On the other hand, since A^{-1} is nonsingular, so that no characteristic root μ of A^{-1} is zero, we have

$$\det(A^{-1} - \mu I) = (-\mu)^n \det A^{-1} \det(A - \mu^{-1}I).$$

For this expression to vanish, we must have

$$\det(A - \mu^{-1}I) = 0.$$

The roots of this equation in μ^{-1} are, of course, $\lambda_1, \lambda_2, \ldots, \lambda_n$. Hence the roots μ of A^{-1} must be $\lambda_1^{-1}, \lambda_2^{-1}, \ldots, \lambda_n^{-1}$, as expected.

We summarize in

Theorem 9.12.2: *If the characteristic roots of A are $\lambda_1, \lambda_2, \ldots, \lambda_n$, those of A^p, where p is any positive integer, are $\lambda_1^p, \lambda_2^p, \ldots, \lambda_n^p$. If A^{-1} exists, this conclusion holds for all negative integers as well.*

Now suppose that A is a real, symmetric matrix with characteristic roots $\lambda_1, \lambda_2, \ldots, \lambda_n$ and with characteristic equation $\phi(\lambda) = 0$. Then $\phi(A)$ is also a real, symmetric matrix and, by Theorem 9.12.1, its characteristic roots are $\phi(\lambda_1), \phi(\lambda_2), \ldots, \phi(\lambda_n)$, all of which are zero. On the other hand, if a real, symmetric matrix has all its roots zero, it must be the zero matrix. (Prove this.) Hence $\phi(A) = 0$. That is, A satisfies its own characteristic equation. This is a special case of the famous **Cayley-Hamilton theorem**:

Theorem 9.12.3: *Every matrix satisfies its own characteristic equation.*

The general proof may be found in F. E. Hohn, *Elementary Matrix Algebra*, second edition, New York, Macmillan, 1964, page 283.

For example, if

$$A = \begin{bmatrix} 1 & 2 & 0 \\ 2 & -1 & 0 \\ 0 & 0 & 1 \end{bmatrix},$$

then

$$\phi(\lambda) = \begin{bmatrix} 1-\lambda & 2 & 0 \\ 2 & -1-\lambda & 0 \\ 0 & 0 & 1-\lambda \end{bmatrix} = -5 + 5\lambda + \lambda^2 - \lambda^3.$$

Hence

$$\begin{aligned}\phi(A) &= -5\begin{bmatrix} 1 & 0 & 0 \\ 0 & 1 & 0 \\ 0 & 0 & 1 \end{bmatrix} + 5\begin{bmatrix} 1 & 2 & 0 \\ 2 & -1 & 0 \\ 0 & 0 & 1 \end{bmatrix} \\ &\quad + \begin{bmatrix} 1 & 2 & 0 \\ 2 & -1 & 0 \\ 0 & 0 & 1 \end{bmatrix}^2 - \begin{bmatrix} 1 & 2 & 0 \\ 2 & -1 & 0 \\ 0 & 0 & 1 \end{bmatrix}^3 \\ &= \begin{bmatrix} -5 & 0 & 0 \\ 0 & -5 & 0 \\ 0 & 0 & -5 \end{bmatrix} + \begin{bmatrix} 5 & 10 & 0 \\ 10 & -5 & 0 \\ 0 & 0 & 5 \end{bmatrix} \\ &\quad + \begin{bmatrix} 5 & 0 & 0 \\ 0 & 5 & 0 \\ 0 & 0 & 1 \end{bmatrix} + \begin{bmatrix} -5 & -10 & 0 \\ -10 & 5 & 0 \\ 0 & 0 & -1 \end{bmatrix} \\ &= 0.\end{aligned}$$

An important application of the Cayley–Hamilton theorem is in the representation of high powers of a matrix. Let us rewrite $\phi(A) = 0$ in the form

$$A^n + a_{n-1}A^{n-1} + \cdots + a_1A + a_0I_n = 0, \tag{9.12.1}$$

so that, if we have already computed $A^2, A^3, \ldots, A^{n-1}$, we can express A^n as a linear combination of these:

$$A^n = -a_0I_n - a_1A - \cdots - a_{n-1}A^{n-1}. \tag{9.12.2}$$

Multiplying through by A and substituting from (9.12.2) for A^n on the right, we obtain

$$\begin{aligned}A^{n+1} = a_{n-1}a_0I_n + (a_{n-1}a_1 - a_0)A + (a_{n-1}a_2 - a_1)A^2 \\ + \cdots + (a_{n-1}^2 - a_{n-2})A^{n-1}.\end{aligned} \tag{9.12.3}$$

By continuing this process, we can express any positive integral power of A as a linear combination of $I, A, A^2, \ldots, A^{n-1}$.

If A^{-1} exists, by multiplying (9.12.1) by A^{-1} and solving for A^{-1}, we obtain

$$A^{-1} = -\frac{a_1}{a_0}I - \frac{a_2}{a_0}A - \cdots - \frac{1}{a_0}A^{n-1}. \tag{9.12.4}$$

Multiplying (9.12.4) by A^{-1} and substituting from (9.12.4) for A^{-1}, we obtain

$$A^{-2} = \left(\frac{a_1^2}{a_0^2} - \frac{a_2}{a_0}\right)I_n + \left(\frac{a_1a_2}{a_0^2} - \frac{a_3}{a_0}\right)A + \cdots + \left(\frac{a_1a_{n-1}}{a_0^2} - \frac{1}{a_0}\right)A^{n-2} + \frac{a_1}{a_0^2}A^{n-1}.$$

By continuation of this process, all negative integral powers of A may also be expressed as linear combinations of $I_n, A, A^2, \ldots, A^{n-1}$.

9.13 EXERCISES

1. Find orthogonal matrices which diagonalize the matrices

(a) $\begin{bmatrix} 1 & 3 \\ 3 & 1 \end{bmatrix}$, (b) $\begin{bmatrix} 1 & -1 & 0 \\ -1 & 1 & 0 \\ 0 & 0 & 1 \end{bmatrix}$, (c) $\begin{bmatrix} 4 & 1 & 0 \\ 1 & 4 & 0 \\ 0 & 0 & 4 \end{bmatrix}$,

(d) $\begin{bmatrix} 0 & \sqrt{2} & -1 \\ \sqrt{2} & 1 & -\sqrt{2} \\ -1 & -\sqrt{2} & 0 \end{bmatrix}$, (e) $\begin{bmatrix} 1 & 3 & 4 \\ 3 & 1 & 0 \\ 4 & 0 & 1 \end{bmatrix}$,

(f) $\begin{bmatrix} 2 & 3 & 0 & 0 \\ 3 & 2 & 0 & 0 \\ 0 & 0 & 3 & 2 \\ 0 & 0 & 2 & 3 \end{bmatrix}$, (g) $\begin{bmatrix} 0 & 0 & 1 & 0 \\ 0 & 1 & 0 & 0 \\ 1 & 0 & 0 & 0 \\ 0 & 0 & 0 & 1 \end{bmatrix}$.

2. Show that if $U_1^{\mathsf{T}}A_1U_1 = D[\lambda_1, \ldots, \lambda_k]$ and $U_2^{\mathsf{T}}A_2U_2 = D[\lambda_{k+1}, \ldots, \lambda_n]$, then

$$\left[\begin{array}{c|c} U_1 & 0 \\ \hline 0 & U_2 \end{array}\right]^{\mathsf{T}}\left[\begin{array}{c|c} A_1 & 0 \\ \hline 0 & A_2 \end{array}\right]\left[\begin{array}{c|c} U_1 & 0 \\ \hline 0 & U_2 \end{array}\right] = D[\lambda_1, \ldots, \lambda_k, \lambda_{k+1}, \ldots, \lambda_n].$$

3. Show that if

$$A = \begin{bmatrix} 1 & 1 \\ 0 & 1 \end{bmatrix},$$

there exists no nonsingular matrix B of order 2 such that $B^{-1}AB = D[d_1, d_2]$. This example shows that *not every square matrix can be diagonalized by a nonsingular transformation of coordinates.*

4. Show that if λ is a *real* characteristic root of an orthogonal matrix U, then λ must be 1 or -1. Show that for every complex root λ, $|\lambda| = 1$.

5. Show that if U is an orthogonal matrix of odd order, then U necessarily has 1 or -1 as a characteristic root according as det U is 1 or -1.

6. Use Theorem 9.12.1 to prove Lemma 9.6.3.

7. Prove that a real, symmetric matrix is also orthogonal if and only if its characteristic roots are all ± 1. Use the argument of the proof to help you construct an example of a real, symmetric, orthogonal matrix of order 3.

***8.** Given that $U^{\mathsf{T}}AU = D[\lambda_1, \lambda_2, \ldots, \lambda_n]$, where U is orthogonal, obtain a simple formula for A^p, where p is any positive integer.

***9.** Let A be a real, symmetric matrix with characteristic roots $\lambda_1, \lambda_2, \ldots, \lambda_r$, $0, 0, \ldots, 0$, where no λ_j is 0. Prove that A has rank r.

10. Let A be a real, symmetric matrix and let $U_1, U_2, \ldots, U_n$ be an orthonormal set of characteristic vectors corresponding to the characteristic roots $\lambda_1, \lambda_2, \ldots, \lambda_n$, respectively. Show that the solution of

$$(A - \lambda I)X = B, \quad \lambda \neq \lambda_j, \qquad j = 1, 2, \ldots, n,$$

is given by

$$X = \sum_{j=1}^{n} \frac{U_j^{\mathsf{T}}B}{\lambda_j - \lambda} U_j.$$

***11.** Prove that the characteristic roots of a real, symmetric matrix A are all equal if and only if A is scalar.

***12.** Prove that all the characteristic roots of a real symmetric matrix A are zero if and only if A is zero. Show by means of an example that a nonzero, nonsymmetric matrix can have all its characteristic roots zero.

13. Verify the Cayley–Hamilton theorem for the matrix

$$A = \begin{bmatrix} 0 & 1 & 0 & 0 \\ 0 & 0 & 1 & 0 \\ 0 & 0 & 0 & 1 \\ 1 & 1 & 1 & 1 \end{bmatrix}$$

and then compute A^5 and A^{-1} by the method of Section 9.12.

14. Given that

$$A = \begin{bmatrix} 1 & \sqrt{3} & 0 \\ \sqrt{3} & -1 & 0 \\ 0 & 0 & 1 \end{bmatrix}, \qquad g(A) = A^2 + A + I_3,$$

find the characteristic roots of A and of $g(A)$.

***15.** Prove that if $AX = \lambda X$, then for all positive integers p, $A^pX = \lambda^pX$. Hence show that if $AX = \lambda X$ and if $g(A)$ is any polynomial function of A, then $g(A)X = g(\lambda)X$.

16. Find the powers and the characteristic roots of the powers of the cyclic permutation matrix

$$P = \begin{bmatrix} 0 & 1 & 0 & 0 & 0 \\ 0 & 0 & 1 & 0 & 0 \\ 0 & 0 & 0 & 1 & 0 \\ 0 & 0 & 0 & 0 & 1 \\ 1 & 0 & 0 & 0 & 0 \end{bmatrix}.$$

***17.** A matrix A is said to be **diagonalizable** if there exists a matrix B such that $B^{-1}AB = D[d_1, d_2, \ldots, d_n]$. Let A be diagonalizable and let $\phi(\lambda) = 0$ be the characteristic equation of A. Show that $\phi(D) = 0$ and hence that $\phi(BDB^{-1}) = 0$, that is, that $\phi(A) = 0$. This proves the Cayley–Hamilton theorem for every diagonalizable matrix.

***18.** Prove that the characteristic roots of a Hermitian matrix are all real.

***19.** Prove that characteristic vectors associated with distinct characteristic roots of a Hermitian matrix H are mutually orthogonal.

***20.** An $n \times n$ matrix U over the complex field is called a **unitary matrix** if and only if $U^*U = I_n$. Prove that the columns of a unitary matrix form an orthonormal basis for unitary n-space $\mathscr{U}^n$.

***21.** Prove that a transformation $X = UY$ of unitary n-space leaves the length of a vector invariant if and only if U is unitary.

***22.** Prove that for each Hermitian matrix H, there exists a unitary matrix U such that

$$U^*HU = D[\lambda_1, \lambda_2, \ldots, \lambda_n],$$

where the λ's are the characteristic roots of H.

***23.** Show how to obtain a unitary matrix U which will diagonalize a given Hermitian matrix H, assuming that its characteristic roots are known.

24. Show that the set of all polynomial functions over a number field $\mathscr{F}$ of a fixed matrix A over $\mathscr{F}$ is a vector space of dimension n over $\mathscr{F}$.

9.14 QUADRATIC FORMS

A homogeneous polynomial q of the type

$$q = X^{\mathsf{T}}AX = \sum_{i,j=1}^{n} a_{ij}x_ix_j, \tag{9.14.1}$$

the coefficients of which are in a field $\mathscr{F}$, is called a **quadratic form** over $\mathscr{F}$. Such forms have many applications in the physical sciences and engineering, in the mathematics of computation, in statistics, in geometry, and so on. In most applications, the field $\mathscr{F}$ is the field of real numbers. From now on we assume that this is the case.

When the matrix product $X^{\mathsf{T}}AX$ is given, the scalar expansion given in (9.14.1) may be written by inspection, since the i,j-entry of A is the coefficient of the product x_ix_j. In this expansion, the similar terms $a_{ij}x_ix_j$ and $a_{ji}x_jx_i$ would naturally be combined into a single term $(a_{ij} + a_{ji})x_ix_j$. It is clear from this fact that distinct $n \times n$ matrices A_1 and A_2 will lead to the same quadratic polynomial provided only that all corresponding sums of the type $a_{ij} + a_{ji}$ are equal. Moreover, given the quadratic form, one could not identify the corresponding matrix A by inspection. It is consistent with the various uses made of quadratic forms to eliminate this ambiguity once and for all by replacing each of the pair of coefficients a_{ij} and a_{ji} of a given form q by their mean, $(a_{ij} + a_{ji})/2$. Then the coefficients of a given quadratic form will define a unique symmetric matrix. In what follows, every quadratic form $X^{\mathsf{T}}AX$ with which we work will therefore be assumed to have a symmetric matrix A.

For example,

$$X^{\mathsf{T}}\begin{bmatrix} 1 & 2 \\ 2 & -3 \end{bmatrix}X = x_1^2 + 4x_1x_2 - 3x_2^2$$

and

$$x_1^2 - 2x_1x_2 + 5x_1x_3 + 2x_3^2 = X^{\mathsf{T}}\begin{bmatrix} 1 & -1 & \frac{5}{2} \\ -1 & 0 & 0 \\ \frac{5}{2} & 0 & 2 \end{bmatrix}X.$$

Note how missing terms lead to zero entries in the coefficient matrix.

If the x's are independent variables, the rank of A is called the **rank of the form** and det A is called the **discriminant of the form**.

A nonsingular linear transformation $X = B\tilde{X}$ maps a quadratic form $X^{\mathsf{T}}AX$ onto the form $\tilde{X}^{\mathsf{T}}(B^{\mathsf{T}}AB)\tilde{X}$, where $B^{\mathsf{T}}AB$ is also symmetric and has the same rank as A. We say that two quadratic forms $X^{\mathsf{T}}A_1X$ and $\tilde{X}^{\mathsf{T}}A_2\tilde{X}$ are **equivalent** if and only if there is a nonsingular transformation $X = B\tilde{X}$ such that

$$X^{\mathsf{T}}A_1X = \tilde{X}^{\mathsf{T}}B^{\mathsf{T}}A_1B\tilde{X} = \tilde{X}^{\mathsf{T}}A_2\tilde{X},$$

that is, if and only if, for a suitable nonsingular matrix B,

$$A_2 = B^{\mathsf{T}}A_1B. \tag{9.14.2}$$

Two matrices A_1 and A_2 related as in (9.14.2), with B nonsingular, are said to be **congruent**.

As an example, consider the form

$$q = 29x_1^2 + 24x_1x_2 + 5x_2^2 = X^{\mathsf{T}}\begin{bmatrix} 29 & 12 \\ 12 & 5 \end{bmatrix}X.$$

This can be rearranged in the form

$$q = (5x_1 + 2x_2)^2 + (2x_1 + x_2)^2.$$

Hence let us apply the transformation

$$\tilde{X} = \begin{bmatrix} 5 & 2 \\ 2 & 1 \end{bmatrix}X \quad \text{or} \quad X = \begin{bmatrix} 1 & -2 \\ -2 & 5 \end{bmatrix}\tilde{X}.$$

Then

$$X^{\mathsf{T}}\begin{bmatrix} 29 & 12 \\ 12 & 5 \end{bmatrix}X = \tilde{X}^{\mathsf{T}}\begin{bmatrix} 1 & -2 \\ -2 & 5 \end{bmatrix}\begin{bmatrix} 29 & 12 \\ 12 & 5 \end{bmatrix}\begin{bmatrix} 1 & -2 \\ -2 & 5 \end{bmatrix}\tilde{X}$$

$$= \tilde{X}^{\mathsf{T}}\begin{bmatrix} 1 & 0 \\ 0 & 1 \end{bmatrix}\tilde{X} = \tilde{x}_1^2 + \tilde{x}_2^2.$$

Thus the forms $29x_1^2 + 24x_1x_2 + 5x_2^2$ and $\tilde{x}_1^2 + \tilde{x}_2^2$ are equivalent and the matrices $\begin{bmatrix} 29 & 12 \\ 12 & 5 \end{bmatrix}$ and $\begin{bmatrix} 1 & 0 \\ 0 & 1 \end{bmatrix}$ are congruent.

9.15 DIAGONALIZATION OF QUADRATIC FORMS

A given quadratic form can be reduced in various ways to equivalent forms which emphasize certain of its basic properties. The most important of these reductions is diagonalization by means of an orthogonal transformation.

We proved in Section 9.10 that for every real symmetric matrix A, there exists an orthogonal matrix U such that

$$U^{\mathsf{T}}AU = D[\lambda_1, \lambda_2, \ldots, \lambda_n],$$

where $\lambda_1, \lambda_2, \ldots, \lambda_n$ are the characteristic roots of A. As a consequence, the transformation $X = U\tilde{X}$ applied to the quadratic form $X^{\mathsf{T}}AX$ gives

$$X^{\mathsf{T}}AX = \tilde{X}^{\mathsf{T}}(U^{\mathsf{T}}AU)\tilde{X} = \lambda_1\tilde{x}_1^2 + \lambda_2\tilde{x}_2^2 + \cdots + \lambda_n\tilde{x}_n^2. \tag{9.15.1}$$

If the rank of A is r, then so is that of $D[\lambda_1, \lambda_2, \ldots, \lambda_n]$. Hence $n - r$ characteristic roots of A must be zero and, if $\lambda_1, \lambda_2, \ldots, \lambda_r$ denote the nonzero characteristic roots, (9.15.1) becomes simply

$$X^{\mathsf{T}}AX = \lambda_1\tilde{x}_1^2 + \lambda_2\tilde{x}_2^2 + \cdots + \lambda_r\tilde{x}_r^2. \tag{9.15.2}$$

The computational aspects of this reduction are, of course, the same as those discussed in Section 9.11 in connection with the determination of the matrix U.

For example, let

$$q = 5x_1^2 + 2x_1x_2 + 5x_2^2 = X^{\mathsf{T}}\begin{bmatrix} 5 & 1 \\ 1 & 5 \end{bmatrix}X.$$

Here A has characteristic roots 6 and 4, with associated normalized characteristic vectors $[1/\sqrt{2}, 1/\sqrt{2}]^{\mathsf{T}}$ and $[1/\sqrt{2}, -1/\sqrt{2}]^{\mathsf{T}}$, respectively. Hence we put

$$X = \begin{bmatrix} \frac{1}{\sqrt{2}} & \frac{1}{\sqrt{2}} \\ \frac{1}{\sqrt{2}} & -\frac{1}{\sqrt{2}} \end{bmatrix}\tilde{X}$$

and get

$$X^{\mathsf{T}}\begin{bmatrix} 5 & 1 \\ 1 & 5 \end{bmatrix}X = \tilde{X}^{\mathsf{T}}\begin{bmatrix} \frac{1}{\sqrt{2}} & \frac{1}{\sqrt{2}} \\ \frac{1}{\sqrt{2}} & -\frac{1}{\sqrt{2}} \end{bmatrix}\begin{bmatrix} 5 & 1 \\ 1 & 5 \end{bmatrix}\begin{bmatrix} \frac{1}{\sqrt{2}} & \frac{1}{\sqrt{2}} \\ \frac{1}{\sqrt{2}} & -\frac{1}{\sqrt{2}} \end{bmatrix}\tilde{X} = \tilde{X}^{\mathsf{T}}\begin{bmatrix} 6 & 0 \\ 0 & 4 \end{bmatrix}\tilde{X}.$$

9.16 DEFINITE FORMS AND MATRICES

Let us take a second look at each of our last two examples:

$$q = (5x_1 + 2x_2)^2 + (2x_1 + x_2)^2$$

and

$$q = 5x_1^2 + 2x_1x_2 + 5x_2^2 = 6\tilde{x}_1^2 + 4\tilde{x}_2^2.$$

Each of these is representable as a sum of squares of linear expressions and hence can never be negative for any real vector X. Moreover, they will take on the value zero only when $X = 0$. This sort of situation is common in applications in both the physical and the social sciences as well as in mathematics. The form typically represents some quantity, such as energy or the square of a distance, that may be zero but cannot be negative. We are thus led to make the following definition: A real quadratic form q, defined by $q(X) = X^{\mathsf{T}}AX$, is **positive definite** if and only if $q(X) > 0$ for all real X except $X = 0$. If $q(X) \geq 0$ for all real X, q is said to be **positive semidefinite** or **nonnegative definite**. Similarly, q is **negative definite** (**negative semidefinite**) if $q(X) < 0$ for all real X except $X = 0$ ($q(X) \leq 0$ for all real X).

The first example cited above illustrates the fact that any quadratic form that can be written as a sum of squares of linear functions is positive definite or positive semidefinite. The second example illustrates

Theorem 9.16.1: *A quadratic form $X^{\mathsf{T}}AX$ is positive definite if and only if the characteristic roots of A are all positive.*

To prove this, assume first that the form is positive definite. Let λ_1, necessarily real since A is real and symmetric, be any characteristic root of A and let U_1 be a normalized, real characteristic vector such that

$$AU_1 = \lambda_1 U_1.$$

Then, since the form is positive definite,

$$0 < U_1^{\mathsf{T}}(AU_1) = U_1^{\mathsf{T}}(\lambda_1 U_1) = \lambda_1 U_1^{\mathsf{T}} U_1 = \lambda_1.$$

Hence all characteristic roots are positive.

Now assume that all characteristic roots of A are positive and let $U_1, U_2, \ldots, U_n$ be an orthonormal set of vectors associated with the roots $\lambda_1, \lambda_2, \ldots, \lambda_n$, respectively. Let X be any real vector. Since the U_j are a basis for $\mathscr{E}^n$, we may write $X = \sum \alpha_i U_i$, where the α's are real. Hence

$$\begin{aligned} X^{\mathsf{T}}AX &= \Big(\sum_i \alpha_i U_i\Big)^{\mathsf{T}} A \Big(\sum_j \alpha_j U_j\Big) = \sum_i \sum_j (\alpha_i \alpha_j U_i^{\mathsf{T}} A U_j) \\ &= \sum_i \sum_j \alpha_i \alpha_j U_i^{\mathsf{T}}(\lambda_j U_j) = \sum_i \sum_j \alpha_i \alpha_j \lambda_j (U_i^{\mathsf{T}} U_j) \\ &= \sum_i \sum_j \alpha_i \alpha_j \lambda_j \delta_{ij} = \sum_i \alpha_i^2 \lambda_i. \end{aligned}$$

Since all the λ_i are positive, and all the α_i are real, $\sum_i \alpha_i^2\lambda_i \geq 0$. Moreover, it can be zero only if all the α_i are zero, that is, only if $X = 0$. Thus the form is seen to be positive definite.

A slight alteration of this proof yields

Theorem 9.16.2: *A quadratic form $X^\mathsf{T}AX$ is positive semidefinite if and only if all the characteristic roots of A are nonnegative.*

Next we establish a necessary condition for positive definiteness.

Theorem 9.16.3: *If $X^\mathsf{T}AX$ is positive definite, then every principal minor determinant of A is positive.*

First, assume that the form is positive definite. Then, since all characteristic roots of A are positive, we have $\det A = \lambda_1\lambda_2\cdots\lambda_n > 0$.

Every principal minor of order r, $0 < r < n$, is the determinant of a submatrix obtainable by deleting symmetrically a suitable set of rows and columns of A. This corresponds to putting $n - r$ of the x_i equal to zero and looking at the quadratic form in the remaining variables only. Since the original form is positive definite, so is this reduced form, for the values of the reduced form will also be values of the original form. Now we apply our first observation to the reduced form and the proof is complete.

In particular, every main diagonal entry a_{ii} of A must be positive.

The reader may prove, in similar fashion,

Theorem 9.16.4: *If $X^\mathsf{T}AX$ is positive semidefinite, then every principal minor determinant of A is nonnegative.*

Whereas the two preceding theorems do not make it possible to prove that a form *is* positive definite or semidefinite, they do make it possible to prove that a form *is not*. For example, the form

$$x_1^2 + 2x_1x_2 + x_1x_3 + 4x_2x_3 - x_2^2 + x_3^2$$

is not positive definite or semidefinite because $a_{22} = -1$. In fact, if $X = [0, 1, 0]^\mathsf{T}$, the form has the value -1. It is not negative definite or semidefinite either.

We can extend Theorem 9.16.4 a bit:

Theorem 9.16.5: *If the quadratic form $X^\mathsf{T}AX$ is positive semidefinite and if x_i actually appears in the form, then $a_{ii} > 0$.*

Since x_i actually appears in the form, we have $a_{ij} \neq 0$ for some j. If $j = i$, then $a_{ii} \neq 0$ and hence $a_{ii} > 0$ by the preceding theorem. If $j \neq i$, then the principal minor

$$\begin{vmatrix} a_{ii} & a_{ij} \\ a_{ji} & a_{jj} \end{vmatrix} \geq 0;$$

that is, since $a_{ij} = a_{ji} \neq 0$,

$$a_{ii}a_{jj} \geq (a_{ij})^2 > 0,$$

so $a_{ii} \neq 0$ and hence again $a_{ii} > 0$ by the preceding theorem. Incidentally, $a_{jj} > 0$ also.

A sort of converse to the last three theorems, and a particularly useful result, is included in

Theorem 9.16.6: *The quadratic form $X^{\mathsf{T}}AX$ is positive definite if and only if all its leading principal minors $p_1, p_2, \ldots, p_n$ are positive.*

The proof involves several pages of computation and we omit it here. Details are given in F. E. Hohn, *Elementary Matrix Algebra*, second edition, New York, Macmillan, 1964.

We illustrate with an example. To establish the positive definiteness of

$$X^{\mathsf{T}}\begin{bmatrix} 2 & 1 & 1 & \cdots & 1 \\ 1 & 2 & 1 & \cdots & 1 \\ \vdots & & & & \\ 1 & 1 & 1 & \cdots & 2 \end{bmatrix}_{n \times n} X$$

note that

$$p_1 = 2,$$

$$p_2 = \begin{vmatrix} 2 & 1 \\ 1 & 2 \end{vmatrix} = 3,$$

$$p_3 = \begin{vmatrix} 2 & 1 & 1 \\ 1 & 2 & 1 \\ 1 & 1 & 2 \end{vmatrix} = \begin{vmatrix} 1 & 0 & -1 \\ 0 & 1 & -1 \\ 1 & 1 & 2 \end{vmatrix} = \begin{vmatrix} 1 & 0 & 0 \\ 0 & 1 & 0 \\ 1 & 1 & 4 \end{vmatrix} = 4.$$

The procedure here is to subtract the bottom row from each of the others, then add the first two columns to the last. The same procedure works on all the principal minors, so that, in fact, $p_r = r + 1$. Since every p_r is greater than 0, the form is positive definite.

We show next that positive definiteness is invariant under transformation:

Theorem 9.16.7: *A quadratic form $X^{\mathsf{T}}AX$ is positive definite if and only if, for every real, nonsingular transformation $X = B\tilde{X}$, the equivalent form $\tilde{X}^{\mathsf{T}}(B^{\mathsf{T}}AB)\tilde{X}$ is positive definite.*

Since B is nonsingular and $X = B\tilde{X}$, $X = 0$ if and only if $\tilde{X} = 0$. Hence, since $X^{\mathsf{T}}AX = \tilde{X}^{\mathsf{T}}(B^{\mathsf{T}}AB)\tilde{X}$, if either member is positive for all nonzero X (or $\tilde{X}$), the other must be also and the theorem is proved.

It is convenient to make the following definition at this point: a real symmetric matrix A is **positive definite** or **semidefinite** if and only if the corresponding quadratic form is positive definite or semidefinite. Hence a real, symmetric matrix A is positive definite if and only if all its characteristic roots are positive, if and only if all its leading principal minors are positive, and if and only if every matrix $B^{\mathsf{T}}AB$, where B is nonsingular, is positive definite. Similarly for semidefinite matrices. At times it is easier to work with matrices rather than with forms. For example:

Theorem 9.16.8: *A real symmetric matrix A is positive definite if and only if A^{-1} exists and is positive definite and symmetric.*

Suppose that A is real, symmetric, and positive definite. Then every $\lambda_i > 0$. Hence $\det A = \lambda_1\lambda_2 \cdots \lambda_n > 0$. Hence A^{-1} exists. From $AA^{-1} = I$ follows $(A^{-1})^{\mathsf{T}}A^{\mathsf{T}} = (A^{-1})^{\mathsf{T}}A = I$, so $(A^{-1})^{\mathsf{T}} = A^{-1}$ and hence A^{-1} is symmetric. The characteristic roots of A^{-1} are $\lambda_1^{-1}, \lambda_2^{-1}, \ldots, \lambda_n^{-1}$, which are all positive. Hence A^{-1} is a positive definite, symmetric matrix.

The converse follows from the same argument, since $A = (A^{-1})^{-1}$.

An extension of the preceding theorem is

Theorem 9.16.9: *A is positive definite and symmetric if and only if A^p is positive definite and symmetric for all integers p.*

This follows principally from the fact that the characteristic roots of A^p are $\lambda_1^p, \lambda_2^p, \ldots, \lambda_n^p$ and from Theorem 9.16.8.

We prove next the often useful

Theorem 9.16.10: *A real matrix A is positive definite and symmetric if and only if there exists a nonsingular matrix B such that $A = B^{\mathsf{T}}B$.*

First assume that $A = B^{\mathsf{T}}B$, where B is nonsingular. Then we have readily that A is symmetric. Moreover, $X^{\mathsf{T}}AX = X^{\mathsf{T}}B^{\mathsf{T}}BX = (BX)^{\mathsf{T}}(BX) = |BX|^2 \geq 0$. Since, therefore, $X^{\mathsf{T}}AX = 0$ if and only if $BX = 0$, and since, because B is nonsingular, $BX = 0$ if and only if $X = 0$, it follows that A is positive definite.

Conversely, assume that A is a positive definite symmetric matrix. Then there exists an orthogonal matrix C such that

$$CAC^{\mathsf{T}} = D[\lambda_1, \lambda_2, \ldots, \lambda_n].$$

Then

$$A = C^{\mathsf{T}}D[\sqrt{\lambda_1}, \sqrt{\lambda_2}, \ldots, \sqrt{\lambda_n}]^2C$$

or

$$A = (D[\sqrt{\lambda_1}, \sqrt{\lambda_2}, \ldots, \sqrt{\lambda_n}]C)^{\mathsf{T}}(D[\sqrt{\lambda_1}, \sqrt{\lambda_2}, \ldots, \sqrt{\lambda_n}]C).$$

Now we put

$$B = D[\sqrt{\lambda_1}, \sqrt{\lambda_2}, \ldots, \sqrt{\lambda_n}]C,$$

which is nonsingular since C and D are, and thus have the stated result.

Finally, we prove

Theorem 9.16.11: *A real, symmetric, positive definite matrix A has a real, symmetric, positive definite pth root for each positive integer p.*

There exists an orthogonal matrix U such that

$$U^{\mathsf{T}}AU = D[\lambda_1, \lambda_2, \ldots, \lambda_n],$$

where all the λ's are positive so that the principal roots $\lambda_1^{1/p}, \lambda_2^{1/p}, \ldots, \lambda_n^{1/p}$ are all positive real numbers. Then

$$\begin{aligned} A &= UD[\lambda_1^{1/p}, \ldots, \lambda_n^{1/p}]U^{\mathsf{T}} \cdot UD[\lambda_1^{1/p}, \ldots, \lambda_n^{1/p}]U^{\mathsf{T}} \cdots UD[\lambda_1^{1/p}, \ldots, \lambda_n^{1/p}]U^{\mathsf{T}} \\ &= (UD[\lambda_1^{1/p}, \ldots, \lambda_n^{1/p}]U^{\mathsf{T}})^p. \end{aligned}$$

Moreover, the product in parentheses is real, symmetric, and positive definite.

9.17 A GEOMETRICAL APPLICATION

The diagonalization of quadratic forms has a ready geometrical interpretation. For example, in $\mathscr{E}^2$, the graph of the quadratic equation

$$a_{11}x_1^2 + 2a_{12}x_1x_2 + a_{22}x_2^2 = b, \tag{9.17.1}$$

where b is a constant, is a proper conic section, a pair of straight lines (not necessarily distinct), a single point, or the empty set. To determine the precise nature of the graph, we first rewrite (9.17.1) in matrix form:

$$X^{\mathsf{T}}\begin{bmatrix} a_{11} & a_{12} \\ a_{12} & a_{22} \end{bmatrix}X = b. \tag{9.17.2}$$

Then there exists an orthogonal matrix U which diagonalizes the symmetric matrix

$$A = \begin{bmatrix} a_{11} & a_{12} \\ a_{12} & a_{22} \end{bmatrix}$$

so that, if we put $X = UY$, we obtain the equation

$$Y^{\mathsf{T}}\begin{bmatrix} \lambda_1 & 0 \\ 0 & \lambda_2 \end{bmatrix}Y = b,$$

where λ_1 and λ_2 are the characteristic roots of A. In scalar form, we now have the familiar equation

$$\lambda_1 y_1^2 + \lambda_2 y_2^2 = b. \tag{9.17.3}$$

If we interpret the transformation $X = UY$ as a transformation of the coordinate system, the figure constituting the graph of (9.17.1) is not changed. Only the reference system is altered. However, this alteration is such that the equation becomes so simple in the new reference system that the nature of the graph can be determined by inspection. In particular, if $\lambda_1 \neq 0$, $\lambda_2 \neq 0$, $b \neq 0$, we can rewrite (9.17.3) in the form

$$\frac{y_1^2}{b/\lambda_1} + \frac{y_2^2}{b/\lambda_2} = 1. \tag{9.17.4}$$

If $b/\lambda_1 > 0$ and $b/\lambda_2 > 0$, equation (9.17.4) represents an **ellipse**. If, in addition, $\lambda_1 = \lambda_2$, the ellipse becomes a **circle**. If b/λ_1 and b/λ_2 are opposite in sign, equation (9.17.4) represents a **hyperbola**. If b/λ_1 and b/λ_2 are both negative, the graph of (9.17.4) is the empty set.

The nature of the graph when one or more of λ_1, λ_2, b is zero is readily determined and is left to the reader to discuss.

The graphs in the three cases cited are illustrated in Figure 9-3, where only the new axes (y_1 and y_2) are shown. These axes are called the **principal axes** of

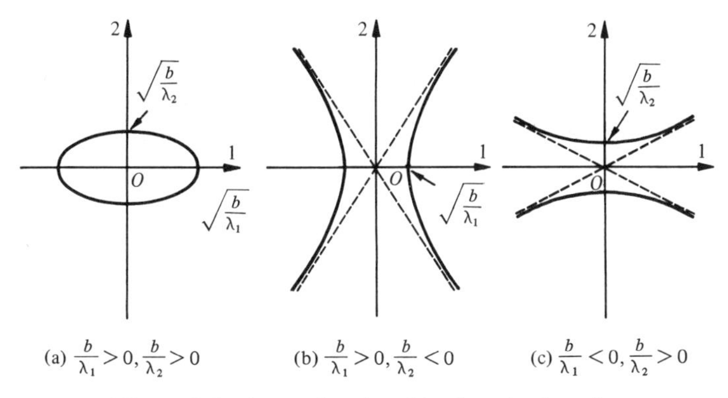

Figure 9-3. *Some Graphs of* $\lambda_1 y_1^2 + \lambda_2 y_2^2 = b$.

the conic and the transformation $X = UY$ is called a **principal axis transformation.**

In $\mathscr{E}^3$, the equation

$$X^{\mathsf{T}}AX = X^{\mathsf{T}}\begin{bmatrix} a_{11} & a_{12} & a_{13} \\ a_{12} & a_{22} & a_{23} \\ a_{13} & a_{23} & a_{33} \end{bmatrix} X = b \tag{9.17.5}$$

represents what is called a **quadric surface**. Again, let U be an orthogonal matrix which diagonalizes the symmetric matrix A. Then the nonsingular mapping $X = UY$ transforms (9.17.5) into the equation

$$\lambda_1 y_1^2 + \lambda_2 y_2^2 + \lambda_3 y_3^2 = b, \tag{9.17.6}$$

where λ_1, λ_2, and λ_3 are the characteristic roots of A. If λ_1, λ_2, λ_3, and b are all different from zero, we may write equation (9.17.6) in the form

$$\frac{y_1^2}{b/\lambda_1} + \frac{y_2^2}{b/\lambda_2} + \frac{y_3^2}{b/\lambda_3} = 1. \tag{9.17.7}$$

If all three denominators are positive, the equation represents an **ellipsoid**. This is the most important case in practice. If one of the denominators is negative, the equation represents a **hyperboloid of one sheet**; if two are negative, it represents a **hyperboloid of two sheets**; if all three are negative, the graph is the empty set.

If $b = 0$, $\lambda_1\lambda_2\lambda_3 \neq 0$, and λ_1, λ_2, and λ_3 all have the same sign, the graph is just the origin but if the signs are not all the same, the graph is a **cone**.

The nontrivial cases mentioned so far are illustrated in Figure 9-4.

The other cases in which one or more of λ_1, λ_2, λ_3, and b are zero should be discussed by the reader.

As in $\mathscr{E}^2$, the columns of the orthogonal matrix U define the **principal axes** of the quadric.

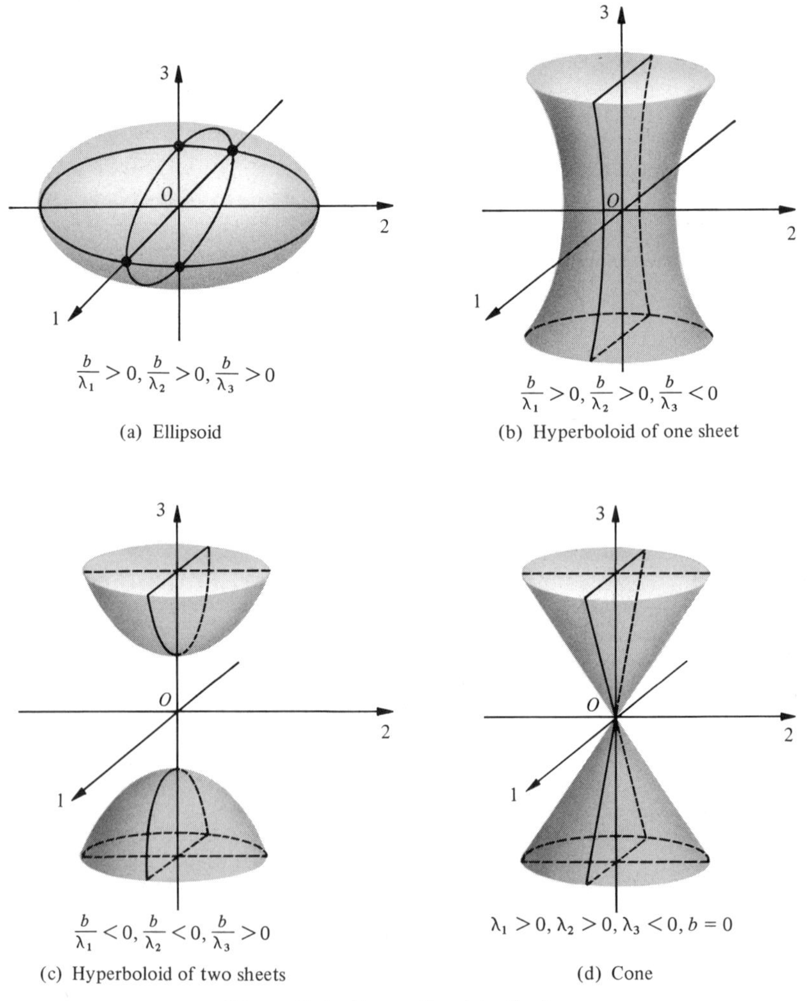

Figure 9-4. *Some Quadric Surfaces.*

In $\mathscr{E}^n$, the equation

$$X^{\mathsf{T}}AX = b$$

represents what is called a **hyperquadric**, an **n-quadric**, or simply a **quadric**. As in $\mathscr{E}^2$ and $\mathscr{E}^3$, we can use an appropriate orthogonal transformation of coordinates $X = UY$ to reduce the equation to the form

$$\sum_{j=1}^{n} \lambda_j y_j^2 = b.$$

The most important special case here is that in which all the λ's are positive. In this event, the locus of the equation is commonly called an **ellipsoid**. As in $\mathscr{E}^2$ and $\mathscr{E}^3$, the columns of U define the **principal axes** of the quadric in $\mathscr{E}^n$.

9.18 BILINEAR FORMS

A function of $m + n$ variables $x_1, x_2, \ldots, x_m$ and $y_1, y_2, \ldots, y_n$ of the type

$$b(X, Y) = \sum_{i=1}^{m} \sum_{j=1}^{n} a_{ij} x_i y_j = X^{\mathsf{T}} A Y, \tag{9.18.1}$$

where $A = [a_{ij}]_{m \times n}$ and the a_{ij} are real, is called a **real bilinear form**.

For example,

$$[x_1, x_2, x_3] \begin{bmatrix} 1 & 0 \\ 0 & 2 \\ -1 & -2 \end{bmatrix} \begin{bmatrix} y_1 \\ y_2 \end{bmatrix} = x_1 y_1 + 2x_2 y_2 - x_3 y_1 - 2x_3 y_2$$

is a real bilinear form.

The $m \times n$ matrix A which contains all the coefficients of a bilinear form is called the **matrix of the form** and, if the x's and y's are independent variables, the rank of A is called the **rank of the form**. In the example just given, the rank of the form is 2. When the matrix of the form is symmetric, the form itself is called a **symmetric bilinear form**. The most important example of a real, symmetric bilinear form is the inner product $X^{\mathsf{T}} Y = X^{\mathsf{T}} I Y$. When $m = n$ and $Y = X$, the bilinear form reduces to a quadratic form.

Since every $m \times n$ matrix may be used as the matrix of a bilinear form in $m + n$ variables, many definitions and theorems about matrices have simple counterparts in the theory of bilinear forms.

It is important to note that in (9.18.1), a_{ij} is the coefficient of the product $x_i y_j$, for by this observation we are enabled to go from the form to its matrix representation and vice versa by inspection. A useful example is provided by a form much used in statistics, the **covariance** of X and Y:

$$\operatorname{cov}(X, Y) = \frac{1}{n-1} \sum_{j=1}^{n} (x_j - m_x)(y_j - m_y), \tag{9.18.2}$$

where

$$m_x = \frac{1}{n}\sum_{j=1}^{n} x_j \qquad \text{and} \qquad m_y = \frac{1}{n}\sum_{j=1}^{n} y_j$$

are the means of the x's and the y's, respectively. This form may be regarded most simply as a bilinear form in the $2n$ "deviations" $x_j - m_x$ and $y_j - m_y$, its matrix being the identity matrix divided by $n - 1$. However, since $\sum (x_j - m_x) = 0$ and $\sum (y_j - m_y) = 0$, the deviations are not independent variables. To determine the rank of the form, we therefore expand the product in (9.18.2), sum the individual terms, and substitute for m_x and m_y, thus obtaining the expansion

$$\text{(9.18.3)} \qquad \operatorname{cov}(X, Y) = \frac{1}{n-1}\left(\frac{n-1}{n}\sum_{j=1}^{n} x_j y_j - \frac{1}{n}\sum_{j\neq k} x_j y_k\right).$$

Hence the matrix of the covariance is

$$\text{(9.18.4)} \qquad \frac{1}{n(n-1)}\begin{bmatrix} n-1 & -1 & \cdots & -1 \\ -1 & n-1 & \cdots & -1 \\ \vdots & & & \\ -1 & -1 & \cdots & n-1 \end{bmatrix}_n,$$

which may be shown to have rank $n - 1$. Assuming the x's and y's to be independent, $n - 1$ is then the rank of the form.

Frequently it is necessary or desirable to introduce new variables into a bilinear form in place of X and Y, that is, to effect linear transformations of coordinates in the spaces $\mathscr{E}^m$ and $\mathscr{E}^n$.

We next investigate the effect of this operation and some of the results which can be accomplished thereby.

Let $X^{\mathsf{T}}AY$ be a real bilinear form. Let $X = B\tilde{X}$ and $Y = C\tilde{Y}$ be nonsingular linear transformations relating X and Y to new variables $\tilde{X}$ and $\tilde{Y}$, the matrices B and C also being real. Then we have

$$X^{\mathsf{T}}AY = (B\tilde{X})^{\mathsf{T}}A(C\tilde{Y}) = \tilde{X}^{\mathsf{T}}(B^{\mathsf{T}}AC)\tilde{Y},$$

which gives the representation of the form in the new reference systems. The matrix $B^{\mathsf{T}}AC$ has the same rank as A, since B^{T} and C are nonsingular.

Alternatively, we can regard the transformations $X = B\tilde{X}$, $Y = C\tilde{Y}$ as nonsingular operators on $\mathscr{E}^m$ and $\mathscr{E}^n$, respectively, the effect of which is to take the bilinear form $X^{\mathsf{T}}AY$ into the new form $\tilde{X}^{\mathsf{T}}(B^{\mathsf{T}}AC)\tilde{Y}$, which always has the same value as the given one at the corresponding pair of vectors.

We have then, in summary,

Theorem 9.18.1: *The rank of a bilinear form is the same in all reference systems. Nonsingular linear operators on $\mathscr{E}^m$ and $\mathscr{E}^n$ take a real bilinear form into a real bilinear form of the same rank. The values of the two forms are always the same at corresponding pairs of vectors (X, Y) and $(\tilde{X}, \tilde{Y})$.*

For example, let us apply the real transformations of coordinates

$$X = \begin{bmatrix} 1 & 1 & 0 \\ 0 & 1 & 0 \\ 0 & 0 & -1 \end{bmatrix} \tilde{X} \quad \text{and} \quad Y = \begin{bmatrix} 1 & 1 & 0 \\ 0 & 1 & 0 \\ 0 & 0 & -1 \end{bmatrix} \tilde{Y}$$

to the symmetric bilinear form

$$X^{\mathsf{T}}AY = X^{\mathsf{T}} \begin{bmatrix} 1 & -1 & 0 \\ -1 & 2 & 0 \\ 0 & 0 & 1 \end{bmatrix} Y = x_1y_1 - x_1y_2 - x_2y_1 + 2x_2y_2 + x_3y_3.$$

We obtain

$$\begin{aligned} X^{\mathsf{T}}AY &= \tilde{X}^{\mathsf{T}} \begin{bmatrix} 1 & 0 & 0 \\ 1 & 1 & 0 \\ 0 & 0 & -1 \end{bmatrix} \begin{bmatrix} 1 & -1 & 0 \\ -1 & 2 & 0 \\ 0 & 0 & 1 \end{bmatrix} \begin{bmatrix} 1 & 1 & 0 \\ 0 & 1 & 0 \\ 0 & 0 & -1 \end{bmatrix} \tilde{Y} \\ &= \tilde{X}^{\mathsf{T}} \begin{bmatrix} 1 & 0 & 0 \\ 0 & 1 & 0 \\ 0 & 0 & 1 \end{bmatrix} \tilde{Y} = \tilde{x}_1\tilde{y}_1 + \tilde{x}_2\tilde{y}_2 + \tilde{x}_3\tilde{y}_3. \end{aligned}$$

In the new reference systems, the bilinear form is diagonal; that is, it has a diagonal matrix.

This example also illustrates the concept of equivalent bilinear forms. We shall say that two real bilinear forms in $m + n$ variables with matrices A_1 and A_2 are **equivalent** if and only if there exist nonsingular real matrices B and C, of orders m and n, respectively, such that $B^{\mathsf{T}}A_1C = A_2$, that is, if and only if the matrices of the two forms are equivalent. If we write the forms as $X^{\mathsf{T}}A_1Y$ and $\tilde{X}^{\mathsf{T}}A_2\tilde{Y}$, the definition amounts to saying that the two forms are equivalent if and only if there exist nonsingular transformations $X = B\tilde{X}$ and $Y = C\tilde{Y}$ which transform the first form into the second. The reader should show that the inverse transformations will then carry the second form into the first, so equivalence is actually symmetric in character, even though it is not symmetrically defined.

If we choose to regard the equivalent forms as being related by linear transformations of coordinates, it follows that, in this case, *distinct but equivalent forms are representations of the same bilinear function but in different reference systems.*

In the event that the transformations are regarded as operators, two forms are equivalent if and only if there exist nonsingular operators on $\mathscr{E}^m$ and $\mathscr{E}^n$ such that the forms always have equal values at corresponding pairs of vectors. Since two real matrices are equivalent if and only if they have the same order and the same rank, we may conclude

Theorem 9.18.2: *Two real bilinear forms, each with an $m \times n$ matrix, are equivalent if and only if they have the same rank.*

In particular, every bilinear form $X^{\mathsf{T}}AY$ in $m + n$ variables and of rank r is equivalent to the canonical form

$$\tilde{X}^{\mathsf{T}}\left[\begin{array}{c|c} I_r & 0 \\ \hline 0 & 0 \end{array}\right]\tilde{Y} = \tilde{x}_1\tilde{y}_1 + \tilde{x}_2\tilde{y}_2 + \cdots + \tilde{x}_r\tilde{y}_r. \tag{9.18.5}$$

In fact, if B^{T} and C are matrices such that $B^{\mathsf{T}}AC$ is the rank normal form of A, then the transformations $X = B\tilde{X}$ and $Y = C\tilde{Y}$ effect the reduction of $X^{\mathsf{T}}AY$ to the canonical form.

By determining first what transformations reduce each of two equivalent bilinear forms to the canonical form, we can determine by what transformations either may be transformed into the other. (Explain how this could be done.)

9.19 EXERCISES

1. Diagonalize each quadratic form by means of an orthogonal transformation ($n = 3$):

$$\text{(a) } q(X) = x_1^2 - 6x_1x_2 + x_2^2,$$
$$\text{(b) } q(X) = 2x_1x_2 + 2x_2x_3,$$
$$\text{(c) } q(X) = 2x_1x_2 + 2x_2x_3 + 2x_3x_1.$$

2. By inspection, write each quadratic form of Exercise 1 as a linear combination of squares of linear expressions.

3. Examine for definiteness:

$$\text{(a) } \begin{bmatrix} 1 & -1 & -1 \\ -1 & 2 & 4 \\ -1 & 4 & 6 \end{bmatrix}, \qquad \text{(b) } \begin{bmatrix} 4 & 2 & -2 \\ 2 & 4 & 2 \\ -2 & 2 & 4 \end{bmatrix}, \qquad \text{(c) } \begin{bmatrix} 1 & 2 & 0 \\ 2 & 1 & 3 \\ 0 & 3 & -3 \end{bmatrix},$$

$$\text{(d) } \begin{bmatrix} 2 & 1 & 1 \\ 1 & 2 & 1 \\ 1 & 1 & \frac{2}{3} \end{bmatrix}, \qquad \text{(e) } \begin{bmatrix} x & 1 & 0 \\ 1 & x & 1 \\ 0 & 1 & x \end{bmatrix}.$$

***4.** Show that for every real matrix A, $A^{\mathsf{T}}A$ is positive semidefinite or positive definite. When is it positive definite?

5. What is the condition that the quadratic form $Ax^2 + 2Bxy + Cy^2$ be positive definite? Test $x^2 + xy + y^2$, $x^2 + 2xy + y^2$, $x^2 + 4xy + y^2$, and $x^2 + 2kxy + my^2$ for definiteness.

6. Show that the quadratic form in $x_1, x_2, \ldots, x_n$ defined by

$$q = x_1^2 + (x_1 + x_2)^2 + \cdots + (x_1 + x_2 + \cdots + x_n)^2$$

is positive definite.

7. Show that the quadratic form in $x_1, x_2, \ldots, x_n$ defined by

$$q = (x_1 - x_2)^2 + (x_2 - x_3)^2 + \cdots + (x_{n-1} - x_n)^2 + (x_n - x_1)^2$$

is positive semidefinite, but not positive definite.

8. Let $f_1(X), f_2(X), \ldots, f_k(X)$ be linear functions of $x_1, x_2, \ldots, x_n$. Under what conditions is the quadratic form defined by

$$q = (f_1(X))^2 + (f_2(X))^2 + \cdots + (f_k(X))^2$$

(a) positive definite, (b) positive semidefinite?

9. Under what conditions on x and a is the real matrix

$$\begin{bmatrix} x & a & a & \cdots & a \\ a & x & a & \cdots & a \\ \vdots \\ a & a & a & \cdots & x \end{bmatrix}_n$$

positive definite?

10. Show that if $\alpha > 1$, the matrix A of order n for which $a_{i,i+k} = \alpha^{(n-1)-k}$ is positive definite.

***11.** Prove that if A and B are real, positive definite, symmetric matrices of order n, then so is $A + B$.

***12.** Show that a real matrix A is negative definite if and only if all its characteristic roots are negative.

13. A real, symmetric matrix A is negative definite if and only if its principal minor determinants alternate in sign thus:

$$p_0 = 1, p_1 < 0, p_2 > 0, p_3 < 0, \ldots.$$

(a) Show that these matrices are negative definite:

$$A = [-2]_{1\times 1}, \qquad B = \begin{bmatrix} -2 & 1 \\ 1 & -2 \end{bmatrix}, \qquad C = \begin{bmatrix} -2 & 1 & -1 \\ 1 & -2 & 1 \\ -1 & 1 & -2 \end{bmatrix}.$$

(b) Given that A is a real, symmetric, negative definite matrix, show how to write $X^{\mathsf{T}}AX$ as a linear combination of squares of linearly independent linear functions of x_1, x_2, and x_3, the coefficients of combination all being negative.

14. Reduce the bilinear form

$$X^{\mathsf{T}} \begin{bmatrix} 1 & 1 & 1 & 2 \\ 2 & 2 & 2 & 3 \\ 3 & 3 & 3 & 4 \end{bmatrix} Y$$

to the canonical form defined in (9.18.5).

15. Show that a real $n \times n$ matrix A is symmetric if and only if $X^{\mathsf{T}}AY = Y^{\mathsf{T}}AX$ for all real n-vectors X and Y.

16. Let A and B be real $m \times n$ matrices. Show that tr $(A^{\mathsf{T}}B)$ is a symmetric bilinear form in the elements of A and B.

17. Identify each of the following conics and sketch a figure:

(a) $x_1^2 + 6x_1x_2 + x_2^2 = 4$,

(b) $x_1^2 + 2x_1x_2 + 4x_2^2 = 6$,

(c) $4x_1^2 - x_1x_2 + 4x_2^2 = 12$.

18. Identify each quadric and sketch a figure:

(a) $X^{\mathsf{T}}\begin{bmatrix} 2 & 1 & 0 \\ 1 & 2 & 0 \\ 0 & 0 & -2 \end{bmatrix}X = 0$, (b) $X^{\mathsf{T}}\begin{bmatrix} 2 & 1 & 0 \\ 1 & 2 & 0 \\ 0 & 0 & 2 \end{bmatrix}X = 24$.

***19.** Prove that if $A_{n \times n}$ is real and symmetric and if $B_{n \times n}$ is real, symmetric, and positive definite, then the roots of det $[A - \lambda B] = 0$ are all real. This generalizes the characteristic value problem.

***20.** Under the assumptions of Exercise 19, prove that if $AX_1 = \lambda_1 BX_1$, $AX_2 = \lambda_2 BX_2$ and $\lambda_1 \neq \lambda_2$, then $X_1^{\mathsf{T}}BX_2 = 0$. This generalizes the concept of orthogonality. Explain.

21. Under the assumptions of Exercise 19, prove that there exists a matrix V such that $V^{\mathsf{T}}AV$ and $V^{\mathsf{T}}BV$ are both diagonal. (See F. E. Hohn, *Elementary Matrix Algebra*, second edition, New York, Macmillan, 1964, pages 347–349, for help on these last three exercises.)

APPENDIX I

The Notations $\sum$ and $\prod$

Through his previous work in mathematics, the reader may already have become somewhat familiar with the notations $\sum$ and $\prod$ for sums and products respectively, but there are operations with these symbols which we use rather frequently in this book and which may well be new to him. For his convenience, we therefore provide a discussion of these symbols.

THE $\sum$ NOTATION

I.1 DEFINITIONS

The $\sum$ notation is simply a shorthand method for designating sums. Thus, for example, we write

$$x_1 + x_2 + x_3 + x_4 + x_5 = \sum_{j=1}^{5} x_j.$$

Here j is a variable ranging over the integers 1, 2, 3, 4, and 5. The symbols $j = 1$ below the $\sum$ sign indicate that 1 is the initial value taken on by j, and the 5 written above the $\sum$ sign indicates that 5 is the terminal value of j. We call j the **index of summation**. The **summand**, x_j, is a function of j which takes on the values

x_1, x_2, x_3, x_4, and x_5, respectively, as j takes on successively the values 1, 2, 3, 4, and 5. Finally, the $\sum$ sign denotes the fact that the values x_1, x_2, x_3, x_4, and x_5 taken on by x_j are to be *added.* The entire symbol $\sum_{j=1}^{5} x_j$ is read, "the summation of x_j as j ranges from 1 to 5."

In the same way, we have

$$x_6 + x_7 + x_8 = \sum_{j=6}^{8} x_j,$$

where now the initial value of j is 6 and the terminal value is 8. Combining these two examples we have

$$x_1 + x_2 + x_3 + x_4 + x_5 + x_6 + x_7 + x_8 = \sum_{j=1}^{5} x_j + \sum_{j=6}^{8} x_j,$$

so

$$\sum_{j=1}^{8} x_j = \sum_{j=1}^{5} x_j + \sum_{j=6}^{8} x_j.$$

Our first example above is an illustration of the basic definition:

(I.1.1) $$\sum_{j=1}^{n} x_j = x_1 + x_2 + \cdots + x_n.$$

Our third example above is an illustration of the theorem:

(I.1.2) $$(x_1 + x_2 + \cdots + x_p) + (x_{p+1} + \cdots + x_n) = \sum_{j=1}^{n} x_j = \sum_{j=1}^{p} x_j + \sum_{j=p+1}^{n} x_j.$$

A familiar function of n quantities $x_1, x_2, \ldots, x_n$ is their "average" or arithmetic mean m_x, namely their sum divided by n. Using the above notation, we can write

$$m_x = \frac{x_1 + x_2 + \cdots + x_n}{n} = \frac{1}{n} \sum_{j=1}^{n} x_j.$$

The compactness of the $\sum$ notation, as here demonstrated, is one indication of its value.

In the above examples, the values actually represented by $x_1, x_2, \ldots, x_n$, of course have to be given before the sums can be evaluated. Sometimes, however, the notation is such as to designate the values of the various terms. An example of such a sum is

$$1^2 + 2^2 + 3^2 + 4^2 + 5^2 = \sum_{k=1}^{5} k^2,$$

or, more generally,

$$1^2 + 2^2 + \cdots + n^2 = \sum_{k=1}^{n} k^2.$$

Here the index of summation k ranges over the values 1, 2, 3, 4, and 5 in the first case, while the summand k^2 ranges over the values 1^2, 2^2, 3^2, 4^2, and 5^2. In the second case the range of the index k is from 1 to n; that of the summand k^2 is from 1^2 to n^2, inclusive, of course.

It should also be pointed out that sometimes the initial value of the summation index is zero or a negative integer. For example,

$$\sum_{j=0}^{k} \frac{1}{2^j} = \frac{1}{2^0} + \frac{1}{2^1} + \cdots + \frac{1}{2^k},$$

and

$$\sum_{j=-n}^{n} a_j x^j = a_{-n}x^{-n} + a_{-n+1}x^{-n+1} + \cdots + a_{-1}x^{-1} + a_0x^0 + a_1x^1 + a_2x^2 + \cdots + a_nx^n \qquad (x \neq 0).$$

As a final illustration, we recall that infinite series are also commonly written with a $\sum$ sign. Thus, for example, we might have

$$\frac{1}{1^p} + \frac{1}{2^p} + \frac{1}{3^p} + \cdots + \frac{1}{n^p} + \cdots = \sum_{n=1}^{\infty} \frac{1}{n^p},$$

or

$$\frac{x}{1+2} + \frac{2x^2}{1+2^2} + \frac{3x^3}{1+2^3} + \cdots + \frac{nx^n}{1+2^n} + \cdots = \sum_{n=1}^{\infty} \frac{nx^n}{1+2^n}.$$

In each case we can obtain the first three terms on the left by substituting $n = 1, 2, 3$, respectively, into the **general term** of the series, namely the term containing the index n which appears on both the left and the right in the appropriate equation. As many more terms as may be desired may, of course, be found in the same way. Here the $\sum$ sign denotes a *purely formal sum* which may or may not represent a number depending on whether the series does or does not converge.

I.2 EXERCISES

1. Given that $x_1 = -2$, $x_2 = 1$, $x_3 = -1$, $x_4 = 3$, $x_5 = 7$, $x_6 = -8$, find

$$\sum_{j=1}^{6} x_j, \ \sum_{j=1}^{6} x_j^2, \ \sum_{j=1}^{6} (2x_j + 3), \text{ and } \sum_{j=1}^{6} (x_j + 2)(x_j - 2).$$

2. Rewrite in the $\sum$ notation:

(a) $2t + 4t^2 + 8t^3 + 16t^4 + 32t^5 + 64t^6$.

(b) $1 + 3 + 5 + \cdots + (2n - 1)$.

(c) $1 \cdot 2 + 2 \cdot 3 + \cdots + n(n + 1)$.

(d) $(x_1 - m_x)(y_1 - m_y) + (x_2 - m_x)(y_2 - m_y) + \cdots + (x_n - m_x)(y_n - m_y)$.

3. Rewrite in the ordinary notation:

(a) $\sum_{k=0}^{5} k(k-1)$; $\sum_{k=2}^{5} k(k-1)$; $\sum_{k=-5}^{5} k(k-1)$.

(b) $\sum_{j=1}^{n} a_j x_j$; $\sum_{j=1}^{n} a_j x^j$.

(c) $\sum_{n=0}^{\infty} \frac{x^n}{n!}$ (0! is defined to be 1, in case you have forgotten, and $x^0 = 1$ here).

(d) $\sum_{n=0}^{\infty} \left(\frac{x^n}{n!} + n(n-1)(n-2)\right)$. [Compare with (c).]

(e) $\sum_{n=0}^{\infty} \left(\frac{x^n}{n!}\right)(1 + \sin n(n-1)(n-2)x)$. [Compare with (d) and (c).]

4. Show that

(a) $\left(\sum_{j=1}^{n} x_j\right) + x_{n+1} = \sum_{j=1}^{n+1} x_j$.

(b) $\sum_{j=p+1}^{n} x_j = \sum_{j=1}^{n} x_j - \sum_{j=1}^{p} x_j$, $n \geq p + 1$.

(c) $\sum_{j=1}^{k} x_j + \sum_{j=1}^{n-k} x_{k+j} = \sum_{j=1}^{n} x_j$.

I.3 BASIC RULES OF OPERATION

In each of the examples given in Section I.1, the symbol used for the index of summation is entirely arbitrary, so it is called a **dummy index**. Thus we have, for example,

$$x_1 + x_2 + \cdots + x_n = \sum_{i=1}^{n} x_i = \sum_{j=1}^{n} x_j = \sum_{p=1}^{n} x_p = \cdots,$$

$$1^2 + 2^2 + \cdots + n^2 = \sum_{j=1}^{n} j^2 = \sum_{k=1}^{n} k^2 = \sum_{v=1}^{n} v^2 = \cdots.$$

There is another kind of arbitrariness in the summation index; this is indicated by the following examples, which the student should examine carefully.

$$\sum_{j=1}^{n} x_j = \sum_{j=0}^{n-1} x_{j+1} = \sum_{j=2}^{n+1} x_{j-1} = \cdots$$

and

$$\sum_{n=0}^{\infty} \frac{x^n}{n!} = \sum_{n=1}^{\infty} \frac{x^{n-1}}{(n-1)!} = \sum_{n=-1}^{\infty} \frac{x^{n+1}}{(n+1)!} = \cdots.$$

Here we have altered the initial value of the index of summation, but we have altered the function being summed in a compensating way, so the net sum remains unaltered. Can you write in words a rule for how this is to be done? This sort of shift in the range of summation is often useful.

Let us consider again the sum

$$\sum_{j=1}^{n} x_j = x_1 + x_2 + \cdots + x_n. \tag{I.3.1}$$

If each of the x's here is equal to the same fixed quantity c, we have

$$\sum_{j=1}^{n} x_n = c + c + \cdots + c = nc,$$

or, as we write it in this case,

$$\sum_{j=1}^{n} c = nc. \tag{I.3.2}$$

For example,

$$\sum_{j=1}^{5} 10 = 50.$$

Equation (I.3.2) is, in fact, a *definition* of the symbol $\sum_{j=1}^{n} c$, which is a priori meaningless since the constant c does not depend on the index of summation j.

Next let us suppose that in (I.3.1) we have $x_j = ky_j$, where k is a constant. Then

$$\sum_{j=1}^{n} x_j = \sum_{j=1}^{n} (ky_j) = ky_1 + ky_2 + \cdots + ky_n = k(y_1 + y_2 + \cdots + y_n)$$
$$= k \sum_{j=1}^{n} y_j.$$

Thus we have our second basic rule,

$$\sum_{j=1}^{n} (ky_j) = k \sum_{j=1}^{n} y_j. \tag{I.3.3}$$

Finally, let us suppose that $x_j = y_j + z_j$ in (I.3.1). Then we have

$$\sum_{j=1}^{n} x_j = \sum_{j=1}^{n} (y_j + z_j) = (y_1 + z_1) + (y_2 + z_2) + \cdots + (y_n + z_n)$$
$$= (y_1 + y_2 + \cdots + y_n) + (z_1 + z_2 + \cdots + z_n) = \sum_{j=1}^{n} y_j + \sum_{j=1}^{n} z_j,$$

which gives our third basic rule,

$$\sum_{j=1}^{n} (y_j + z_j) = \sum_{j=1}^{n} y_j + \sum_{j=1}^{n} z_j. \tag{I.3.4}$$

We make one more observation in this section. When there is no possible misinterpretation, the index of summation is often omitted. Thus we write simply $\sum x$ in place of $\sum_{j=1}^{n} x_j$ if the range of summation is clearly indicated by the context. Similarly, we could write $\sum x^2 - (\sum x)^2$ in place of $\sum_{j=1}^{n} x_j^2 - (\sum_{j=1}^{n} x_j)^2$, and so on.

I.4 EXERCISES

1. Use (I.3.3) and (I.3.4) to show that

$$\sum_{j=1}^{n} (ax_j + by_j) = a \sum_{j=1}^{n} x_j + b \sum_{j=1}^{n} y_j.$$

2. Show that

$$\sum_{j=1}^{n} x_j(x_j - 1) = \sum_{j=1}^{n} x_j^2 - \sum_{j=1}^{n} x_j,$$

and that

$$\sum_{j=1}^{n} (x_j - 1)(x_j + 1) = \left(\sum_{j=1}^{n} x_j^2\right) - n.$$

3. Using the fact that $m_x = (\sum x)/n$, (I.3.2), and Exercise 1, show that $\sum (x_j - m_x) = 0$. (Fill in the missing indices of summation first. The differences $x_j - m_x$ are called the *deviations of the* x's *from their mean*. You are thus to prove that the sum of the deviations of a set of quantities from their mean is zero.)

4. Show that

(a) $\displaystyle\sum_{j=1}^{k} (x_j + 1)^2 f_j = \sum_{j=1}^{k} x_j^2 f_j + 2 \sum_{j=1}^{k} x_j f_j + \sum_{j=1}^{k} f_j,$

(b) $\displaystyle\sum_{j=1}^{n} (x_j - m_x)^2 = \left(\sum_{j=1}^{n} x_j^2\right) - nm_x^2.$

These two results are used in deriving various formulas in statistics.

I.5 FINITE DOUBLE SUMS

We shall now consider the matter of **double sums**. Let us suppose that we have a set of nm quantities U_{ij}, where $i = 1, 2, \ldots, n$ and $j = 1, 2, \ldots, m$. We arrange these in a rectangular pattern, thus:

$$\begin{array}{cccc} U_{11} & U_{12} & \cdots & U_{1m} \\ U_{21} & U_{22} & \cdots & U_{2m} \\ \vdots & & & \\ U_{n1} & U_{n2} & \cdots & U_{nm}. \end{array}$$

If we wish to add all the U's, we may add first the various rows and then add the row totals to get the desired result:

$$\sum_{j=1}^{m} U_{1j} + \sum_{j=1}^{m} U_{2j} + \cdots + \sum_{j=1}^{m} U_{nj},$$

which may be written more compactly by using a second summation sign thus:

$$\sum_{i=1}^{n} \left(\sum_{j=1}^{m} U_{ij} \right).$$

If we had found the column totals first instead of the row totals, we would have obtained in the same way the result

$$\sum_{j=1}^{m} \left(\sum_{i=1}^{n} U_{ij} \right).$$

Since the sum will be the same in either case, we have

$$\text{(I.5.1)} \qquad \sum_{i=1}^{n} \left(\sum_{j=1}^{m} U_{ij} \right) = \sum_{j=1}^{m} \left(\sum_{i=1}^{n} U_{ij} \right),$$

which says that *in a finite double sum, the order of summation is immaterial.* This result does not necessarily hold for sums of infinitely many terms.

Such double sums are usually written without parentheses:

$$\sum_{i=1}^{n} \sum_{j=1}^{m} U_{ij} = \sum_{j=1}^{m} \sum_{i=1}^{n} U_{ij}.$$

The indices of summation here are, of course, dummy indices, just as in the case of simple sums.

An important kind of double sum is obtained when we put

$$U_{ij} = a_{ij}x_i y_j \begin{cases} i = 1, 2, \ldots, n, \\ j = 1, 2, \ldots, m, \end{cases}$$

and obtain

$$\sum_{i=1}^{n} \sum_{j=1}^{m} a_{ij}x_i y_j.$$

The expanded form of this sum is a polynomial in the $m + n$ variables $x_1, x_2, \ldots, x_n, y_1, y_2, \ldots, y_m$. Since each term of this polynomial is of the first degree in the x variables as well as in the y variables, we call it a *bilinear form* in these variables.

If we had, for example, $n = 2, m = 3$ and $a_{11} = a_{12} = a_{13} = 1, a_{21} = a_{22} = a_{23} = -1$, the bilinear form would be

$$\begin{aligned} \sum_{i=1}^{2} \sum_{j=1}^{3} a_{ij}x_i y_j &= a_{11}x_1y_1 + a_{12}x_1y_2 + a_{13}x_1y_3 + a_{21}x_2y_1 + a_{22}x_2y_2 + a_{23}x_2y_3 \\ &= x_1y_1 + x_1y_2 + x_1y_3 - x_2y_1 - x_2y_2 - x_2y_3. \end{aligned}$$

Another special situation of prime importance is obtained when $m = n$ and $U_{ij} = a_{ij}x_ix_j$, $i, j = 1, 2, \ldots, n$. We have then

$$\sum_{i=1}^{n} \sum_{j=1}^{n} a_{ij}x_ix_j,$$

or, as it is more commonly written,

$$\sum_{i,j=1}^{n} a_{ij}x_ix_j.$$

(When several indices of summation have the same range, as here, it is convenient to write them on one summation sign. This is permissible because the order of summation is irrelevant in a finite sum.) If, for example, $n = 2$, the expanded form of the sum is

$$a_{11}x_1x_1 + a_{12}x_1x_2 + a_{21}x_2x_1 + a_{22}x_2x_2 = a_{11}x_1^2 + (a_{12} + a_{21})x_1x_2 + a_{22}x_2^2.$$

A polynomial of this kind is called a *quadratic form* in $x_1, x_2, \ldots, x_n$, since every term in it is of the second degree in those variables. In most applications, the requirement $a_{ij} = a_{ji}$, $i, j = 1, 2, \ldots, n$, is useful. In this case we call the quadratic form *symmetric*.

I.6 EXERCISES

1. Write out in full the *trilinear* form

$$\sum_{i=1}^{2} \sum_{j=1}^{2} \sum_{k=1}^{3} a_{ijk}x_iy_jz_k.$$

2. Show in Exercise 1 that the same result is obtained independently of the order in which the various summations are carried out.

3. Write out the quadratic form for which $a_{ij} = 0$, $i \neq j$, and $a_{ii} = 1$, $i, j = 1, 2, \ldots, n$.

4. In how many different orders may the summation in

$$\sum_{i_1=1}^{n_1} \sum_{i_2=1}^{n_2} \cdots \sum_{i_k=1}^{n_k} U_{i_1 i_2 \cdots i_k}$$

be carried out? Are the results all equal? By what method of proof would you establish your answer to this last question?

5. Show that

$$\left(\sum_{j=1}^{n} x_j\right)^2 - \sum_{j=1}^{n} x_j^2 = \sum_{\substack{i,j=1 \\ i \neq j}}^{n} x_ix_j = 2 \sum_{\substack{i,j=1 \\ i<j}}^{n} x_ix_j.$$

(When, as here, a restriction is imposed on a summation process, the intention is that the summation should proceed as usual except that only those terms satisfying the restriction are to be written.)

6. Show that, if $\sum_{j=1}^{n} x_j = 0$, then

$$\sum_{j=1}^{n} x_j^2 = -\sum_{\substack{i,j=1\\ i\neq j}}^{n} x_i x_j.$$

7. Given that $n_i m_{x_i} = \sum_{j=1}^{n_i} x_{ij}$, $i = 1, 2, \ldots, k$ and that

$$\left(\sum_{i=1}^{k} n_i\right) m_x = \sum_{i=1}^{k}\sum_{j=1}^{n_i} x_{ij},$$

show that

$$\sum_{i=1}^{k} n_i(m_{x_i} - m_x) = 0$$

and

$$\sum_{i=1}^{k}\sum_{j=1}^{n_i} (x_{ij} - m_x)^2 = \sum_{i=1}^{k}\sum_{j=1}^{n_i} (x_{ij} - m_{x_i})^2 + \sum_{i=1}^{k} n_i(m_{x_i} - m_x)^2.$$

These are more formulas useful in statistics.

8. Write out in full the "triangular" sums

(a) $\displaystyle\sum_{\substack{i,j=1\\ i<j}}^{6} U_{ij}$ or $\displaystyle\sum_{1\leqslant i<j\leqslant 6} U_{ij}$,

and

(b) $\displaystyle\sum_{\substack{i,j=1\\ i\leqslant j}}^{6} U_{ij}$.

9. Show that the sum of the elements in the triangular array

$$\begin{matrix} a_{11} \\ a_{12} & a_{22} \\ a_{13} & a_{23} & a_{33} \\ \vdots \\ a_{1n} & a_{2n} & a_{3n} & \cdots & a_{nn} \end{matrix}$$

may be represented as either

$$\sum_{i=1}^{n}\left(\sum_{j=i}^{n} a_{ij}\right) \quad\text{or}\quad \sum_{j=1}^{n}\left(\sum_{i=1}^{j} a_{ij}\right).$$

It is instructive to compare this with the change of order in a double integration (Figure I-1):

$$\int_0^a \int_x^a f(x, y)\, dy\, dx = \int_0^a \int_0^y f(x, y)\, dx\, dy.$$

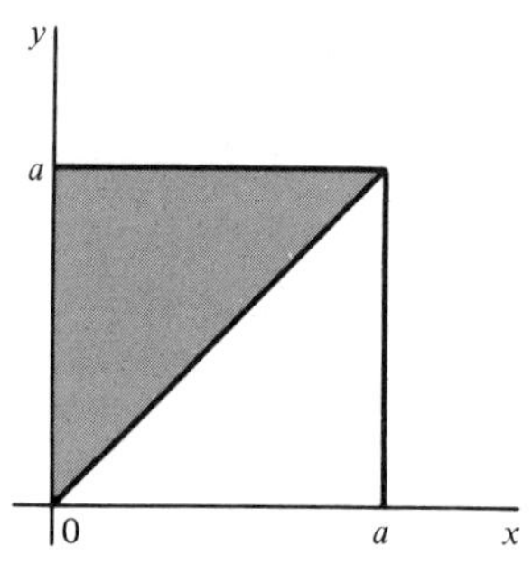

Figure I-1.

10. Show that, if

$$\frac{1}{N}\left(\sum_{i=1}^{N} \alpha_{ik}\right) = \mu, \qquad k = 1, 2, \ldots, M,$$

and

$$\frac{1}{M}\left(\sum_{k=1}^{M} \beta_{kj}\right) = \nu, \qquad j = 1, 2, \ldots, R,$$

then

$$\frac{1}{NRM} \sum_{i=1}^{N} \sum_{j=1}^{R} \sum_{k=1}^{M} \alpha_{ik}\beta_{kj} = \mu\nu.$$

THE $\prod$ NOTATION

I.7 DEFINITIONS AND BASIC PROPERTIES

We have for **products** a notation, analogous to the $\sum$ notation for sums, the definition of which is contained in the equation

$$\prod_{j=1}^{n} x_j = x_1 x_2 \cdots x_n. \tag{I.7.1}$$

Here again, j is an index whose range is indicated by the notations on the $\prod$ symbol, and x_j is a function of j, just as before. The values taken on by x_j are, however, multiplied in this case, as the symbol $\prod$ (for "product") is intended to imply.

We have the further definition

$$\prod_{j=1}^{n} c = c^n, \tag{I.7.2}$$

and the properties

$$\prod_{j=1}^{n} (kx_j) = k^n\left(\prod_{j=1}^{n} x_j\right), \tag{I.7.3}$$

$$\prod_{j=1}^{n} x_j y_j = \left(\prod_{j=1}^{n} x_j\right)\left(\prod_{j=1}^{n} y_j\right), \tag{I.7.4}$$

$$\prod_{i=1}^{n}\left(\prod_{j=1}^{m} U_{ij}\right) = \prod_{j=1}^{m}\left(\prod_{i=1}^{n} U_{ij}\right). \tag{I.7.5}$$

We also have **triangular products**, just as we have triangular sums. (See Exercises 8 and 9, Section I.6.) An important example of this type of product is

$$\prod_{1 \leqslant i < j \leqslant n} (x_i - x_j),$$

or, as it is also written,

$$\prod_{\substack{i=1,\ldots,n-1 \\ j=i+1,\ldots,n}} (x_i - x_j).$$

The notations under the $\prod$'s here mean that we are to use all factors of the form $(x_i - x_j)$ as i and j range over the values $1, 2, \ldots, n$, subject to the restriction that i be always less than j. We have, therefore,

$$\begin{aligned}\prod_{1 \leqslant i < j \leqslant n} (x_i - x_j) = (x_1 - x_2)(x_1 - x_3)&\cdots(x_1 - x_n)\\ (x_2 - x_3)&\cdots(x_2 - x_n)\\ &\vdots\\ &(x_{n-1} - x_n).\end{aligned}$$

How many factors are there in this product?

This function of $x_1, x_2, \ldots, x_n$ is known as the **alternating function**. It is zero unless the x's are all distinct, and when it is not zero, it changes its sign if any two x's are interchanged. (Verify this last statement for x_1 and x_2.)

For other ways of writing triangular products, see Exercise 5 in the next section.

I.8 EXERCISES

1. Prove rules (I.7.3), (I.7.4), and (I.7.5).

2. Show that

(a) $\left(\prod_{j=1}^{n} x_j\right) x_{n+1} = \prod_{j=1}^{n+1} x_j.$

(b) $\prod_{j=1}^{k} x_j \cdot \prod_{j=k+1}^{n} x_j = \prod_{j=1}^{n} x_j, \qquad n > k.$

(c) $\prod_{i=1}^{k} x_i \cdot \prod_{j=1}^{n-k} x_{k+j} = \prod_{p=1}^{n} x_p, \qquad n > k.$

3. Show by examples that the index in the $\prod$ notation is a dummy index and that the range of the index may be shifted if desired.

4. Simplify

$$\left(\prod_{k=1}^{n} (a^{1/2^k} + b^{1/2^k})\right)(a^{1/2^n} - b^{1/2^n}), \quad a, b > 0.$$

5. Show that

$$\prod_{j=1}^{n}\left(\prod_{i=1}^{j} a_{ij}\right) = \prod_{i=1}^{n}\left(\prod_{j=i}^{n} a_{ij}\right).$$

6. Write in expanded notation:

(a) $\displaystyle\sum_{i=1}^{n}\left(\prod_{j=1}^{n} x_{ij}\right)$.

(b) $\displaystyle\prod_{j=1}^{n}\left(\sum_{i=1}^{n} x_{ij}\right)$.

(c) $\displaystyle\sum_{i=1}^{n}\left\{\frac{y_i}{x_0 - x_i}\prod_{\substack{1\leq j\leq n\\ j\neq i}}\left(\frac{x - x_j}{x_0 - x_j}\right)\right\}$.

7. Show that

$$\prod_{\substack{i,j=1,2,\ldots,n\\ i\neq j}} (x_i - x_j) = (-1)^{n(n-1)/2}\left[\prod_{\substack{i,j=1,2,\ldots,n\\ i<j}} (x_i - x_j)\right]^2.$$

8. Show that

$$\prod_{i=1}^{n}\left(\sum_{j=1}^{m} x_{ij}\right) = \sum_{j_1,\ldots,j_n=1}^{m}\left(\prod_{i=1}^{n} x_{ij_i}\right).$$

APPENDIX II

The Algebra of Complex Numbers

Since complex numbers are used rather extensively in this book, and since some readers may have only a passing acquaintance with them, we give in this appendix a brief review of their more important algebraic properties.

II.1 DEFINITIONS AND FUNDAMENTAL OPERATIONS

If a and b are real numbers, symbols of the form $a + bi$ (subject to the rules of operation listed below) are called **complex numbers**. The real number a is called the **real part** of $a + bi$, and the real number b is called its **imaginary part**. It is often convenient to denote complex numbers by single letters from the end of the alphabet: $a + bi = z$, etc.

Two complex numbers $a + bi$ and $c + di$ are defined to be equal if and only if the real and imaginary parts of one are respectively equal to the real and imaginary parts of the other, that is, if and only if $a = c$ and $b = d$.

The complex numbers which are the **sum** and the **product** of two complex numbers are defined as follows:

$$(a + bi) + (c + di) = (a + c) + (b + d)i, \tag{II.1.1}$$

$$(a + bi)(c + di) = (ac - bd) + (ad + bc)i. \tag{II.1.2}$$

From the preceding equations, we note that

$$(a + 0i) + (c + 0i) = (a + c) + 0i,$$
$$(a + 0i)(c + 0i) = (ac) + 0i.$$

Thus complex numbers of the form $a + 0i$ behave just like the corresponding real numbers a with respect to addition and multiplication. This fact leads us to redefine the symbol $a + 0i$ as the real number a:

(II.1.3) $$a + 0i = a$$

for every real number a. In particular, we have

$$0 + 0i = 0 \qquad \text{and} \qquad 1 + 0i = 1.$$

The first of these special cases leads to:

Theorem II.1.1: *A complex number is zero if and only if its real and imaginary parts are both zero.*

The preceding definitions have the consequence that the set of real numbers is contained in the set of complex numbers: Every real number a is a complex number of the special form $a + 0i$. Every statement true for complex numbers in general is thus true in particular for real numbers, but not conversely, of course.

Since

$$(a + bi) + 0 = (a + bi) + (0 + 0i) = (a + 0) + (b + 0)i = a + bi,$$

we see that 0 is an **identity element for addition**.

We define the **negative** of $a + bi$ to be the complex number

$$-(a + bi) = (-a) + (-b)i,$$

so

$$(a + bi) + (-(a + bi)) = (a + bi) + ((-a) + (-b)i) = 0.$$

Subtraction is next defined thus:

$$(a + bi) - (c + di) = (a + bi) + (-(c + di)),$$

so

(II.1.4) $$(a + bi) - (c + di) = (a - c) + (b - d)i.$$

This makes subtraction the operation inverse to addition; that is, we will have $(z + w) - w = z$ for all complex numbers z and w.

It is consistent with these definitions of addition and subtraction to define also

(II.1.5) $$a + (-b)i = a - bi.$$

We could establish easily, but not in little space, that the associative, commutative, and distributive laws, the laws of signs and parentheses, and the laws of positive integral exponents apply in the above operations with complex numbers just as in the case of real numbers. We shall simply assume these results and proceed.

The equation

$$(a + bi)1 = (a + bi)(1 + 0i) = a + bi,$$

which follows from (II.1.2), shows that 1 is an **identity element for multiplication**.

For economy of representation, we define

$$0 + bi = bi. \tag{II.1.6}$$

A complex number of this form is called a **pure imaginary number**. The pure imaginary number $1i = i$ is called the **imaginary unit**. From (II.1.2), by putting $a = c = 0$, $b = d = 1$, we obtain the equation

$$i^2 = -1.$$

Thus, when multiplying complex numbers, we proceed as though we were multiplying polynomials of the form $a + bx$, except that now we replace i^2 by -1 whenever it appears. It is helpful in this connection to observe that $i^3 = -i$, $i^4 = 1$, etc. For some purposes it is helpful to rewrite the relation $i^2 = -1$ in the form $i = \sqrt{-1}$, in which case $-\sqrt{-1} = -i$.

Using (II.1.2) we may now verify that

$$(c + di)\left(\left(\frac{c}{c^2 + d^2}\right) + \left(\frac{-d}{c^2 + d^2}\right)i\right) = 1,$$

provided that $c^2 + d^2 \neq 0$. Hence we define the **inverse** or **reciprocal** of $c + di$ as follows:

$$(c + di)^{-1} = \left(\frac{c}{c^2 + d^2}\right) + \left(\frac{-d}{c^2 + d^2}\right)i = \frac{c - di}{c^2 + d^2}. \tag{II.1.7}$$

Since c^2 and d^2 are both nonnegative, $c^2 + d^2 = 0$ if and only if $c = d = 0$. Thus *the only complex number $c + di$ which has no reciprocal is zero.*

This definition of the reciprocal leads us to define **division** of complex numbers thus:

$$\frac{a + bi}{c + di} = (a + bi)(c + di)^{-1}, \qquad c^2 + d^2 \neq 0. \tag{II.1.8}$$

Substituting and expanding, we obtain

$$\frac{a + bi}{c + di} = \frac{(a + bi)(c - di)}{c^2 + d^2} = \left(\frac{ac + bd}{c^2 + d^2}\right) + \left(\frac{bc - ad}{c^2 + d^2}\right)i,$$

which is again a complex number. With this definition, we have $(zw) \div w = z$ for all z and w except $w = 0$, so division as defined here is indeed the inverse of multiplication.

Any collection $\mathscr{F}$ of complex numbers which has the property that the sum, difference, product, and quotient (division by zero excepted) of any two numbers of $\mathscr{F}$ also belong to $\mathscr{F}$ is called a **number field**. In particular, the set of all complex numbers is a field since, as has been shown above, the four operations applied to any two complex numbers each yield a complex number. The set of all complex numbers of the form $a + 0i$, namely the set of all real numbers, is likewise a number field, since

$$(a + 0i) \pm (b + 0i) = (a \pm b) + 0i,$$
$$(a + 0i)(b + 0i) = (ab) + 0i,$$
$$(a + 0i) \div (b + 0i) = \frac{a}{b} + 0i, \qquad b \neq 0;$$

that is, the four operations applied to two real numbers again result in real numbers. The real field is a **subfield** of the complex number field. A third example of a number field is the set of all rational real numbers, that is, the set of all real numbers of the form a/b, where a and b are integers ($b \neq 0$). There are many other examples of number fields.

II.2 EXERCISES

1. Simplify $(1 - i)^3 - (1 + i)^3$.

2. If x and y are real, what can we conclude about them from the equation

$$(x - y + 2) + (2x + y)i = 3 + 5i?$$

3. Let n be any integer, so that $n = 4q + r$, where q is an integer and $r = 0$, 1, 2, or 3. Give a rule for evaluating i^n.

4. Show that the complex numbers $1, -1, i, -i$ form a group with respect to the operation of multiplication.

5. Use (II.1.1) and (II.1.2) to show that, for arbitrary complex numbers $z_j = x_j + iy_j$, $j = 1, 2, 3$, the commutative law

$$z_1 z_2 = z_2 z_1$$

and the distributive law

$$z_1(z_2 + z_3) = z_1 z_2 + z_1 z_3$$

hold.

6. Show that, for all complex numbers z, $z \cdot 0 = 0$. [Use (II.1.2) and the fact that $0 = 0 + 0i$.]

7. Prove that, if $z_1 + w = z_2 + w$, then $z_1 = z_2$, and that, if $z_1 w = z_2 w$ and $w \neq 0$, then $z_1 = z_2$. (**Cancellation laws** for addition and multiplication.)

8. Given that

$$\omega = \frac{-1 + \sqrt{3}i}{2},$$

show that

$$\omega^2 = \frac{-1 - \sqrt{3}i}{2}$$

and that

$$\omega^3 = 1.$$

Show also that

$$\omega^2 + \omega + 1 = 0.$$

9. Show that, for every complex number $x_1 + x_2i$, there exist unique real numbers y_1 and y_2 such that

$$x_1 + x_2i = y_1\frac{1 + i}{\sqrt{2}} + y_2\frac{1 - i}{\sqrt{2}}.$$

II.3 CONJUGATE COMPLEX NUMBERS

The complex numbers $a + bi$ and $a - bi$ are called **conjugate complex numbers**, each being the conjugate of the other. A real number is its own conjugate. The conjugate of a complex number z is denoted by $\bar{z}$.

If $z = a + bi$, then $\bar{z} = a - bi$, and

$$z + \bar{z} = 2a, \qquad z - \bar{z} = 2bi, \qquad z\bar{z} = a^2 + b^2.$$

We have, therefore,

Theorem II.3.1: *The sum, difference, and product of conjugate complex numbers are respectively a real number, a pure imaginary number, and a nonnegative real number.*

Again, by the definition of equality, if $z = \bar{z}$, we have $b = -b$, so $b = 0$ and $z = a$. If $z = -\bar{z}$, then $a = -a$ so $a = 0$ and $z = bi$. This yields

Theorem II.3.2: *If a complex number equals its conjugate it is a real number, but if it equals the negative of its conjugate it is a pure imaginary number.*

An important use of the conjugate of a complex number is in the evaluation of a quotient according to the process

$$\frac{a + bi}{c + di} = \frac{(a + bi)(c - di)}{(c + di)(c - di)} = \frac{(ac + bd) + (bc - ad)i}{c^2 + d^2}.$$

The rule "multiply numerator and denominator by the conjugate of the denominator" is easier to use and remember than is the formula for the quotient.

By direct computation, we can show readily that for any two complex numbers w and z, we have

(II.3.1)
$$\begin{aligned} \overline{(\bar{z})} &= z, \\ \overline{w + z} &= \bar{w} + \bar{z}, \\ \overline{wz} &= \bar{w}\bar{z}, \\ \overline{\left(\frac{w}{z}\right)} &= \frac{\bar{w}}{\bar{z}}, \qquad z \neq 0. \end{aligned}$$

For example, if

$$w = a + bi, \qquad z = c + di,$$

then

$$\overline{wz} = \overline{(ac - bd) + (ad + bc)i} = (ac - bd) - (ad + bc)i$$

and

$$\bar{w}\bar{z} = (a - bi)(c - di) = (ac - bd) + (-ad - bc)i,$$

so

$$\overline{wz} = \bar{w}\bar{z}.$$

Similar procedures apply in the other cases.

The **absolute value** of the complex number $z = a + bi$, denoted by $|z|$, is by definition the nonnegative real number $\sqrt{z\bar{z}} = \sqrt{a^2 + b^2}$. The square of the absolute value, namely $z\bar{z}$ or $a^2 + b^2$, appears in the process of division:

$$z^{-1} = \frac{1}{z} = \frac{\bar{z}}{z\bar{z}}, \qquad z \neq 0,$$

and similarly

$$\frac{w}{z} = \frac{w\bar{z}}{z\bar{z}}, \qquad z \neq 0.$$

In particular, the reciprocal of a complex number is its conjugate divided by the square of its absolute value.

Concerning the absolute value, we may show, again by direct computation, that

(II.3.2)
$$\begin{aligned} |\bar{z}| &= |z|, \\ |wz| &= |w| \cdot |z|, \\ \left|\frac{w}{z}\right| &= \frac{|w|}{|z|}. \end{aligned}$$

For example, if $w = a + bi$, $z = c + di$, we have

$$\begin{aligned} |wz| &= |(ac - bd) + (ad + bc)i| \\ &= \sqrt{(ac - bd)^2 + (ad + bc)^2} = \sqrt{(a^2 + b^2)(c^2 + d^2)} \\ &= |w| \cdot |z|. \end{aligned}$$

Alternatively,

$$(wz) \cdot (\overline{wz}) = wz\bar{w}\bar{z} = (w\bar{w})(z\bar{z}).$$

Hence, taking positive square roots on both sides,

$$\sqrt{(wz)(\overline{wz})} = \sqrt{w\bar{w}} \cdot \sqrt{z\bar{z}},$$

or

$$|wz| = |w| \cdot |z|.$$

Similar proofs may be developed for the other rules.

The absolute value of a complex number also appears in some important inequalities:

(II.3.3) If $z = a + bi$, then $|a| \leq |z|$ and $|b| \leq |z|$.

$$|w + z| \leq |w| + |z|.$$

The first two of these are left to the reader to prove. To prove the last one, we note that

$$(w + z)\overline{(w + z)} = (w + z)(\bar{w} + \bar{z}) = w\bar{w} + z\bar{z} + \bar{w}z + w\bar{z},$$

or

$$|w + z|^2 = |w|^2 + |z|^2 + (\bar{w}z + \overline{\bar{w}z}).$$

The quantity in parentheses is twice the real part of $\bar{w}z$ and hence, by the first of (II.3.3), we have

$$\bar{w}z + \overline{\bar{w}z} \leq |\bar{w}z + \overline{\bar{w}z}| \leq 2|\bar{w}z| = 2|w|\,|z|.$$

Therefore,

$$|w + z|^2 \leq |w|^2 + 2|w|\,|z| + |z|^2 = (|w| + |z|)^2.$$

Taking the positive square root on both sides, we then have the desired result:

$$|w + z| \leq |w| + |z|.$$

II.4 EXERCISES

1. Complete the proof of (II.3.1) and (II.3.2).

***2.** Prove that, if f is a polynomial with *real coefficients*, $f(\bar{z}) = \overline{f(z)}$.

3. Give an example of a function g such that $g(\bar{z}) \neq \overline{g(z)}$.

***4.** Prove that, if f is a polynomial with real coefficients and if $f(z) = 0$, then $f(\bar{z}) = 0$ also. In words, this says that in the case of polynomial equations with real coefficients, complex roots occur in conjugate pairs.

5. Prove that $|w + z| \geq ||w| - |z||$.

6. For what complex numbers do we have $|z| = z$? $|z| = -z$?

7. Simplify $(1 + i)^3/(1 - i)^3$.

***8.** Show that $\sum_{j=1}^{n} |w_j|^2 \geq 0$ for arbitrary complex numbers w_j, and that, if

$$\sum_{j=1}^{n} |w_j|^2 = 0,$$

then

$$w_j = 0, \qquad j = 1, 2, \ldots, n.$$

9. Show that, if $wz = 1$ and $|w| = 1$, where w and z are complex numbers, then $|z| = 1$ and $w^{-1} = \bar{w}$.

INDEX

D